# 14
# Topics in Medicinal Chemistry

## Aims and Scope

Drug research requires interdisciplinary team-work at the interface between chemistry, biology and medicine. Therefore, the new topic-related series Topics in Medicinal Chemistry will cover all relevant aspects of drug research, e.g. pathobiochemistry of diseases, identification and validation of (emerging) drug targets, structural biology, drugability of targets, drug design approaches, chemogenomics, synthetic chemistry including combinatorial methods, bioorganic chemistry, natural compounds, high-throughput screening, pharmacological in vitro and in vivo investigations, drug-receptor interactions on the molecular level, structure-activity relationships, drug absorption, distribution, metabolism, elimination, toxicology and pharmacogenomics.

In general, special volumes are edited by well known guest editors.

In references Topics in Medicinal Chemistry is abbreviated Top Med Chem and is cited as a journal.

More information about this series at
http://www.springer.com/series/7355

Nuska Tschammer
Editor

# Chemokines

## Chemokines and Their Receptors in Drug Discovery

With contributions by

R. Abagyan · A. Christopoulos · I.J.P. de Esch · C. de Graaf · T.M. Handel · R. Horuk · A. Junker · A.K. Kokornaczyk · A.K. Strunz · T. Kenakin · I. Kufareva · R. Leurs · W. Mooij · J.E. Pease · D. Scholten · M.J. Smit · N. Tschammer · M. Wijtmans · B. Wünsch

Springer

*Editor*
Nuska Tschammer
Dept. of Chemistry and Pharmacy
Friedrich Alexander University
Erlangen, Germany

ISSN 1862-2461 ISSN 1862-247X (electronic)
Topics in Medicinal Chemistry
ISBN 978-3-319-14059-9 ISBN 978-3-319-14060-5 (eBook)
DOI 10.1007/978-3-319-14060-5

Library of Congress Control Number: 2015930521

Springer Cham Heidelberg New York Dordrecht London


Printed on acid-free paper

Springer International Publishing AG Switzerland is part of Springer Science+Business Media (www.springer.com)

# Preface

Sophisticated homeostatic and inflammatory actions of our immune system are orchestrated by a myriad of different proteins and other signaling molecules. One of the crucial molecular components of the immune system is a complex network of small soluble proteins named chemokines and their G protein coupled receptors. The GPCR superfamily is the largest family of transmembrane receptors, which transmit signals from outside of the cell across the membrane to signaling pathways within the cell. A misbalance in the functions of chemokines and their receptors often leads to severe pathologies like autoimmune diseases (e.g., rheumatoid arthritis, psoriasis, and multiple sclerosis), asthma, and cancer. Furthermore, the chemokine receptors CCR5 and CXCR4 are also hijacked by HIV as co-receptors needed for the viral entry in the $CD4^+$ T cells. In accordance with the overall importance of chemokines and their receptors in pathologies, multiple pharmaceutical companies initiated screening campaigns dedicated to the development of chemokine receptor antagonists in a recent decade. After all, GPCRs are the site of action of about 30% current drugs, making the GPCR superfamily the largest and single most important family of drug targets. More than 40 antagonists of chemokine receptors entered the clinical trials that unfortunately largely failed. Only two candidates for the therapy of noninflammatory diseases progressed successfully on the market, one for the treatment of HIV (the CCR5 antagonist maraviroc, Selzentry®, Pfizer) and one for the hematopoietic stem cell transplantation in patients with lymphoma and multiple myeloma (the CXCR4 antagonist, perixafor, Mozobil®, Genzyme). The reasons for a tremendous failure of drug candidates in clinical trials were largely attributed to the redundancy of chemokine system, inappropriate target selection, suboptimal dosing regimen, off-target effects, and in some cases even poor drug-like properties of a drug candidate. Despite the higher-than-average failure rate, the quest for successful drug candidates, which would modulate the function of chemokine receptors, continues.

In this book the current development, opportunities, and challenges in the field of drug discovery related to chemokine receptors are presented and debated. As an example for the role of chemokines in autoimmunity and inflammation, their functions in the pathophysiology of asthma, multiple sclerosis, and rheumatoid arthritis are illustrated in the chapter "Chemokine Receptors in Allergy, Inflammation, and Infectious Disease", written by James Pease and Richard Horuk. The authors describe various strategies that pharmaceutical companies have come up with to block the effect of chemokines in driving these disease processes, and how they have progressed in the clinic. Chapter "Role of 3D Structures in Understanding, Predicting, and Designing Molecular Interactions in the Chemokine Receptor Family", written by Irina Kufareva, Ruben Abagyan, and Tracy M. Handel, provides an excellent overview of pre- and post-structure efforts in understanding, predicting, and designing chemokine receptor interactions with small molecules and peptides, chemokines, and HIV gp120 proteins, as well as structure-guided insights regarding chemokine receptor dimerization and the impact of structures on rational molecular design initiatives. The efficient symbiosis of computational approaches with experimental structure determination is discussed in depth. The concept that GPCRs are natural allosteric proteins led to chapter "Allosteric Modulation of Chemokine Receptors", written by Arthur Christopoulos, Terry Kenakin, and myself, which discusses complex allosteric mechanisms by which the functions of chemokines and their receptors are fine-tuned and presents their impact on preclinical drug discovery. The opportunities and challenges of bench-to-clinic approaches are elucidated. Although allosteric modulation of chemokine receptors adds a level of complexity to analyses and approaches to drug discovery, it also introduces a tremendous capacity for pharmacologic control of this physiological system for therapeutic advantage. Chapter "Exploring the CXCR3 Chemokine Receptor with Small-Molecule Antagonists and Agonists", written by Rob Leurs and colleagues, illustrates on the example of the chemokine receptor CXCR3 nicely, how the combination of chemical, computational, and pharmacological tools and techniques increases our understanding of the molecular mechanisms by which small-molecule antagonists and agonists bind to the chemokine receptors compared to the relatively large chemokines. This knowledge potentially opens up novel therapeutic opportunities in the area of inflammation. In the last chapter " Selective and Dual Targeting of CCR2 and CCR5 Receptors: A Current Overview" Bernhard Wünsch and his colleagues present classical approaches in medicinal chemistry that fueled the development of antagonists for the chemokine receptors CCR2 and CCR5. These efforts led to the discovery of the CCR5 targeting drug that is used for the treatment of HIV-1 (maraviroc, Selzentry®, Pfizer). The reasons of failure of other promising clinical candidates are critically discussed.

I thank the authors for their valuable contributions to this volume. With their assistance this book provides profound insights into the failure-rich past, exciting present developments and promising future opportunities and challenges in the field of drug discovery dedicated to the manipulations of chemokine receptor network.

Erlangen, Germany Nuska Tschammer

# Contents

Top Med Chem (2015) 14: 1–40
DOI: 10.1007/7355_2014_39

Published online: 30 March 2014

# Chemokine Receptors in Allergy, Inflammation, and Infectious Disease

**James E. Pease and Richard Horuk**

**Abstract** Chemokines play an important role in disease by virtue of their effects on immune cells. They mediate their biological effects by acting on G-protein-coupled receptors, which represent one of the most druggable classes of proteins. In this review we will examine the role of chemokines in autoimmunity and inflammation by concentrating on the part they play in the pathophysiology of several diseases including asthma, multiple sclerosis, and rheumatoid arthritis. We will describe the various strategies that pharmaceutical companies have come up with to block the effect of chemokines in driving these disease processes and how they have fared in the clinic. We will also briefly discuss the repurposing of chemokine receptor antagonists in new indications.

**Keywords** Allergy, Antagonists, Chemokine receptors, Chemokines, Infection, Inflammation

## Contents



J.E. Pease
Leukocyte Biology Section, Faculty of Medicine, National Heart and Lung Institute,
MRC & Asthma UK Centre in Allergic Mechanisms of Asthma, Imperial College of Science, Technology and Medicine, London, SW7 2AZ, UK
e-mail: j.pease@imperial.ac.uk

R. Horuk (✉)
Department of Pharmacology, UC Davis, Davis, CA 95616, USA
e-mail: Horuk@pacbell.net

## 1 Introduction

Chemokines belong to a small family of chemoattractant proteins that orchestrate the directed trafficking of immune cells; thus, they play an important role in host defense. Inappropriate activation of the immune response by chemokines can also lead to autoimmunity, giving rise to a number of devastating diseases that include asthma, multiple sclerosis, and rheumatoid arthritis. These and other autoimmune diseases pose an ever-increasing health burden on our society and affect millions of individuals each year. Consequently, pharmaceutical companies have poured billions of dollars into research and development to identify safe and effective drugs to treat these diseases. The chemokines, because of their central role in coordinating the immune system have provided an enticing target for pharmacological intervention. In this chapter we will describe some of the key chemokines that are believed to be responsible for the leukocyte recruitment and underlying pathology in asthma, multiple sclerosis, and rheumatoid arthritis. In addition, we will describe strategies directed at inhibiting specific chemokine receptors that have been identified as being important in driving disease processes. Finally, we will discuss the progress that chemokine receptor antagonists have made in the clinic and we will conclude by looking at potential new therapeutic uses for them.

## 2 The Role of Chemokines in Asthma

Asthma is a complex, multifactorial, heterogeneous disease, which has reached epidemic levels in the Western world, with over 5 million people in the UK alone receiving some type of asthma treatment. Asthma broadly describes a variety of patients with hyperactive airways, which when trigged by antigen results in compromised lung function. This so-called early phase asthmatic reaction is triggered by antigen crosslinking the high affinity IgE receptor on mast cells and results in their degranulation and the release of preformed mediators such as histamine and newly synthesized lipid mediators such as the leukotriene $LTB_4$ and the prostaglandin $PGD_2$. These molecules exert a variety of actions on the airways including bronchoconstriction, increased microvascular permeability and increased mucus production, all of which contribute to the asthmatic phenotype of breathlessness [1]. The mediators also influence the recruitment of leukocytes to the allergic lung, either by directly serving as leukocyte chemoattractants themselves ($LTB_4$ and $PGD_2$) or directing the structural cells of the lung to produce chemokines. This chemokine production is highly dependent upon the interplay of structural and immune cells, notably dendritic cells and $T_H2$ cells and results in the late-phase reaction. This typically occurs several hours after the early phase reaction and is notable for the recruitment of a variety of leukocytes to the lung. Originally thought of as a chiefly eosinophilic disease, the identification of different subgroups or asthma phenotypes over recent years suggests that novel strategies to target the trafficking of disparate cell types may provide additional therapeutic benefit [2].

## 3 Chemokine-Driven Eosinophil Recruitment in Asthma

The recruitment of eosinophils to the respiratory system is considered a characteristic hallmark of asthma. As long ago as the late 1800s, Paul Ehrlich, who is credited with the discovery of the eosinophil, postulated that eosinophil recruitment to specific tissue sites required a stimulus that induced their "chemotactic irritability" [3]. A little over a century later, the group of Tim Williams at Imperial College London proved this hypothesis, with the discovery of a CC chemokine that was produced in the guinea pig lung following allergen challenge [4]. They named this chemokine "eotaxin," and the identification of the mouse orthologue and the eotaxin receptor, CCR3, quickly followed [5]. Subsequent identification of the additional CCR3 ligands, Eotaxin-2/CCL24, and Eotaxin-3/CCL26 suggested further means by which eosinophils could be recruited. In mouse models of allergic airways disease, CCL11 expression is induced following allergen challenge [6]. In humans, both CCL11 mRNA and CCL11 protein levels have been observed indirectly in the allergic lung tissue of both atopic asthmatics (individuals suffering from allergic conditions, e.g., hay fever or allergic dermatitis) and also that of non-atopic asthmatics [7]. Similarly, CCL11 levels have been reported to be elevated in the plasma of acute asthmatics compared with stable asthmatic patients [8]. Histamine release from degranulating mast cells can trigger the localized production of CCL11 by endothelial cells [9], as can the action of TNF-α and IL-4 on lung fibroblasts and human airway epithelial cells [10].

More recently, the role of the type 2 innate lymphoid cell in eosinophil homeostasis has begun to be appreciated. These are long-lived tissue resident cells and co-express IL-13 at sites of allergic inflammation, resulting in the expression of CCL11 and the recruitment of eosinophils [11]. At present, little is known about how these cells traffic to the lungs, but targeting this process may provide an alternative angle for intervention in asthma. Likewise, the specific roles of the other two eotaxins, CCL24 and CCL26, in asthma pathogenesis are not fully appreciated. Mice differ from humans in lacking a functional orthologue of CCL26, although CCL11 and CCL24 both seem to be important for eosinophil recruitment to the allergic lung, with deletion of both chemokines needed for ablation recruitment in an ovalbumin sensitization model [12]. A recent study by Provost and coworkers suggested that CCL26 was a particularly potent in vitro recruiter of eosinophils from asthmatic individuals and that complete blockade was not achieved with a CCR3-specific antibody leading to the notion that an additional receptor for CCL26 may exist in humans [13]. Further work is needed to test this intriguing finding.

## 4 Targeting Eosinophil Chemokine Receptors

CCR3 is the principal chemokine receptor expressed by human eosinophil and was therefore an obvious target for eosinophil-directed drug development. Blockade of CCR3 by a specific monoclonal antibody by Heath and colleagues showed that the majority of eosinophil responses to CC chemokines could be inhibited by targeting of CCR3, establishing it as a key therapeutic target for the treatment of asthma [14]. A mouse CCR3-specific antibody was also developed by scientists at DNAX to validate CCR3 in vivo. The antibody, which was functional, had the unexpected property of depleting eosinophils from the circulation [15]. Other protein-based therapeutics aimed at the blockade of CCR3 on eosinophils were chemokine based such as Met-RANTES, a modified version of CCL5/RANTES with an N-terminal methionine extension which can bind but not activate CCR3 [16]. Likewise, a modified version of the chemokine CCL18 similarly extended at the N-terminus by a single methionine residue was shown to act as an antagonist [17].

The first description of a small-molecule antagonist of CCR3 came from one of our own groups, with the compound UCB 35625 shown to block CCR3 at low nanomolar concentrations (Table 1 and **1** Fig. 1) [25]. Intriguingly, despite blocking CCR3 in a number of different assays, UCB 35625 did not displace radiolabeled CCL11 from CCR3-transfectants in contrast to small-molecule antagonists of other chemokine receptors described at the time. This led to what was then a controversial hypothesis; that is, these compounds did not antagonize the chemokine-binding site directly, but instead altered the receptor conformation such that signaling could not take place. Since the compound also had significant activity at CCR1 (which shares excellent homology with CCR3 in the transmembrane helices), we postulated that the compound resided in the intrahelical bundle of either receptor, which was formally proven in subsequent studies [26, 27].

Since these initial studies, several different CCR3 antagonists have been described in the literature with typically low nanomolar affinity despite quite diverse chemical structures [18–20, 22–24, 28–30]. Some notable examples and their effects in vitro and in vivo are summarized in Table 1. However despite often demonstrating quite

**Table 1** Preclinical CCR3 antagonists to treat asthma

| Compound | In vitro activity | In vivo activity | References |
|---|---|---|---|
| A-122057 A-122058 (Abbott Laboratories) | Inhibition of CCL11 binding ($IC_{50}$ values of 600 and 975 nM) | Reduction in CCL11-induced peritoneal eosinophilia in mice (10 mg/kg) | [18] |
| GW701897B (GlaxoSmithKline) | None published | Prevention of antigen-induced clustering of eosinophils along the vagus nerves and hyperresponsiveness to vagal stimulation following antigen inhalation in guinea pigs | [19] |
| 14n (Schering-Plough) | Inhibitor of CCL11 binding ($K_i = 4$ nM) and chemotaxis ($IC_{50} = 160$ nM) of human eosinophils | | [20] |
| LH31407 (Boehringer Mannheim) | Blockade of CCL11 binding to human eosinophils ($K_i = 14$ nM). Inhibition of CCL11-induced $Ca^{2+}$ influx in human eosinophils ($IC_{50} = 11$ nM) | Diminished infiltration of eosinophils into the airway lumen at 30 mg/kg | [21] |
| YM-344031 (Yamanouchi Pharmaceutical Co.) | Inhibition of chemotaxis of human CCR3-expressing cells ($IC_{50} = 20$ nM) | Oral administration to macaques (1–10 mg/kg) significantly inhibited CCL11-induced eosinophil shape change in whole blood. Oral administration to mice (100 mg/kg) prevented both immediate- and late-phase allergic skin reactions | [22] |
| YM-355179 (Yamanouchi Pharmaceutical Co.) | Inhibition of intracellular $Ca^{2+}$ influx, chemotaxis, and eosinophil degranulation (respective $IC_{50}$ values of 8, 24, and 29 nM) | Oral administration (1 mg/kg) inhibited CCL11-induced shape change of eosinophils in macaques. Intravenous injection (1 mg/kg) also inhibited eosinophil infiltration into macaque airways following segmental bronchoprovocation with CCL11 | [23] |
| Ki19003 (Gifu Pharmaceutical University) | Inhibition of CCL11-induced murine eosinophil migration ($IC_{50} = 200$ nM) | Blockade of eosinophil levels in mouse BALF at 3 or 10 mg/kg | [24] |

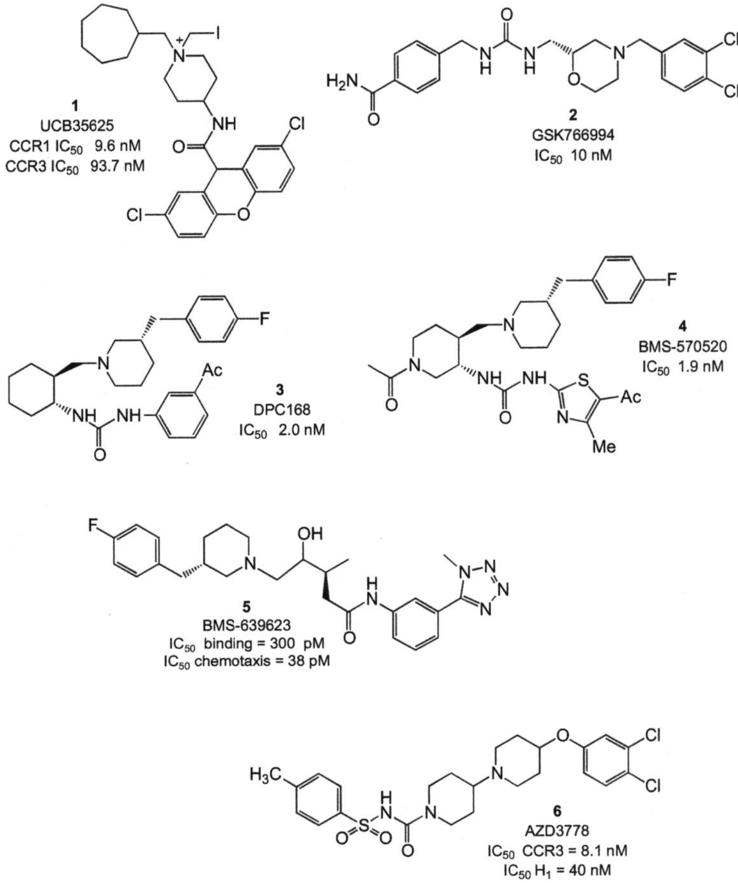


**Fig. 1** CCR3 antagonists in asthma and allergic disease (unless otherwise noted all kinetic data are inhibition of receptor binding)

convincing in vitro and in vivo data, none have so far progressed beyond phase II clinical trials (Table 2).

GlaxoSmithKline (GSK) have been active in the search for CCR3 antagonists and have identified several clinical candidates (Table 2). Although very little published data on the potency of these compounds is available, the company has published a number of patents claiming various acyl and urea derivatives of 2-aminomethyl-4-benzylmorpholine [21]. One of the compounds GSK766994 (Table 2 and **2** Fig. 1) demonstrated excellent pharmacokinetics in preclinical studies [31] and also showed efficacy in a mouse model of age-related macular degeneration [32]. Although the drug showed no safety concerns, it failed to show efficacy in a phase II clinical trial for the treatment of allergic rhinitis [31]. Despite this setback the development of the antagonist was continued and it was tested in a phase II clinical trial for asthma in 53 patients [33]. Unfortunately, the compound

**Table 2** Summary of clinical development of chemokine receptor antagonists to treat asthma, allergic rhinitis, multiple sclerosis, and rheumatoid arthritis

| Receptor | Company | Compound | Affinity (nM) | Indication | Clinical phase | Status |
|---|---|---|---|---|---|---|
| CCR1 | Schering AG (Berlex) | BX 471 | 1.0 | MS, psoriasis, endometriosis | II | No efficacy |
| CCR1 | Millennium | MLN 3701 | | MS, multiple myeloma | II | Not reported |
| CCR1 | Millennium | MLN 3897 | 2.3 | RA | II | No efficacy |
| CCR1 | Pfizer | CP-481,715 | 64 | RA | II | No efficacy |
| CCR1 | GSK | CCX354 | 1.5 | RA | II | Ongoing |
| CCR1 | BMS | BMS-817399 | | RA | II | Ongoing |
| CCR2 | Millennium | MLN 1202[a] | | RA | II | No efficacy |
| | | | | Atherosclerosis, MS | II | Ongoing |
| | | | | | II | Uncertain |
| CCR2 | CCX | CCX915 | | MS | I | Terminated |
| CCR2 | Merck | MK-0812 | 5.0 | RA, MS | II | No efficacy |
| CCR2 | Incyte | INCB8696 | | MS, lupus | I | Not reported |
| CCR2 | Incyte | INCB3284 | 3.7 | RA, type II diabetes | II | Not reported |
| CCR3 | Pharmaxis | ASM8[b] | | Asthma | II | Ongoing |
| CCR3 | GSK | GSK766994 | 10.0 | Asthma and allergic rhinitis | II | No efficacy |
| CCR3 | GSK | GSK766904 | | Asthma | II | Ongoing |
| CCR3 | GSK | GW824575 | | Asthma | I | Terminated |
| CCR3 | DuPont | DPC168 | 2.0 | Asthma | I | Terminated |
| CCR3 | BMS | BMS-639623 | 0.3 | Asthma | I | Ongoing |
| CCR3 | Novartis | QAP-642 | | Allergic rhinitis | I | Terminated |
| CCR3 | AstraZeneca | AZD3778 | 8.1 | Allergic rhinitis | II | Not reported |
| CCR4 | Amgen | KW-0761[a]/Mogamulizumab | | Oncology | II | Ongoing |
| | | | | Asthma | I | Ongoing |
| CCR4 | GSK | GSK2239633 | 10.0 | Asthma | I | Terminated |
| CCR5 | Pfizer | UK-427,857 (maraviroc) | 3.0 | RA | II | No efficacy |
| | | | | AIDS | Approved | Registered drug |

(continued)

**Table 2** (continued)

| Receptor | Company | Compound | Affinity (nM) | Indication | Clinical phase | Status |
|---|---|---|---|---|---|---|
| CCR5 | Schering-Plough | SCH-C | 2.0 | AIDS | I | Terminated |
| | | | | RA | II | No efficacy |
| CCR5 | GSK | GSK-873140 (aplaviroc) | | AIDS | II | Terminated |
| CCR5 | AstraZeneca | AZD5672 | 0.26 | RA | II | No efficacy |

[a]Neutralizing monoclonal antibodies
[b]Antisense oligonucleotide
*BMS*, Bristol–Myers Squibb, *GSK* GlaxoSmithKline, *CCX* ChemoCentryx, *MS* multiple sclerosis, *RA* rheumatoid arthritis

failed to meet its clinical end points ([34], and so its future development is uncertain.

Bristol–Myers Squibb (BMS) took a lead compound from DuPont-Merck, DPC168 (Table 2 and **3** Fig. 1), which was a potent CCR3 antagonist, but had cytochrome P450 and cardiovascular liabilities [30], as their starting point for the development of CCR3 antagonists. Extensive structural modifications allowed them to overcome the cytochrome P450 and cardiovascular liabilities with BMS-570520 (**4** Fig. 1) [35] which after further optimization gave rise to their clinical development compound BMS-639623 that was reported to be in phase I clinical development for asthma (Table 2 and **5** Fig. 1) [36].

AZD3778 is a novel low molecular weight dual CCR3 and histamine $H_1$ receptor antagonist developed by AstraZeneca. The compound has an $IC_{50}$ of 8.1 nM for the inhibition of eotaxin binding to CCR3 and an $IC_{50}$ of 40 nM for the inhibition of binding to the $H_1$ histamine receptor (Table 2 and **6** Fig. 1) [37]. A phase II clinical trial in patients with allergic rhinitis revealed that AZD3778 exerted moderately anti-eosinophilic and symptom-reducing effects thought to be through inhibition of CCR3 rather than through its effects on the histaminergic receptor [37]. Since the effects of the compound were only modest, no further development of has been reported.

Novartis had a CCR3 antagonist program and identified the compound QAP 642 (structure not disclosed) as the clinical lead, but have not disclosed any structural or potency data. A human clinical pharmacodynamic study reporting the effects of QAP642 on cutaneous eosinophil migration in the skin following subcutaneous injection of eotaxin in human volunteers has been reported [38]. At the highest dose the compound caused a modest increase in the QTc prolongation. The compound was able to inhibit eosinophil migration in this human pharmacodynamic study; however, it subsequently failed in clinical trials for asthma, and its development was discontinued [39].

A novel approach to inhibiting CCR3, ASM8, has been recently described by scientists at Pharmaxis (Table 2) [40]. ASM8 contains two modified phosphorothioate antisense oligonucleotides designed to inhibit allergic inflammation by downregulating human CCR3 and the common beta chain of the IL-3, IL-5, and GMCSF receptors. In a small clinical study with patients with mild asthma, the drug was safe and well tolerated. It attenuated the allergen-induced increase in target gene mRNA, allergen-induced sputum eosinophils, and the early and late asthmatic responses [40, 41]. It also reduced the number of CD34(+) CCR3(+) cells and CD34 (+) IL-5Rα(+) cells and the proportion of CD34(+) cells expressing IL-5Rα. Currently ASM8 is being evaluated in larger phase II clinical trials for asthma.

## 5 Chemokine-Driven Lymphocyte Recruitment in Asthma

As mentioned previously, the recruitment of the $T_H2$ type lymphocyte in asthma is thought to be pivotal to asthma pathogenesis and strategies to block $T_H2$ cell recruitment are of great interest in asthma. $T_H2$ cells are notable for their expression of a number of chemokine receptors including CCR3, CCR4, and CCR8

[42–44]. CCR4 is activated by the chemokines CCL22 and CCL17 [45] which are produced by dendritic cells in response to allergen [46, 47].

Several studies have identified CCR4 as being preferentially expressed by $T_H2$ cells [42, 44], regulatory T cells [48], and mast cells [49] suggestive of a role in allergic disease. High levels of CCR4 expression on specific subpopulations of T cells, including skin-homing cutaneous lymphocyte antigen $(CLA)^+$ T cells [50], implicate the receptor in the pathology of atopic dermatitis (AD) [51]. In vivo studies suggest that CCR4 is expressed by the majority of murine $T_H2$ lymphocytes and facilitates CCL17- and CCL22-mediated chemotaxis [52]. While deletion of CCR4 has no effect on either $T_H2$ lymphocyte differentiation in vitro or on a $T_H2$-dependent model of allergic airway inflammation [53], the CCR4/CCL17/CCL22 axes have been shown to play a pivotal role in the late phase of allergic airway inflammation, in studies employing treatment with blocking antibodies specific for the murine orthologues of CCL22 and CCL17 [6, 54]. Moreover, in clinical studies of allergen-challenged atopic asthmatics and rhinitis, the majority of T lymphocytes present in bronchial biopsies were found to be CCR4 positive [55]. Consequently, CCR4 arouses much interest as a potential therapeutic target for the treatment of allergic disease [1].

As stated earlier, CCR8 is also expressed on lymphocytes of the $T_H2$ lineage and is therefore perceived to play a role in adaptive immunity. CCR8 is best known as the sole receptor for the chemokine CCL1 [56, 57] that has been reported to be upregulated in the allergic lung [55, 58, 59]. The level of infiltrating CCR8-expressing $T_H2$ cells has been shown to correlate with the severity of asthmatic responses following allergen challenge [55]. More recently, CCL18 has been identified as an additional CCR8 ligand [60]. CCL18 is highly expressed in the human lung [61] and has been reported to be upregulated in the BAL of allergic asthmatics [62] and to be chemotactic for $T_H2$ cells [60]. Interestingly, no direct equivalent of CCL18 exists in the mouse; instead, CCL8 acts as a functional orthologue of CCL18, activating mouse CCR8 despite sharing little identity with CCL18 [63].

## 6 Targeting Lymphocyte Chemokine Receptors in Asthma

The role of CCR4 in T-cell trafficking to the allergic lung was initially supported by studies in mice in which the CCR4 ligands CCL22 and CCL17 were neutralized by antibody [54, 64] and also adoptive studies in which $T_H2$ cells from CCR4-deficient were introduced into wild-type allergic mice [65]. In each case, perturbation of CCR4 signaling resulted in reduced $T_H2$ recruitment and associated inflammation. Likewise, blockade of human CCR4 by a monoclonal antibody was reported to abolish many of the features of inflammation in a mouse model in which human peripheral blood mononuclear cells were used to reconstitute a SCID mouse [66]. Consequently, many pharmaceutical companies pursued small-molecule antagonists of CCR4.

**Table 3** Preclinical CCR4 antagonists in allergic disease

| Compound | In vitro activity | In vivo activity | References |
|---|---|---|---|
| 22 (Bristol–Myers Squibb) | Inhibition of CCL22-induced chemotaxis ($IC_{50} = 3$ nM) | A 30 mg/kg dose effective in reducing eosinophil numbers into murine BALF | [67] |
| 8c (Astellas Pharma Inc) | Inhibition of CCL22-induced chemotaxis ($IC_{50} = 23$ nM) | A 30 mg/kg dose resulted in inhibition of ear swelling in a murine contact hypersensitivity model | [69] |
| Compound 1 (AstraZeneca) | Inhibition of CCL22-induced $Ca^{2+}$ influx ($K_i = 10$ nM) | | [70] |
| RS-1154 (Daiichi Sankyo Co.) | Inhibition of CCL17-induced chemotaxis ($IC_{50} = 5.5$ nM) | Effective at reducing ovalbumin-induced ear swelling at 30 mg/kg in mice | [76] |
| RS-1748 (Daiichi Sankyo Co.) | $IC_{50}$ of 60nM in CCL17 binding assay and $^{35}$S-GTPγS | Effective at ovalbumin lung inflammation at 10 mg/kg in a guinea pigs | [73] |
| K327 (Kyowa Hakko Kirin Co.) | Inhibition of CCL17 binding ($IC_{50} = 72$ nM) | Inhibited the of $CCR4^+$ $CD4^+$ T-cell recruitment to the murine lung in an ovalbumin-challenge model (44 mg/kg, twice daily) | [77] |
| K777 (Kyowa Hakko Kirin Co.) | Inhibition of CCL17 binding ($IC_{50}$: 57 nM) and CCL17-induced chemotaxis (9 nM) | | [72] |

The high level of conservation between human and rodent CCR4 species has meant that many of the compounds developed against human CCR4 generally possess good potency at the murine counterpart. Table 3 lists some published data regarding the in vitro and in vivo activities of preclinical CCR4 antagonists [67, 69, 70, 72, 73, 76, 77]. AstraZeneca has been active in the CCR4 antagonist area and has identified *N*-pyrazin-2-yl-arylsulphonamides as potent CCR4 antagonists [70]. One of these molecules is AstraZeneca compound 1 (Table 3 and **7** Fig. 2) which appears to have a different mode of action compared to other chemokine receptor antagonists. Instead of binding to an intrahelical site composed of the transmembrane regions, compound 1 appears to require access to an intracellular site within the CCR4 C-terminus to exert its antagonistic effects [70]. Another CCR4 antagonist (Table 3 and **8** Fig. 2) from this series was able to dose-responsively inhibit CCR4 ligand-induced actin polymerization in T lymphocytes, which is a measure of T-cell function [68].

To date, only one small-molecule CCR4 antagonist, the GSK compound GSK2239633, has been reported in clinical trials (Table 2 and **9** Fig. 2). Although generally well tolerated, less than 80% receptor occupancy was achieved at doses of 1,500 mg 1 h following administration, which dropped to 50% occupancy by 4 h

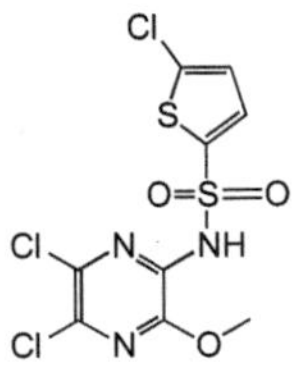

**7**
Compound 1
$IC_{50}$ CCR4 = 1.2 nM

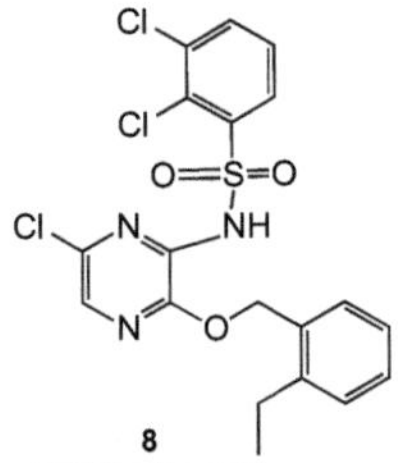

**8**
CCR4 antagonist
$IC_{50}$ CCR4 = 1.2 nM

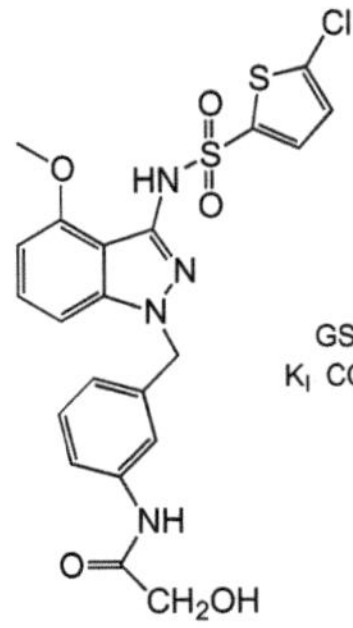

**9**
GSK2239633
$K_I$ CCR4 = 10 nM

**Fig. 2** CCR4 antagonists in asthma and allergic disease (unless otherwise noted all kinetic data are inhibition of receptor binding)

post-dose. As a consequence, GSK2239633 is not being developed further at this time [71]. The antibody Mogamulizumab (KW-0761; AMG-761) a defucosylated humanized IgG1 mAb specific for CCR4 is being developed by Kyowa Hakko Kirin and Amgen for the intravenous treatment of T-cell lymphoma. At the time of writing, Mogamulizumab has been approved in Japan for the treatment of relapsed or refractory adult T-cell leukemia-lymphoma, while Amgen is currently conducting a phase I asthma trial for the use of Mogamulizumab in asthma (Table 2) [78].

In terms of targeting CCR8, supportive data have been slow in emerging, with the reporting of small-molecule CCR8 antagonists slower still (Table 4). An initial study of allergen-challenged CCR8-deficient mice [85] appeared to support a role in allergic airway disease, although proved controversial, with subsequent in vivo studies failing to support such a role [86, 87]. In humans, the lack of reliable CCR8-specifc antibodies has been one obstacle [88] although the generation of the 433 H mAb, by ICOS scientists, provided a work-around [89]. Using this antibody, Mutalithas and colleagues were able to show that greater percentages of $CCR8^+$ T cells were found in PBMCs isolated from the venous blood of asthmatics compared with those of control subjects (4.7% c.f. 3.0%), suggesting a role for CCR8 in asthma and the use of CCR8 as a biomarker of disease progression. A handful of small-molecule CCR8 antagonists have subsequently been described in the literature (Table 4) with both in vitro and in vivo efficacy.

Against this backdrop, scientists from Medlmmune recently reported the results of a study using the CCR8 antagonist ML604086 (Table 4 and **10** Fig. 3), in an *Ascaris suum* airway challenge model in cynomolgus monkeys [83]. Despite almost complete occupancy of CCR8 with the drug during the study, no significant effects on any marker of airway inflammation were observed, leading the authors to conclude that CCR8 plays a dispensable role in asthma, certainly in the primate model employed. One potential caveat of targeting both CCR4 and CCR8 is their expression on regulatory T cells [48], which may have undesired proinflammatory effects.

**Table 4** Preclinical CCR8 antagonists in allergic disease

| Compound | In vitro activity | In vivo activity | References |
|---|---|---|---|
| AZ6 (AstraZeneca) | Inhibition of CCL1-induced chemotaxis ($IC_{50}$ = 300 nM) and $Ca^{2+}$ ($IC_{50}$ = 630 nM) | | [80] |
| AZ084 (AstraZeneca) | Inhibition of CCL1-induced chemotaxis ($IC_{50}$ = 1.3 nM) | Well tolerated in rats and dogs exposed for 7 consecutive days at doses up to 650 mg/kg/day and 8.7 mg/kg/day, respectively | [82] |
| ML604086 (MedImmune) | Inhibition of CCL1-induced chemotaxis ($IC_{50}$ = 1.3 μM) and $Ca^{2+}$ $IC_{50}$ = 1 μM) | In a primate model of asthma, no significant effect on Ag-induced BAL eosinophilia, mucus production, or $T_H2$ cytokine production, despite >98% coverage on T cells | [83] |

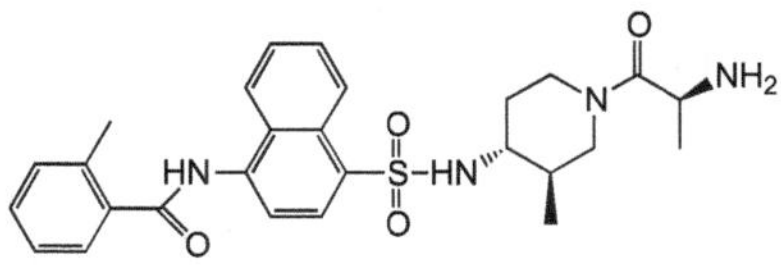


**10**
ML604086
$IC_{50}$ CCR8 in whole blood = 1.0 uM

**Fig. 3** CCR8 antagonists in asthma and allergic disease (unless otherwise noted all kinetic data are inhibition of receptor binding)

# 7 Multiple Sclerosis

Multiple sclerosis is a chronic autoimmune disease in which the immune system attacks and destroys the myelin sheath that surrounds nerve cells. Although the cause of the disease is unknown, it is thought to involve a combination of toxicological, viral, bacterial, and genetic factors. Multiple sclerosis affects close to half a million individuals in the USA alone and is the most common form of paralysis in young adults in the developed world. A conservative annual cost for patients with multiple sclerosis in the USA has recently been estimated at anywhere from 3 to 8 billion dollars [90] and this has attracted massive investment from the pharmaceutical industry in the development of new therapeutic approaches. Chemokine receptors have been considered an attractive target for the treatment of multiple sclerosis. The major rationale for targeting these proteins has been based both on the pathophysiology of the disease and also from animal models of disease.

Multiple sclerosis appears to be induced when T Helper 1 cells ($T_H1$) recognize components of the myelin sheath. Activated, autoreactive T cells within the lesions are believed to drive the chronic inflammatory process and activate local or hematogenous macrophages that destroy myelin. This inflammatory cascade leads to large focal lesions of primary demyelination with relative axonal preservation. Recent research suggests that the pathogenetic scheme described above is oversimplified and cannot explain lesion formation. It is known that T-cell populations like $T_H17$ cells can also contribute to inflammation in multiple sclerosis [81] and that amplification of demyelination in a chronic inflammatory reaction in the brain requires additional factors. Furthermore, the patterns of demyelination are different between different subgroups of multiple sclerosis patients, which suggests that the disease is heterogeneous [79, 84].

Evidence from a variety of studies has implicated chemokines in the pathophysiology of multiple sclerosis. Early studies by Godiska demonstrated that mRNAs encoding a variety of chemokines including CCL3 and CCL5 were induced in the spinal cord 1–2 days before the clinical signs of disease were apparent [91]. This was followed by studies that showed that a neutralizing antibody to CCL3 ameliorated disease in an animal model of multiple sclerosis in the mouse [92]. The link to the human disease was soon established from studies in which demyelinating plaques from the brains of multiple sclerosis patients were shown to express a variety of inflammatory chemokines and their receptors [93]. Microglial activation is thought to contribute directly to myelin destruction in multiple sclerosis through mechanisms that include the production of proinflammatory cytokines and chemokines [94]. In chronic active human multiple sclerosis lesions, the chemokine receptors CCR2, CCR3, and CCR5 have been shown to be present on infiltrating macrophages and activated microglia, while CCR2 and CCR5 were also present on large numbers of infiltrating lymphocytes [95]. In addition, macrophages derived from blood-borne monocytes and microglia have been shown to express CCR1 and its ligand CCL3 [79]. T cells expressing the chemokine receptors CCR5 and CXCR3 and their ligands CCL3 and CXCL10 are expressed in demyelinating brain lesions [96]. Interestingly the cytokine interferon beta which is used to treat multiple sclerosis reduced the expression of CXCR3 on $CD4^+$ and $CD8^+$ T cells [97]. The authors concluded that since CXCR3 cells are enriched in cerebrospinal fluid and are detected in lesion material in multiple sclerosis, this might represent one important means of interferon-beta action in treating multiple sclerosis.

As outlined above animal models of disease, particularly the experimental autoimmune encephalomyelitis (EAE) models carried out in rodents, have provided valuable insight into the role of chemokines in the human disease [98]. However, although these models have led to an understanding of the pathogenesis of the human disease, they need to be interpreted with some caution especially because they do not recapitulate the complex spectrum of the human disease. A prime example is the animal studies that showed that blocking the TNF receptor was effective in decreasing disease in a rodent model [99]. In contrast, when this approach was translated to human clinical trials in patients suffering from multiple sclerosis, the trials had to be halted because the TNF receptor blockers actually made the disease worse [100]. Also it is clear that many aspects of the human disease, in particular, the contributions of B lymphocytes and $CD8^+$ cells in disease pathology, are not captured by these models. Finally, selection of the appropriate

EAE model is important in determining the validity of a disease target. For example, the acute EAE model in rats is driven by a cell type, neutrophils, that does not really figure in the human disease [101].

# 8 Targeting Chemokine Receptors in Multiple Sclerosis

CCR1 and CCR2 are the major chemokine receptors that have been targeted pharmacologically in human clinical trials for the treatment of multiple sclerosis (Table 2). These include two CCR1 receptor antagonists BX 471 and MLN 3701 and four CCR2 antagonists MLN1202, INCB8696, CCX915, and MK-0812.

The evidence for a role of CCR1 in the pathophysiology of multiple sclerosis was based on a number of studies. First, neutralizing antibodies to one of the CCR1 ligands, CCL3, prevented the development of both acute and relapsing paralytic disease as well as infiltration of mononuclear cells into the CNS initiated by the transfer of activated T cells [102]. Second, deletion of CCR1 was protective in a myelin oligodendrocyte glycoprotein (MOG) model of multiple sclerosis in mice decreasing the disease score by around half compared to their wild-type littermates [103]. Finally, CCR1 is expressed in human multiple sclerosis lesions associated with hematogenous macrophages usually coexpressed with CCR5 [79].

Based on these data Berlex initiated a CCR1 antagonist program and identified BX 471 (Table 2 and **11** Fig. 4), a potent diacyl piperazine, as its clinical candidate. The antagonist had a reported $K_D$ of 1.0 nM for human CCR1 and was more than 1,000-fold selective for CCR1 [104]. Although the antagonist was poorly cross-reactive with rat and mouse, CCR1, it had sufficient affinity to be tested in animal models and it was efficacious in an acute rat EAE model of multiple sclerosis [104]. Based on these data the antagonist entered human clinical trials. The drug was well tolerated and had no safety issues in phase I; however, its development was stopped after the phase II study failed to demonstrate a positive clinical end point, a reduction in the number of new inflammatory CNS lesions [105].

Millennium has reported a CCR1 antagonist MLN3701 in phase II clinical trials for multiple sclerosis. This compound was being codeveloped with its partner Sanofi-Aventis (AVE9897), but no structures or data were ever published [106].

The evidence for a role of CCR2 in the pathophysiology of multiple sclerosis was based on a number of studies.

First, CCL2 is one of the major chemokines responsible for the recruitment and activation of monocytes in the blood and macrophages in the tissues [107]. These cells play a central role in the disease pathology involved in multiple sclerosis and thus their modulation might be of benefit in treating multiple sclerosis.

Second, Karpus [92] demonstrated that the production of the CCR2 ligand CCL2 correlated with the relapse induced in an EAE model of multiple sclerosis in the mouse. Furthermore, neutralizing antibodies to CCL2 significantly reduced the severity of relapsing EAE and significantly inhibited the adoptive transfer of EAE when included in in vitro activation cultures, suggesting a regulatory anti-inflammatory property.

**Fig. 4** Chemokine receptor antagonists in multiple sclerosis (unless otherwise noted all kinetic data are inhibition of receptor binding)

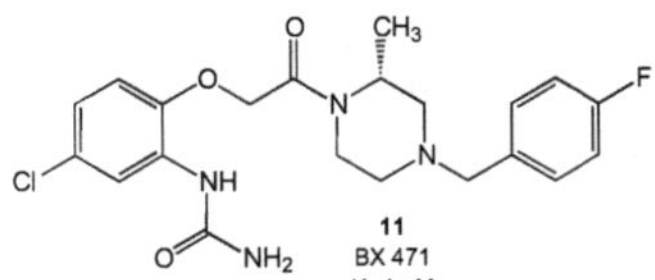


**11**
BX 471
$K_i$ 1 nM

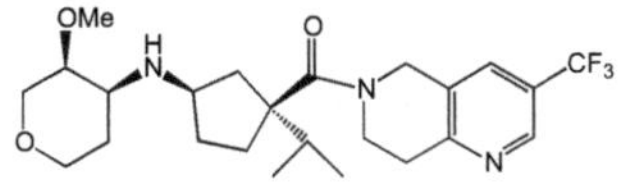


**12**
MK-0812
$IC_{50}$ 5 nM

Finally, animals genetically deficient in the receptor for CCL2 (CCR2) were found to be resistant to disease induction in an EAE model of disease [108]. These animals failed to develop mononuclear cell inflammatory infiltrates in the CNS and failed to increase CNS levels of the chemokines CCL5, CCL2, and CXCL10 as well as the chemokine receptors CCR1, CCR2, and CCR5.

Based on these studies a number of companies felt encouraged to pursue CCR2 antagonists as therapeutics for treating multiple sclerosis. One of the first to be described was Merck's MK-0812, a pyridine-substituted piperidine (Table 2 and **12** Fig. 4) [109]. MK-812 is a potent CCR2 antagonist, $IC_{50}$ of 5.0 nM, and was tested in phase II clinical trials for multiple sclerosis and rheumatoid arthritis (see later) [105]. The multiple sclerosis trial was a randomized, double-blind, placebo-controlled study with a 12-week protocol and 120 patients. The primary end point was for the compound to decrease the presence of new gadolinium-enhancing lesions as measured by MRI [110]. Unfortunately, the compound increased rather than decreased the presence of gadolinium-enhancing lesions and the development of the compound for multiple sclerosis was terminated by Merck [111].

Millennium has also described the development of MLN 1202, which is a blocking antibody to CCR2, as a potential therapeutic for multiple sclerosis and rheumatoid arthritis (see later). The antibody was reported to have positive results in a phase II trial for multiple sclerosis [112]. Millennium announced at the American Neurological Association meeting in 2007 that MLN1202 reduced gadolinium-enhancing lesions on magnetic resonance images of the brain in a multicenter phase II clinical trial involving 50 patients with relapsing-remitting multiple sclerosis [111, 112]. However beyond these data there have been no further reports of activity in multiple sclerosis and currently this molecule is reported to be in a phase II clinical trial for the treatment of bone metastases [113].

Incyte had reported that INCB3344 a tool compound that they developed for target validation was efficacious in a mouse model of multiple sclerosis [114]. Based on these studies they developed a clinical CCR2 receptor antagonist INCB8696 (structure not known) that they reported was in phase I clinical trial for multiple sclerosis [115]. However beyond this initial communication by the company in 2007 there have been no further reports of any activity of this molecule in multiple sclerosis and we are left to conclude that the program was discontinued.

ChemoCentryx identified a CCR2 antagonist CCX915 (structure not disclosed) as a clinical candidate for the treatment of multiple sclerosis [116, 117]. Unfortunately the development of CCX915 was terminated due to its poor pharmacokinetic properties in phase I clinical trials.

## 9 Reasons for the Clinical Failure of Chemokine Receptor Antagonists in Multiple Sclerosis

The data from the chemokine receptor antagonists in the various clinical trials for multiple sclerosis has been hugely disappointing. None of the compounds have advanced further than phase II clinical trials and the failures have cost the pharmaceutical companies developing these compounds multiple millions of dollars. It is difficult to determine the reasons for the failures of these drugs as therapeutics for multiple sclerosis, because in the majority of cases the companies developing these compounds have chosen not to reveal any clinical data. One can guess that the primary reason for companies to take such a negative approach is that by doing this they will not provide any potential advantage to their competitors. However, this approach certainly does a major disservice to patients suffering from multiple sclerosis, since valuable insights into the potential mechanisms of the disease that could provide new approaches for the development of potentially useful novel drugs are lost.

Although there has been very little clinical data published to help explain the failures of chemokine receptor antagonists in treating multiple sclerosis, one obvious reason that comes to mind is that simply blocking one receptor to treat such a heterogeneous disease is way too simplistic an approach. Recall that multiple sclerosis is a complex disease; not only are there four clinical subtypes – relapsing remitting, secondary progressive, primary progressive, and progressive relapsing (of which relapsing remitting is the most common form of the disease) – but also recent work by several groups have revealed a further level of complexity based on the patterns of demyelination that exists between patients [84]. These data suggest that the disease is even more heterogeneous than simply classifying it according to clinical subtypes [79].

Since we do not yet have specific clinical markers to be able to stratify patients into chemokine receptor-specific subpopulations, then the selection of specific responders in a clinical trial is exceedingly difficult and could account for some of the observed clinical failures. In support of this idea is that the mechanism of action of most of the clinically approved multiple sclerosis treatments is relatively broad mechanistically. For example, the antibody Natiluzumab targets adhesion molecules that block the migration of all activated T cells [118], while the small-molecule SIP-1 receptor agonist Fingolomid causes the retention of activated

T cells in the lymph nodes where they accumulate, thus strongly intervening in the pathophysiology of the disease [119]. In contrast, blocking individual chemokine receptors on cells that can respond to more than one receptor will clearly not be as effective as either of these two therapies. Thus, the clinical failure of chemokine receptor antagonists to treat multiple sclerosis may be partly ascribed to the issues discussed above.

In line with this argument is a recent study with the cytokine interferon beta [120]. This study could help to provide an explanation why, although the cytokine is widely prescribed for the treatment of relapsing-remitting multiple sclerosis, only about 50% of patients respond to treatment with interferon beta and that in some individuals the cytokine can actually make the disease worse. In this study the investigators initiated disease in an animal model of multiple sclerosis either with $T_H1$ or $T_H17$ cells [120]. Then they compared the treatment of the disease in control versus interferon-beta-treated animals. They found that the $T_H1$-induced disease was reduced by interferon beta, but the $T_H17$-induced disease was made worse by interferon beta. Furthermore patients with high IL-17 in which the disease is $T_H17$ driven do not respond to treatment with interferon beta and their symptoms are exacerbated. In contrast patients in whom the disease is $T_H1$ driven respond well to treatment with interferon beta. Thus in this case it may be possible to predict therapeutic success in multiple sclerosis by determining a patient's cytokine profile and if this approach is successful, then it could be an important step forward in designing a personalized therapy for multiple sclerosis. Extrapolating from these studies, if it were possible to classify patients with multiple sclerosis into chemokine receptor responsive subgroups, we might have a more rationalized approach to successfully treat the disease with chemokine receptor antagonists.

## 10 Rheumatoid Arthritis

Rheumatoid arthritis is a chronic autoimmune disease in which the immune system targets the destruction of the cells that line the joints of the body. The disease is initiated by the migration of immune cells from the vasculature into the synovial tissue, notably monocytes and $T_H1$ cells, where they initiate an autoimmune reaction leading to destruction of the cartilage. There is currently no cure for rheumatoid arthritis and consequently, there has been much interest in the chemokines and receptors responsible for leukocyte trafficking to the synovium as potential drug targets. The evidence for a role of chemokines in the pathophysiology of rheumatoid arthritis is provided from a variety of sources both in animal studies and from the human disease.

The presence of chemokines in the inflamed joints and tissues of patients with rheumatoid arthritis has been well described and although both CC and CXC chemokines appear to be present, the former are thought to be more involved in driving the disease than the latter. For example, the CC chemokines, CCL3 and CCL5, are readily detectable in synovial fluid of rheumatoid arthritis patients with increased levels correlating with disease severity [121, 122]. Furthermore, studies of material from rheumatoid arthritis patients suggest that CD68$^+$ macrophages which express CCR1, CCR2, and CCR5 are recruited to the synovium by these chemokines [123]. These clinical data have been supported by studies in murine models of rheumatoid arthritis. For example, in the collagen-induced arthritis (CIA) model [124] and in a rat adjuvant-induced arthritis model CCL3 levels are elevated [125]. Moreover, disease severity was significantly reduced by neutralizing anti-CCL3 and anti-CCL5 antibodies establishing a proof of principle for antagonism of CCR1 signaling in rheumatoid arthritis [126].

Based on the studies described above, the major effort in targeting chemokine receptors in rheumatoid arthritis has centered on the receptors CCR1, CCR2, and CCR5, and antagonists to these three receptors have all undergone human clinical trials (Table 2).

Several lines of evidence implicate CCR1 in the pathophysiology of rheumatoid arthritis. First, CCR1 is expressed in macrophages in rodent models of rheumatoid arthritis [127]. Second, CCR1 and its ligands showed significant expression in peripheral mononuclear cells obtained from biopsied synovial tissue from patients with rheumatoid arthritis [128]. Finally, a nonpeptide antagonist of murine CCR1 was efficacious in a collagen-induced arthritis model in the mouse [129].

Based on these data, several pharmaceutical companies focused on CCR1 as a target for rheumatoid arthritis and five different CCR1 antagonists have been reported in human clinical trials (Table 2). One of the first was a substituted pyridylbenzoxepine from Millennium, MLN3897 (Table 2 and **13** Fig. 5), which was a potent CCR1 antagonist with a $K_D$ of 2.3 nM [130]. The compound was effective in vivo and inhibited CCL3-induced immune cell recruitment in a guinea pig skin sensitization model [130]. In 2004 Millennium announced that they were in phase I clinical trials with MLN3897 and the major indications appeared to be rheumatoid arthritis and multiple myeloma. However, in November 2007 Millennium announced that they were terminating the development of the antagonist for rheumatoid arthritis because it had failed to reach its clinical end point in a phase II trial [131].

Pfizer identified CP-481715 a quinoxaline-2-carboxylic acid derivative as a potent CCR1 inhibitor, $K_D$ 64 nM (Table 2 and **14** Fig. 5). The compound is a competitive and reversible antagonist and is more than 100-fold selective for CCR1 [132]. Unfortunately, the antagonist is species specific for human CCR1 precluding its evaluation in classical animal models of disease. To circumvent this problem, Pfizer researchers generated transgenic mice expressing human CCR1 and demonstrated efficacy in models of inflammation in these animals [133]. The molecule successfully completed phase I safety studies and demonstrated efficacy in a 16-patient phase Ib clinical trial [132]. Based on these data CP-481715 entered phase II studies but the trial was stopped after 6 weeks because the compound did not demonstrate any efficacy [132].

GSK is currently in phase II clinical trials with a compound acquired from ChemoCentryx, CCX354 (Table 2) [134]; although the structure of this compound has not been formally disclosed, it is possible that it belongs to a series of

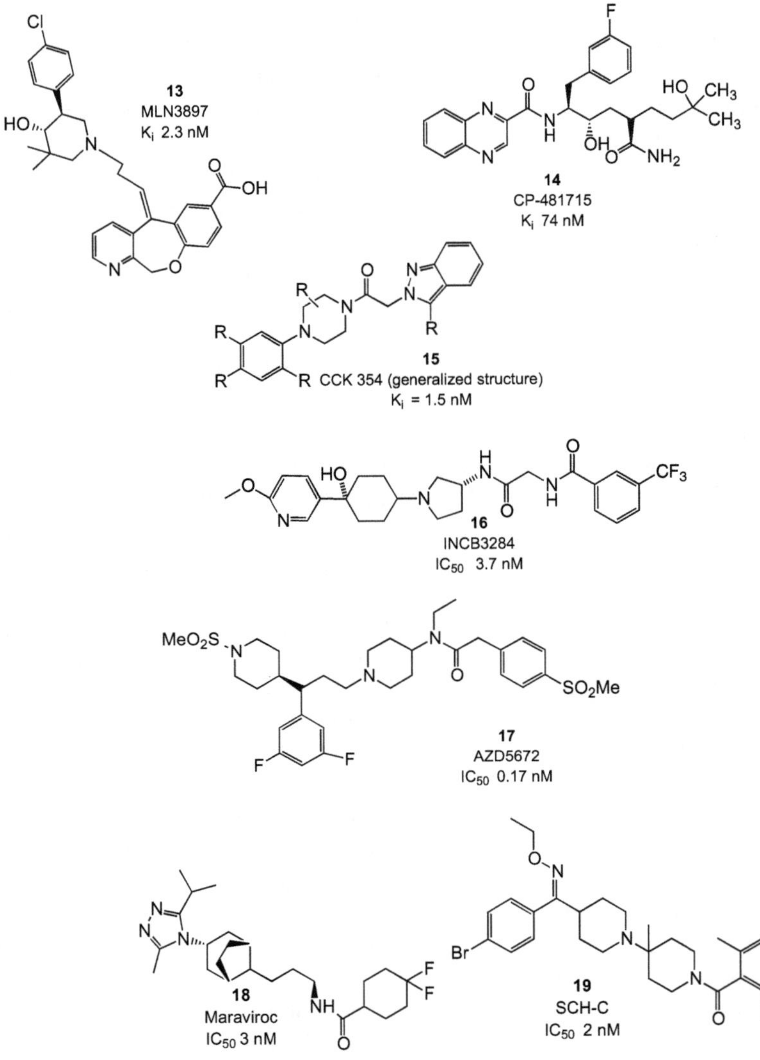


**Fig. 5** Chemokine receptor antagonists in rheumatoid arthritis (unless otherwise noted all kinetic data are inhibition of receptor binding)

azaindazoles which the company has been actively pursuing [135] and a generalized structure is shown (**15** Fig. 5). The compound was potent, $K_i$ of 1.5 nM, and was specific for CCR1 [134]. In addition, the compound blocked the chemotaxis of THP-1 cells that were induced with synovial fluid from rheumatoid arthritis patients. The antagonist was active in two animal models, blocking leukocyte trafficking in a thioglycollate-induced peritonitis model in rats and an LPS-induced synovitis model in rabbits. Based on these favorable animal data, the compound was evaluated in a human phase I clinical trial and found to be well tolerated with no serious adverse events. Phase II clinical trial data recently reported at the American College of Rheumatology meetings showed that at a once daily dose of 200 mg, the antagonist was safe and well tolerated by patients with rheumatoid arthritis. Furthermore the compound reached its clinical end points in the study (reduction in disease score and in the levels of proinflammatory markers) [136].

BMS has long had an interest in CCR1 antagonists for treating rheumatoid arthritis [137] and recently described their efforts in identifying potent CCR1 inhibitors [138, 139]. The preclinical candidate identified from these studies BMS-457 had excellent potency ($K_i$ 0.8 nM) and was specific for CCR1 but exhibited significant QT prolongation in both rabbits and dogs during advanced safety studies; thus, further development of the antagonist was halted [139]. Despite these setbacks the company is currently in phase II clinical trials with another CCR1 antagonist BMS-817399 (Table 2) [140]. Although the structure of BMS-817399 is not reported the company had previously filed a number of patents claiming a series of hydroxy-piperidine derivatives that are strikingly similar to Millennium's CCR1 antagonists, for example, MLN3897 [130].

A number of studies have suggested a role of CCR2 in the pathophysiology of rheumatoid arthritis. First, CCL2 the ligand for CCR2 is highly expressed in the synovial fluid and in macrophages from synovial fluid of patients with rheumatoid arthritis [141]. Second, neutralizing antibodies to CCL2 were beneficial in reducing inflammation in a collagen arthritis model of rheumatoid arthritis [142]. Third, an N-terminal truncated CCL2 antagonist, CCL2 [9–76], prevented the onset of arthritis as monitored by measuring joint swelling and by histopathological evaluation of the joints [143]. Finally, INCB3344 a small-molecule antagonist of CCR2 blocked disease in an adjuvant-induced arthritis model of rheumatoid arthritis [114].

Based on these and a number of other studies several pharmaceutical companies initiated programs to identify CCR2 antagonists for the treatment of rheumatoid arthritis. Foremost amongst these was Millenium, which developed a blocking antibody to CCR2 as a potential therapeutic for multiple sclerosis (see earlier) and rheumatoid arthritis. Although as discussed above MLN1202 showed benefit in phase II studies with multiple sclerosis and atherosclerosis it disappointed in a phase IIa trial in patients with active rheumatoid arthritis [131, 144]. Thirty-two patients received three infusions, over a period of 6 weeks, with either placebo or MLN1202 at three doses. Treatment with MLN1202 reduced the levels of free CCR2 on $CD14^+$ monocytes demonstrating the biological activity of the compound. However, there was no reduction in the levels or expression of any of the synovial biomarkers and no clinical improvement was observed. These findings were in line with recent phase II clinical trials with a neutralizing antibody to the CCR2 chemokine CCL2 [145]. In this study in which 33 patients received the inhibitor, there was no detectable clinical benefit of ABN912 compared with placebo, nor did treatment with the study drug result in a significant change in the levels of biomarkers in synovial tissue and peripheral blood.

Incyte had reported that INCB3344, a potent 3-aminopyrrolidine derivative, reduced inflammation in a rat adjuvant-induced model of arthritis [114]. SAR of this molecule resulted in the identification of INCB3284, a potent, 3.7 nM, CCR2 antagonist (Table 2 and **16** Fig. 5) [146]. This molecule showed minimal inhibition of the hERG potassium current, high selectivity over other G-protein-coupled receptors, and acceptable oral bioavailability in rodents and primates and was reported to have a pharmacokinetic profile suitable for once-a-day dosing [146]. The company stated in 2005 that "We recently initiated a one-month double-blind, placebo-controlled Phase IIa trial of INCB3284 in patients with rheumatoid arthritis. The primary goal of this first Phase IIa trial is to determine the safety and pharmacokinetics of INCB3284 in approximately 48 patients with active disease who are also receiving methotrexate" [147]. However, since INCB3284 is now commercially available from chemical suppliers and it is no longer mentioned in the company's pipeline, this either suggests that the clinical trials were not successful or that the company is no longer pursuing these projects for business reasons.

As described earlier Merck identified a series of pyridine-substituted piperidines as potent CCR2 inhibitors. The clinical candidate was MK-0812 (Table 2 and **12** Fig. 4) [109, 148]. This compound was reported to be in a phase II clinical trial for rheumatoid arthritis [149]. However, MK-0812 failed to show any significant improvement compared to placebo for any of the end points studied and the program was discontinued [149].

Although there is limited evidence suggesting a potential role of the chemokine receptor CCR5 in the pathophysiology of rheumatoid arthritis, three companies nevertheless pursued CCR5 antagonists as potential therapeutics for this indication (Table 2). Interestingly two of these CCR5 programs, those from Pfizer and Schering-Plough, were initially developed as therapeutics to treat AIDS since CCR5 is one of the major co-receptors for HIV invasion (more of this later under chemokine receptors as vehicles of entry for HIV).

In the first program, AstraZeneca had identified AZD5672 (Table 2) as a potent CCR5 antagonist, $K_i = 0.17$ nM (Table 2 and **17** Fig. 5). The molecule exhibited excellent pharmacokinetic properties that merited its further development as a clinical compound. AZD5672 was tested in a phase IIb study in methotrexate-refractory rheumatoid arthritis patients [150]. Although the compound had excellent, once daily oral pharmacokinetic properties and exhibited high levels of receptor occupancy and maximal inhibition of CCR5 as confirmed by an ex vivo pharmacodynamic assay, it had no efficacy [150].

The second approach was initiated by Pfizer, which tested its potent CCR5 antagonist maraviroc (Table 2 and **18** Fig. 5), which is a registered drug in the treatment of AIDS (see later under chemokine receptors as vehicles of entry for HIV), in a phase II trial in patients with rheumatoid arthritis [151]. Although maraviroc had an acceptable safety profile and was well tolerated, it was not clinically efficacious in patients with active rheumatoid arthritis who had shown inadequate responses to methotrexate [151].

The third approach was initiated by Schering-Plough who tested their CCR5 inhibitor SCH-C (Table 2 and **19** Fig. 5), which was originally developed for the treatment of AIDS (see later under chemokine receptors as vehicles of entry for HIV), in rheumatoid arthritis patients [152]. The rheumatoid arthritis phase II study involved 32 patients; 20 received SCH-C and 12 placebo. Three patients who received SCH-C did not complete the study due to adverse events; none of these were serious. No improvement was observed in the active treatment group compared to placebo and this proof of concept study does not support the use of CCR5 blockade as a therapeutic strategy in patients with active rheumatoid arthritis [152].

## 11 Reasons for the Clinical Failure of Chemokine Receptor Antagonists in Rheumatoid Arthritis

As with the chemokine receptor antagonist programs for the treatment of multiple sclerosis described above the results for the clinical trials in rheumatoid arthritis have been largely unsuccessful. Of the 11 programs described in the literature, only two are still active. What could account for the failures described above? It will be largely instructive at this point to use the CCR5 programs, all three of which failed (one with a registered CCR5 antagonist, maraviroc) as examples.

The evidence for a role of CCR5 in rheumatoid arthritis was based on the following data. First, synovial tissue from patients with rheumatoid arthritis shows abundant expression of CCR5 and its ligands [121, 122]. Second, a protein antagonist of CCL5 known as Met-RANTES, which blocks both CCR1 and CCR5, decreased disease in an adjuvant-induced arthritis model in the rat [153]. Third, two small-molecule CCR5 inhibitors SCH-X and SCH-C were efficacious in collagen-induced arthritis models of disease [154, 155]. Finally, individuals who have a 32-base pair deletion in the gene for CCR5 (CCR5-Δ32 allele), which abolishes receptor expression in homozygotes, appear to be protected from developing rheumatoid arthritis since this gene mutation was significantly lower in rheumatoid arthritis patients than in healthy individuals [156].

Clearly based on the studies presented above it would appear that there was sufficient evidence for targeting CCR5 in rheumatoid arthritis. However, the following points need to be considered. First, the expression of CCR5 on inflamed synovial tissue is not by itself evidence for a role of the protein in the disease process; these data are merely guilt by association. Furthermore an equally convincing case could also be made for other chemokine receptors such as CCR1, CCR2, and CXCR3, which have all been shown to be expressed in rheumatoid tissue. Second, the therapeutic effect of the antagonist Met-RANTES in the adjuvant-induced arthritis model of rheumatoid arthritis could equally well have been due to the inhibition of CCR1 which is also antagonized by this protein. Third, the collagen-induced arthritis studies with the CCR5 inhibitors SCH-X and SCH-C involved small numbers of animals, 5 in the monkey study and 10 in the mouse study, and the antagonists were given prophylactically not therapeutically. In addition the induction of arthritis in the mouse collagen-induced arthritis model resulted in a massive leukocyte infiltration into the joints consisting mainly of neutrophils, which is clearly not consistent with the pathophysiology of the human disease. Finally, part of the rationale for treating rheumatoid arthritis patients with a CCR5 inhibitor is based on the finding that individuals expressing the Δ32 mutation of CCR5 appear to be protected from the disease. However, it is possible that the genetic deletion of CCR5 has quite different effects on the immune response than that induced by simply blocking the receptor with a small-molecule inhibitor and this may account for the failure of the antagonist in treating rheumatoid arthritis patients.

These points aside there has recently been a quite considerable discussion on what constitutes sufficient receptor blockade for a chemokine receptor antagonist to be effective therapeutically. It has been argued that to be effective a chemokine receptor antagonist would have to be present at levels to result in the blockade of more than 95% of receptors [157]. This is a tall order and if true, could well account for some of the clinical failures discussed above. For example, a recent study determined that at the doses used in the clinical studies with the CCR1 antagonist MLN3897, only around 74–83% receptor blockade was achieved; in contrast, the CCR1 antagonist CCX354 that was effective in a phase II study for rheumatoid arthritis showed greater than 95% CCR1 blockade [134]. These data suggest that almost complete antagonism is required for 24 h a day for clinical efficacy. Thus successful agents will necessarily require excellent human pharmacokinetics and/or a very slow receptor off rate to achieve complete pharmacodynamic blockade of the receptor system being antagonized.

## 12 Chemokine Receptors as Vehicles of Entry for HIV

As a primary means of recruiting leukocytes to sites of infection, it is perhaps unsurprising that the chemokine system has been a major target for microbial subversion [158, 159]. There are numerous examples of microbial products aimed at disrupting host defense including poxvirus-encoded chemokine scavenging proteins and viral orthologues of chemokines and their receptors. Perhaps one of the more notable examples is the exploitation of chemokine receptors such as CCR5 and CXCR4 by human immunodeficiency virus type 1 (HIV-1). The virus uses these receptors to gain cellular entry, a process, which has fuelled the hunt for small-molecule antagonists which block viral entry.

The first link made between chemokines and HIV-1 pathogenesis was the discovery by Cocchi and colleagues that the chemokines CCL3, CCL4, and CCL5 had a profound antiviral effect [160]. Around the same time, several groups in independent laboratories identified CCR5 as a receptor for CCL3, CCL4, and CCL5 [161–163] leading to the notion that CCR5 might be utilized by HIV-1 to gain cellular entry. Before this was formally proven, the group of Edward Berger utilized an ingenious reporter system to identify an elusive leukocyte co-receptor that was needed in conjunction with CD4 to allow recognition of the HIV-1 viral envelope protein gp120 and permit fusion with the host cell membrane [164]. This co-receptor turned out to be another chemokine receptor, CXCR4, which binds the ligand CXCL12 [165]. Occupation of CXCR4 by CXCL12 was shown to block viral entry [166] and CCR5 was subsequently shown by several groups to also function as a portal for HIV-1 [167–170] with the process again being blocked by occupation of CCR5 by chemokine [171].

The expression of CCR5 by monocytes/macrophages and dendritic cells suggested a primary route by which macrophage-tropic strains of HIV-1 gained entry into the host. The critical importance of CCR5 in this process was emphatically shown with the discovery that approximately 1% of Northern European Caucasians are homozygous carriers of an allele encoding a 32-base pair deletion in the CCR5 gene (*CCR5 Δ32*), which results in premature truncation of the CCR5 protein and an absence of CCR5 upon the cell surface [172, 173]. Such individuals are greatly protected from HIV-1 infection, despite often being exposed multiple times to the virus [172, 174, 175].

CXCR4 is typically utilized by the virus to enter T cells during later stages of infection, following evolution of the gp120 protein to become T cell tropic and recognize CXCR4. The resulting infection of $CD4^+$ T cells via CXCR4 is responsible for the ensuing loss in host T-cell count associated with AIDS and opportunistic infections associated with the condition. In addition to CCR5 and CXCR4, additional chemokine receptors including CCR3 [168, 176] and CXCR6 [177] and the human cytomegalovirus-encoded receptor US28 [178] can function with CD4 as HIV-1 as co-receptors, although their importance in disease progression is not fully understood. Several examples of CCR5-deficient people becoming HIV positive have been described in the literature, suggesting that in certain circumstances, additional co-receptors may be a route of viral entry [179–181].

## 13 Inhibiting Viral Entry via CCR5 and CXCR4

Initial strategies aimed at blocking viral entry via the receptors came in the form of chemokine derivatives. AOP-RANTES an N-terminally modified form of the chemokine CCL5/RANTES was shown to inhibit the entry of macrophage-tropic HIV-1 strains at nanomolar concentrations [182]. Similar modification of the chemokine CCL3L1 produced another CCR5-blocking molecule with around 10 times more activity than AOP-RANTES with respect to the inhibition of HIV-1 entry [183]. The mechanism of action of these molecules appears to revolve around their ability to induce the rapid endocytosis of CCR5 [184]. Likewise, CXCL12 was shown to inhibit HIV-1 entry by inducing CXCR4 internalization [185].

The first specific small-molecule antagonist of CCR5 to be described was the Takeda compound TAK-779, which at low nanomolar concentrations potently blocked both the binding of CCL5 to CCR5 transfectants and the entry of M-tropic HIV-1 strains [186]. Subsequent mutagenesis studies showed that TAK-779 bound to a hydrophobic binding pocket composed of the transmembrane

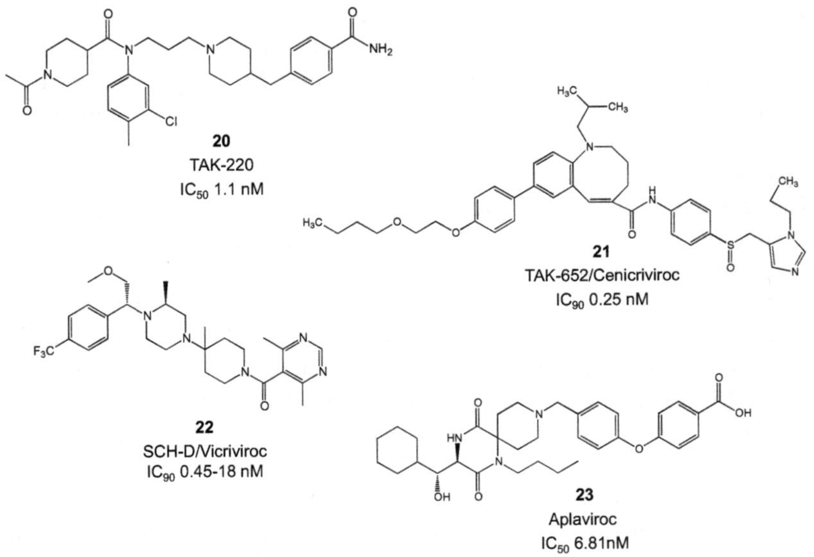


**Fig. 6** CCR5 antagonists in AIDS (unless otherwise noted all kinetic data are inhibition of receptor binding)

helices [187]. Further studies by Takeda scientists led to the discovery of the piperidine-4-carboxamide TAK-220 (**20** Fig. 6), which had nanomolar affinity for CCR5 and was a potent inhibitor of the replication of clinical isolates of HIV-1 [188]. Subsequent development of TAK-220 leads to the derivative TAK-652 (**21** Fig. 6) with improved pharmacokinetics and subnanomolar inhibitory activity in HIV-1 entry assays [189]. Renamed cenicriviroc, TAK-652 is currently being developed by Tobira Therapeutics Inc and encouraging efficacy results have been reported from phase II clinical trials in patients with CCR5-tropic HIV-1 [190].

Many promising CCR5-specific antagonists emerged from in vitro studies only to fail in phase I trials due to off-target activity at the human ether-a-go-go related gene (hERG) and subsequent undesirable cardiac effects. Novartis scientists discovered a lead CCR5-specific compound via a ligand-binding screen, which upon SAR was optimized to give a compound with low nanomolar affinity. However, the modifications resulted in unwanted activity for hERG, despite good pharmacokinetic (PK) properties [191]. Similarly, scientists at Schering-Plough also identified CCR5 antagonists via ligand-binding assays and then tested promising candidates in viral entry assays. The most promising compound identified thus inhibited the replication of a primary HIV-1 with an $IC_{50}$ of 8 nM [192]. Subsequent SAR optimization led to the development of SCH-350634, which exhibited subnanomolar activity in cell-based viral entry assays [193]. Side by side, SCH-351125 (also known as SCH-C,**19** Fig. 5) was developed and, since it had more favorable pharmacokinetics, was advanced as Schering's lead compound. Unfortunately, phase I studies were suspended, due chiefly to hERG, resulting in QTc prolongation in patients [194]. Consequently, an earlier lead compound was chosen for SAR studies, resulting in the discovery of SCH-D also known as vicriviroc (**22** Fig. 6). This had improved receptor selectivity and notably a reduced affinity for hERG [195]. Vicriviroc entered clinical trials for the treatment of AIDS but unfortunately experienced several problems, including increased incidence of liver malignancies [196].

Prominent amongst the CCR5-selective compounds without hERG activity were the GSK compound aplaviroc (**23** Fig. 6) [197] and the Pfizer compound maraviroc (**18** Fig. 5) [198], both with subnanomolar inhibitory activity in HIV-1 entry assays. Unfortunately aplaviroc fell by the wayside due to serious liver toxicity issues [199], leaving maraviroc as the front-runner. Happily, maraviroc cleared the many obstacles for clinical licensing and to date is the major success story in the development of chemokine receptor antagonists [200]. As a clinically validated compound, it is also receiving attention in additional disease areas where CCR5 may play a role in associated inflammation, for example, in atherosclerosis. Since an increased burden of cardiovascular disease has been reported in HIV-infected individuals, often associated with the use of antiretroviral therapies, it may merit usage amongst these individuals. Supportive of this, in a murine model of atherosclerosis exacerbated by the protease inhibitor ritonavir, maraviroc was seen to attenuate plaque progression [201]. A recent crystal structure of CCR5 with maraviroc bound has been published [202]. Within the structure, maraviroc lies in a deep hydrophobic pocket formed by several helices. A salt bridge between the protonated nitrogen of the tropane group and Glu-283 of helix 7 was observed as predicted by previous mutagenesis studies [203].

The first small-molecule antagonist of CXCR4 to be described was the bicyclam AMD3100, developed by scientists at AnorMED [204]. The identification of AMD3100 followed on from an earlier program in which bicyclams were discovered to inhibit HIV-1 entry by what was then described as an unknown viral uncoating event [205]. AMD3100 caries an overall positive charge (**24** Fig. 7) and is dependent upon a cluster of transmembrane glutamate residues for its binding to CXCR4 [206]. Mutation of these residues impairs both the co-receptor activity of CXCR4 and also the antagonistic activity of the compound. This suggests that AMD3100 inhibits HIV-1 entry by directly competing with the gp120 protein of HIV-1 for binding to key regions of CXCR4 [207]. Starting with the pharmacophore of AMD3100, other groups have identified the compound WZ811 with subnanomolar potency at CXCR4 in ligand-binding assays, but no disclosure of activity in viral entry assays [208].

Although possessing excellent in vitro potency, AMD3100 suffers from a marked lack of oral bioavailability, limiting its use as an HIV-1 therapeutic [209]. However, the compound has happily found another clinical application, namely, the mobilization of hematopoietic stem cells (HSCs) from the bone marrow in the treatment of lymphoma and multiple myeloma. The mechanism of action lies with its ability to interrupt the CXCR4:CXCL12 axis within the bone marrow, which is involved in the retention of bone marrow progenitors. Known in the clinic as plerixafor, AMD3100 received FDA approval in 2008 for use in combination with granulocyte colony-stimulating factor (G-CSF) for HSC mobilization into the peripheral blood for collection and subsequent autologous transplantation.

**Fig. 7** CXCR4 antagonists in AIDS (unless otherwise noted all kinetic data are inhibition of receptor binding)

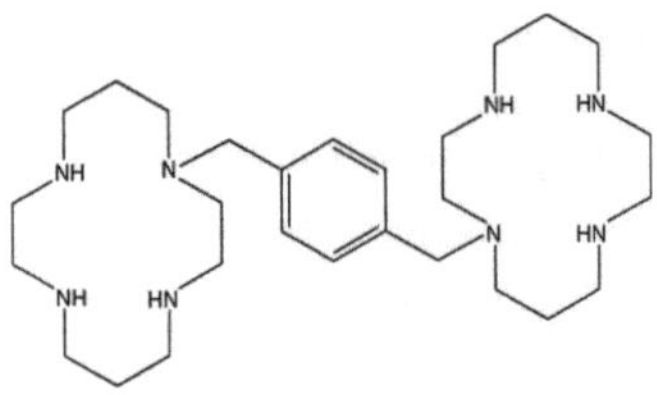


**24**
AMD3100
$IC_{50}$ 74nM

Current clinical studies are employing plerixafor in the treatment of WHIM syndrome (warts, hypogammaglobulinemia, infections, and myelokathexis), a rare immunodeficiency syndrome. Mutations in the C-terminus of CXCR4 are responsible for increased CXCR4 signaling and the resultant myelokathexis [210] and in vitro studies showed that this signaling could be blocked by plerixafor [211]. Recently published studies of WHIM syndrome patients treated with plerixafor showed increased peripheral blood leukocyte counts, suggesting that the panleukopenia can be successfully treated via CXCR4 antagonism [212, 213].

Several other CXCR4 antagonists with pharmacophores distinct from AMD3100 and its relatives have been described in the literature, notably the 18-mer peptide T22 an analogue of polyphemusin II, a protein made by the American horseshoe crab, *Limulus polyphemus*, with an $IC_{50}$ of 8 ng/ml with respect to inhibition of viral entry [214, 215]. CVX15, a 16-residue cyclic peptide analogue of T22, was recently co-crystallized with a thermostable variant of CXCR4 and shown to occupy the bulk of the intrahelical binding pocket [216]. Residues Asp171 and Asp187 (helix 4) and Asp262 (helix 6) were all shown to contact CVX15, these latter two residues also important for the interaction of AMD3100 with CXCR4 [206]. Thus, quite dissimilar antagonists can possess similar modes of binding.

## 14 New Tricks for Old Drugs?

As covered in the previous section, many chemokine receptor antagonists have failed at the final hurdle, that is to say showing efficacy in a clinical inflammatory setting. The reasons for this are likely to be complex. In many cases it has been presumed that the target chemokine receptor did not have the presumed importance in the disease process and neutralization of the chemokine:receptor axis was therefore ineffective. Alternatively, the receptor coverage afforded by either a suboptimal compound or the dosing regimen used may have been insufficient to achieve efficacy. Examples of both scenarios come from failed trials of chemokine receptor antagonists in the treatment of RA.

Antagonists of CCR1 from Millennium Pharmaceuticals and Pfizer both failed to show efficacy in trials [131, 132]. However, subsequent in vitro analysis of the compound's ability to inhibit the monocyte recruitment activity of synovial fluid from arthritic patients suggested that CCR1 was a valid target and that receptor coverage was the main issue [217]. In the case of the Millennium compound, more optimal dosing may provide an efficacious treatment. In contrast, in the same study, inhibition of the chemokine receptors CCR2 and CCR5 was found to be without effect, suggesting that these targets, although validated in in vivo RA studies, do not translate to bona fide roles in monocyte recruitment in the human setting [217].

As we have journeyed deeper into our understanding of chemokine biology, the discovery of unexpected roles for chemokine receptors in disease may provide additional avenues for antagonists that had been assigned to the laboratory shelf. One such example is the role of the chemokine receptor CCR3 in the pathogenesis of age-related macular degeneration (AMD). In a landmark paper, Takeda and colleagues observed expression of CCR3 and the eotaxins CCL11, CCL24, and CCL26 in human AMD tissue [218]. Intriguingly, there were no evidence of infiltrating leukocytes such as eosinophils or mast cell, suggesting the process is not driven by typical inflammatory pathways. In vitro assays showed that the eotaxins could drive the migration of human choroidal epithelial cells (CECs) leading to the hypothesis that this migration process drove the choroidal neovascularization (CNV) of the retina and subsequent macular degeneration. A combination of CCR3-deficient and CCL11/CCL22 double knockout mice supported this hypothesis in a murine model of laser-induced retinal injury. Again this took place in the absence of inflammatory cells such as eosinophils and mast cells, and notably, blockade of CCR3 via the CCR3 antagonist SB328437 was more efficacious than a blocking antibody directed against vascular endothelial growth factor A (VEGF-A), the current gold standard for treatment. The initial induction of CCR3 in AMD is thought to be due to the oxidative stresses associated with aging [219]. In rat CECs, blockade of the canonical Notch signaling pathway following laser-induced retinal injury resulted in the upregulation of the CCR3 gene and increased CNV lesions severity, suggesting a role for notch signaling in maintaining ocular homeostasis [220].

A recent follow-up study by and coworkers used an identical mouse model of AMD and the CCR3 antagonist YM-344031 [221]. Oral administration (50 mg/kg before and after laser treatment) or intravitreous injection of YM-344031 (0.1–10.0 μg dose immediately after injury) resulted in a significant reduction in the lesion volume. Similarly, in a model of alkali-induced CNV, topical application of SB328437 at a concentration of 500 μg/ml was found to have efficacy in terms of reduced CNV lesion size. In contrast, use of the same compound in a matrigel-induced model of AMD was without effect [222], which raises the question as to which animal models of AMD are relevant for study of the human condition.

A potential role for CCR3 in the mechanisms underlying cognitive dysfunction during the aging process has also been recently proposed [223]. In the search for soluble factors associated with this process, Villeda and coworkers employed a classical heterochronic parabiosis system coupling the circulation of old and young mice [223]. Proteomic analysis of the plasma from the mice identified 17 proteins whose levels increased during aging and whose expression correlated with decreased neurogenesis. Most prominent of these was the CCR3 ligand CCL11. Plasma levels of CCL11 were shown to increase during normal aging and also in the younger mice during the heterochronic parabiosis. Intravenous injection of recombinant CCL11 was shown to inhibit neurogenesis and impair the learning and memory functions of young mice. The authors suggest that this process is conserved across species since an age-related increase in CCL11 in both plasma and cerebrospinal fluid of healthy humans was also observed. A recent study also suggested that regular cannabis use can increase plasma levels of CCL11 which may suggest a potential mechanism for the deleterious effect of cannabis on brain function [224]. It should be pointed out that at present it remains to be proven that these deleterious functions of CCL11 are directly attributable to CCR3 activation, although the eotaxins are generally thought of as signaling only via CCR3. If this is indeed the case, then blockade of CCR3 might find a niche in a disease far removed from the original sphere of action envisaged for these drugs.

## 15 Summary

In conclusion, since their discovery a little more than two decades ago, chemokines and their receptors have been the subject of intensive study. Indeed, at the time of writing, a PubMed search will uncover over 70,000 articles referring to chemokines. There have been notable successes in translating this knowledge to the clinic. Targeting of CCR5 and CXCR4 is currently achievable with the licensed small molecules maraviroc and plerixafor. However, in the treatment of allergy and inflammation, no chemokine receptor antagonist has been successful, despite exhaustive efforts. As we glean more knowledge about the requirements for success in this area, for example, the validity of the target and the coverage required from a drug, then perhaps we can move forward with optimism. Moreover, the discovery of a role for chemokines in aging-associated diseases such as AMD and neurodegeneration, suggests that additional therapeutic avenues may be available to both old and new drugs alike. All in all, there is cause for increased optimism.

## References

1. Pease JE (2011) Biochem J 434:11–24
2. Wenzel SE (2012) Nat Med 18:716–725
3. Lee J, Rosenberg H (2013) Eosinophils in health and disease. Elsevier, Waltham MA, USA
4. Jose PJ, Griffiths-Johnson DA, Collins PD, Walsh DT, Moqbel R, Totty NF, Truong O, Hsuan JJ, Williams TJ (1994) J Exp Med 179:881–887
5. Ponath PD, Qin S, Post TW, Wang J, Wu L, Gerard NP, Newman W, Gerard C, Mackay CR (1996) J Exp Med 183:2437–2448
6. Gonzalo J-A, Lloyd CM, Wen D, Albar JP, Wells TNC, Proudfoot A, Martinez-AC, Dorf M, Bjerke T, Coyle AJ, Gutierrez-Ramos J-C (1998) J Exp Med 188:157–167

7. Ying S, Robinson DS, Meng Q, Rottman J, Kennedy R, Ringler DJ, Mackay CR, Daugherty BL, Springer MS, Durham SR, Williams TJ, Kay AB (1997) Eur J Immunol 27:3507–3516
8. Lilly CM, Woodruff PG, Camargo CA Jr, Nakamura H, Drazen JM, Nadel ES, Hanrahan JP (1999) J Allergy Clin Immunol 104:786–790
9. Menzies-Gow A, Ying S, Phipps S, Kay AB (2004) Clinical and experimental allergy. J Br Soc Allergy Clin Immunol 34:1276–1282
10. Teran LM, Mochizuki M, Bartels J, Valencia EL, Nakajima T, Hirai K, Schroder JM (1999) Am J Respir Cell Mol Biol 20:777–786
11. Nussbaum JC, Van Dyken SJ, von Moltke J, Cheng LE, Mohapatra A, Molofsky AB, Thornton EE, Krummel MF, Chawla A, Liang HE, Locksley RM (2013) Nature 502:245–248
12. Pope SM, Zimmermann N, Stringer KF, Karow ML, Rothenberg ME (2005) J Immunol 175:5341–5350
13. Provost V, Larose MC, Langlois A, Rola-Pleszczynski M, Flamand N, Laviolette M (2013) J Leukoc Biol 94:213–222
14. Heath H, Qin S, Wu L, LaRosa G, Kassam N, Ponath PD, Mackay CR (1997) J Clin Invest 99:178–184
15. Grimaldi JC, Yu NX, Grunig G, Seymour BW, Cottrez F, Robinson DS, Hosken N, Ferlin WG, Wu X, Soto H, O'Garra A, Howard MC, Coffman RL (1999) J Leukoc Biol 65:846–853
16. Elsner J, Petering H, Hochstetter R, Kimmig D, Wells TN, Kapp A, Proudfoot AE (1997) Eur J Immunol 27:2892–2898
17. Nibbs RJ, Salcedo TW, Campbell JD, Yao XT, Li Y, Nardelli B, Olsen HS, Morris TS, Proudfoot AE, Patel VP, Graham GJ (2000) J Immunol 164:1488–1497
18. Warrior U, McKeegan EM, Rottinghaus SM, Garcia L, Traphagen L, Grayson G, Komater V, McNally T, Helfrich R, Harris RR, Bell RL, Burns DJ (2003) J Biomol Screen 8:324–331
19. Fryer AD, Stein LH, Nie Z, Curtis DE, Evans CM, Hodgson ST, Jose PJ, Belmonte KE, Fitch E, Jacoby DB (2006) J Clin Invest 116:228–236
20. Ting PC, Lee JF, Wu J, Umland SP, Aslanian R, Cao J, Dong Y, Garlisi CG, Gilbert EJ, Huang Y, Jakway J, Kelly J, Liu Z, McCombie S, Shah H, Tian F, Wan Y, Shih NY (2005) Bioorg Med Chem Lett 15:1375–1378
21. (2004) Exp Op Therap Patents 14:577–582
22. Suzuki K, Morokata T, Morihira K, Sato I, Takizawa S, Kaneko M, Takahashi K, Shimizu Y (2006) Biochem Biophys Res Commun 339:1217–1223
23. Morokata T, Suzuki K, Masunaga Y, Taguchi K, Morihira K, Sato I, Fujii M, Takizawa S, Torii Y, Yamamoto N, Kaneko M, Yamada T, Takahashi K, Shimizu Y (2006) J Pharmacol Exp Ther 317:244–250
24. Komai M, Tanaka H, Nagao K, Ishizaki M, Kajiwara D, Miura T, Ohashi H, Haba T, Kawakami K, Sawa E, Yoshie O, Inagaki N, Nagai H (2010) J Pharmacol Sci 112:203–213
25. Sabroe I, Peck MJ, Jan Van Keulen B, Jorritsma A, Simmons G, Clapham PR, Williams TJ, Pease JE (2000) J Biol Chem 275:25985–25992
26. de Mendonca FL, da Fonseca PC, Phillips RM, Saldanha JW, Williams TJ, Pease JE (2005) J Biol Chem 280:4808–4816
27. Wise EL, Duchesnes C, da Fonseca PC, Allen RA, Williams TJ, Pease JE (2007) J Biol Chem 282:27935–27943
28. Hodgson S, Charlton S, Warne P (2004) Drug News Perspect 17:335–338
29. De Lucca GV, Kim UT, Johnson C, Vargo BJ, Welch PK, Covington M, Davies P, Solomon KA, Newton RC, Trainor GL, Decicco CP, Ko SS (2002) J Med Chem 45:3794–3804
30. De Lucca GV, Kim UT, Vargo BJ, Duncia JV, Santella JB 3rd, Gardner DS, Zheng C, Liauw A, Wang Z, Emmett G, Wacker DA, Welch PK, Covington M, Stowell NC, Wadman EA, Das AM, Davies P, Yeleswaram S, Graden DM, Solomon KA, Newton RC, Trainor GL, Decicco CP, Ko SS (2005) J Med Chem 48:2194–2211

31. Murdoch RD (2006) The challenges of drug discovery & development. Oral presentation, 4th James Black Conference, University of Hertfordshire, 13th September
32. Adamson P, Shima D, Eric YS (2013) Methods of treatment and prevention of eye diseases WO 2013,079,696 A1
33. (2010) Oral GW766944 (Oral CCR3 Antagonist) http://clinicaltrials.gov/ct2/show/NCT01160224
34. Neighbour H, Boulet L-P, Lemiere C, Sehmi R, Leigh R, Sousa AR, Martin JGN, Dallow N, Gilbert J, Allen A, David H, Nair P, Hamilton P (2013) Am J Respir Crit Care Med 187: A3862
35. Pruitt JR, Batt DG, Wacker DA, Bostrom LL, Booker SK, McLaughlin E, Houghton GC, Varnes JG, Christ DD, Covington M, Das AM, Davies P, Graden D, Kariv I, Orlovsky Y, Stowell NC, Vaddi KG, Wadman EA, Welch PK, Yeleswaram S, Solomon KA, Newton RC, Decicco CP, Carter PH, Ko SS (2007) Bioorg Med Chem Lett 17:2992–2997
36. Santella JB 3rd, Gardner DS, Yao W, Shi C, Reddy P, Tebben AJ, DeLucca GV, Wacker DA, Watson PS, Welch PK, Wadman EA, Davies P, Solomon KA, Graden DM, Yeleswaram S, Mandlekar S, Kariv I, Decicco CP, Ko SS, Carter PH, Duncia JV (2008) Bioorg Med Chem Lett 18:576–585
37. Greiff L, Ahlstrom-Emanuelsson C, Bahl A, Bengtsson T, Dahlstrom K, Erjefalt J, Widegren H, Andersson M (2010) Respir Res 11:1–9
38. Pascoe SJ, Bangert C, Bartlett M, Stary G, Stingl G (2007) Am J Resp Crit Care Med 175(Abstracts issue):A484
39. Toda M, Nakamura T, Ohbayashi M, Ikeda Y, Dawson M, Aye CC, Miyazaki D, Ono SJ (2007) Expert Rev Clin Immunol 3:351–364
40. Gauvreau GM, Boulet LP, Cockcroft DW, Baatjes A, Cote J, Deschesnes F, Davis B, Strinich T, Howie K, Duong M, Watson RM, Renzi PM, O'Byrne PM (2008) Am J Respir Crit Care Med 177:952–958
41. Imaoka H, Campbell H, Babirad I, Watson RM, Mistry M, Sehmi R, Gauvreau GM (2011) Clin Exp Allergy 41:1740–1746
42. Bonecchi R, Bianchi G, Bordignon PP, D'Ambrosio D, Lang R, Borsatti A, Sozzani S, Allavena P, Gray PA, Mantovani A, Sinigaglia F (1998) J Exp Med 187:129–134
43. Sallusto F, Mackay CR, Lanzavecchia A (1997) Science 277:2005–2007
44. Sallusto F, Lenig D, Mackay CR, Lanzavecchia A (1998) J Exp Med 187:875–883
45. Imai T, Baba M, Nishimura M, Kakizaki M, Takagi S, Yoshie O (1997) J Biol Chem 272:15036–15042
46. Katou F, Ohtani H, Nakayama T, Ono K, Matsushima K, Saaristo A, Nagura H, Yoshie O, Motegi K (2001) Am J Pathol 158:1263–1270
47. Pilette C, Francis JN, Till SJ, Durham SR (2004) Eur Respir J 23:876–884
48. Iellem A, Mariani M, Lang R, Recalde H, Panina-Bordignon P, Sinigaglia F, D'Ambrosio D (2001) J Exp Med 194:847–853
49. Juremalm M, Olsson N, Nilsson G (2002) Biochem Biophys Res Commun 297:480–485
50. Campbell JJ, Haraldsen G, Pan J, Rottman J, Qin S, Ponath P, Andrew DP, Warnke R, Ruffing N, Kassam N, Wu L, Butcher EC (1999) Nature 400:776–780
51. Vestergaard C, Bang K, Gesser B, Yoneyama H, Matsushima K, Larsen CG (2000) J Invest Dermatol 115:640–646
52. Andrew DP, Ruffing N, Kim CH, Miao W, Heath H, Li Y, Murphy K, Campbell JJ, Butcher EC, Wu L (2001) J Immunol 166:103–111
53. Chvatchko Y, Hoogewerf AJ, Meyer A, Alouani S, Juillard P, Buser R, Conquet F, Proudfoot AE, Wells TN, Power CA (2000) J Exp Med 191:1755–1764
54. Lloyd CM, Delaney T, Nguyen T, Tian J, Martinez-A C, Coyle AJ, Gutierrez-Ramos J-C (2000) J Exp Med 191:265–273
55. Panina-Bordignon P, Papi A, Mariani M, Di Lucia P, Casoni G, Bellettato C, Buonsanti C, Miotto D, Mapp C, Villa A, Arrigoni G, Fabbri LM, Sinigaglia F (2001) J Clin Invest 107:1357–1364

56. Roos RS, Loetscher M, Legler DF, Clark-Lewis I, Baggiolini M, Moser B (1997) J Biol Chem 272:17251–17254
57. Tiffany HL, Lautens LL, Gao J-L, Pease J, Locati M, Combadiere C, Modi W, Bonner TI, Murphy PM (1997) J Exp Med 186:165–170
58. Ying S, O'Connor B, Ratoff J, Meng Q, Mallett K, Cousins D, Robinson D, Zhang G, Zhao J, Lee TH, Corrigan C (2005) J Immunol 174:8183–8190
59. Montes-Vizuet R, Vega-Miranda A, Valencia-Maqueda E, Negrete-Garcia MC, Velasquez JR, Teran LM (2006) Eur Respir J 28:59–67
60. Islam SA, Ling MF, Leung J, Shreffler WG, Luster AD (2013) J Exp Med 210:1889–1898
61. Hieshima K, Imai T, Opdenakker G, Van Damme J, Kusuda J, Tei H, Sakaki Y, Takatsuki K, Miura R, Yoshie O, Nomiyama H (1997) J Biol Chem 272:5846–5853
62. de Nadai P, Charbonnier AS, Chenivesse C, Senechal S, Fournier C, Gilet J, Vorng H, Chang Y, Gosset P, Wallaert B, Tonnel AB, Lassalle P, Tsicopoulos A (2006) J Immunol 176:6286–6293
63. Islam SA, Chang DS, Colvin RA, Byrne MH, McCully ML, Moser B, Lira SA, Charo IF, Luster AD (2011) Nat Immunol 12:167–177
64. Kawasaki S, Takizawa H, Yoneyama H, Nakayama T, Fujisawa R, Izumizaki M, Imai T, Yoshie O, Homma I, Yamamoto K, Matsushima K (2001) J Immunol 166:2055–2062
65. Mikhak Z, Fukui M, Farsidjani A, Medoff BD, Tager AM, Luster AD (2009) J Allergy. Clin Immunol 123(67–73):e63
66. Perros F, Hoogsteden HC, Coyle AJ, Lambrecht BN, Hammad H (2009) Allergy 64:995–1002
67. Purandare AV, Gao A, Wan H, Somerville J, Burke C, Seachord C, Vaccaro W, Wityak J, Poss MA (2005) Bioorg Med Chem Lett 15:2669–2672
68. Banfield G, Watanabe H, Scadding G, Jacobson MR, Till SJ, Hall DA, Robinson DS, Lloyd CM, Nouri-Aria KT, Durham SR (2010) Allergy 65:1126–1133
69. Yokoyama K, Ishikawa N, Igarashi S, Kawano N, Masuda N, Hattori K, Miyazaki T, Ogino SI, Orita M, Matsumoto Y, Takeuchi M, Ohta M (2008) Bioorg Med Chem 16:7968–7974
70. Andrews G, Jones C, Wreggett KA (2008) Mol Pharmacol 73:855–867
71. Cahn A, Hodgson S, Wilson R, Robertson J, Watson J, Beerahee M, Hughes SC, Young G, Graves R, Hall D, van Marle S, Solari R (2013) BMC Pharmacol Toxicol 14:14
72. Sato T, Iwase M, Miyama M, Komai M, Ohshima E, Asai A, Yano H, Miki I (2013) Pharmacology 91:305–313
73. Nakagami Y, Kawashima K, Etori M, Yonekubo K, Suzuki C, Jojima T, Kuribayashi T, Nara F, Yamashita M (2010) Basic Clin Pharmacol Toxicol 107:793–797
74. Nakagami Y, Kawase Y, Yonekubo K, Nosaka E, Etori M, Takahashi S, Takagi N, Fukuda T, Kuribayashi T, Nara F, Yamashita M (2010) Biol Pharm Bull 33:1067–1069
75. Procopiou PA, Ford AJ, Graves RH, Hall DA, Hodgson ST, Lacroix YM, Needham D, Slack RJ (2012) Bioorg Med Chem Lett 22:2730–2733
76. Nakagami Y, Kawashima K, Yonekubo K, Etori M, Jojima T, Miyazaki S, Sawamura R, Hirahara K, Nara F, Yamashita M (2009) Eur J Pharmacol 624(1–3):38–44
77. Sato T, Komai M, Iwase M, Kobayashi K, Tahara H, Ohshima E, Arai H, Miki I (2009) Pharmacology 84:171–182
78. Subramaniam JM, Whiteside G, McKeage K, Croxtall JC (2013) AMG 761 in adults with asthma. Durgs 72(9):1293–8
79. Trebst C, Sorensen TL, Kivisakk P, Cathcart MK, Hesselgesser J, Horuk R, Sellebjerg F, Lassmann H, Ransohoff RM (2001) Am J Pathol 159:1701–1710
80. Karlsson AKC, Walles K, Bladh H, Connolly S, Skrinjar M, Rosendahl A (2011) Small molecule antagonists of CCR8 inhibit eosinophil and T cell migration. Biochem Biophys Res Commun 407(4):764–771
81. Langrish CL, Chen Y, Blumenschein WM, Mattson J, Basham B, Sedgwick JD, McClanahan T, Kastelein RA, Cua DJ (2005) J Exp Med 201:233–240

82. Connolly S, Skrinjar M, Rosendahl A (2012) Orally bioavailable allosteric CCR8 antagonists inhibit dendritic cell, T cell and eosinophil migration. Biochem Pharmacol 83(6):778–787
83. Wang L, Jenkins TJ, Dai M, Yin W, Pulido JC, Lamantia-Martin E, Hodge MR, Ocain T, Kolbeck R (2013) Thorax 68:506–512
84. Lassmann H, Bruck W, Lucchinetti C (2001) Trends Mol Med 7:115–121
85. Chensue SW, Lukacs NW, Yang TY, Shang X, Frait KA, Kunkel SL, Kung T, Wiekowski MT, Hedrick JA, Cook DN, Zingoni A, Narula SK, Zlotnik A, Barrat FJ, O'Garra A, Napolitano M, Lira SA (2001) J Exp Med 193:573–584
86. Chung CD, Kuo F, Kumer J, Motani AS, Lawrence CE, Henderson WR Jr, Venkataraman C (2003) J Immunol 170:581–587
87. Goya I, Villares R, Zaballos A, Gutierrez J, Kremer L, Gonzalo JA, Varona R, Carramolino L, Serrano A, Pallares P, Criado LM, Kolbeck R, Torres M, Coyle AJ, Gutierrez-Ramos JC, Martinez AC, Marquez G (2003) J Immunol 170:2138–2146
88. Pease JE (2010) Clin Exp Allergy 40:1110–1112
89. Mutalithas K, Guillen C, Raport C, Kolbeck R, Soler D, Brightling CE, Pavord ID, Wardlaw AJ (2010) Clin Exp Allergy 40(8):1175–1185
90. Adelman G, Rane SG, Villa KF (2013) J Med Econ 16:639–647
91. Godiska R, Chantry D, Dietsch GN, Gray PW (1995) J Neuroimmunol 58:167–176
92. Karpus WJ, Kennedy KJ (1997) J Leukoc Biol 62:681–687
93. Charo IF, Ransohoff RM (2006) N Engl J Med 354:610–621
94. Benveniste EN (1997) J Mol Med 75:165–173
95. Simpson J, Rezaie P, Newcombe J, Cuzner ML, Male D, Woodroofe MN (2000) J Neuroimmunol 108:192–200
96. Balashov KE, Rottman JB, Weiner HL, Hancock WW (1999) Proc Natl Acad Sci U S A 96:6873–6878
97. Sorensen TL, Sellebjerg F (2002) Mult Scler 8:104–107
98. Steinman L, Zamvil SS (2005) Trends Immunol 26:565–571
99. Ruddle NH, Bergman CM, McGrath KM, Lingenheld EG, Grunnet ML, Padula SJ, Clark RB (1990) J Exp Med 172:1193–1200
100. (1999) Neurology 53:457–465
101. Levine S, Hoenig EM (1971) Am J Pathol 64:13–30
102. Karpus WJ, Lukacs NW, McRae BL, Strieter RM, Kunkel SL, Miller SD (1995) J Immunol 155:5003–5010
103. Rottman JB, Slavin AJ, Silva R, Weiner HL, Gerard CG, Hancock WW (2000) Eur J Immunol 30:2372–2377
104. Liang M, Mallari C, Rosser M, Ng HP, May K, Monahan S, Bauman JG, Islam I, Ghannam A, Buckman B, Shaw K, Wei GP, Xu W, Zhao Z, Ho E, Shen J, Oanh H, Subramanyam B, Vergona R, Taub D, Dunning L, Harvey S, Snider RM, Hesselgesser J, Morrissey MM, Perez HD (2000) J Biol Chem 275:19000–19008
105. Horuk R (2009) Nat Rev Drug Discov 8:23–33
106. R & D Focus Drug News (2007) MLN 3701 Sanofi-Aventis phase change I. Europe
107. Leonard EJ, Yoshimura T (1990) Immunol Today 11:97–101
108. Izikson L, Klein RS, Charo IF, Weiner HL, Luster AD (2000) J Exp Med 192:1075–1080
109. Yang L, Jiao RX, Moyes C, Morriello G, Butora G, Shankaran K, Pasternak A, Goble SD, Zhou C, MacCoss M, Cumiskey AM, Peterson L, Forrest M, Ayala J, Jin H, DeMartino J, Mills SG (2007) The discovery of MK-0812, a potent and selective CCR2 antagonist. In: American Chemical Society Meeting, Chicago. www.acsmedchem.org/mediabstracts2007.pdf
110. Zipp F, Hartung HP, Hillert J, Schimrigk S, Trebst C, Stangel M, Infante-Duarte C, Jakobs P, Wolf C, Sandbrink R, Pohl C, Filippi M (2006) Neurology 67:1880–1883
111. Kalinowska A, Losy J (2008) Expert Opin Investig Drugs 17:1267–1279
112. Sharrack B, Leach T, Jacobson E, Donaldson DD, Xu X, Hu M (2007) Ann Neurol 62(Suppl 11):S74

113. Datamabs (2013) Designing the drugs of tomorrow http://www.datamabs.com/
114. Brodmerkel CM, Huber R, Covington M, Diamond S, Hall L, Collins R, Leffet L, Gallagher K, Feldman P, Collier P, Stow M, Gu X, Baribaud F, Shin N, Thomas B, Burn T, Hollis G, Yeleswaram S, Solomon K, Friedman S, Wang A, Xue CB, Newton RC, Scherle P, Vaddi K (2005) J Immunol 175:5370–5378
115. (2007) Incyte announces first quarter financial results and provides update on drug discovery and development programs http://investor.incyte.com/phoenix.zhtml?c=69764&p=irol-newsArticle_print&ID=963494&highlight
116. Charvat TT, Hu C, Jin J, Li Y, Melikian A, Pennell AMK, Punna S, Ungashe S, Zeng Y (2008) US 20080,039,504
117. Krasinski A, Punna S, Ungashe S, Wang Q, Zeng Y (2011) US 7,884,110
118. Ransohoff RM (2007) N Engl J Med 356:2622–2629
119. Aktas O, Kury P, Kieseier B, Hartung HP (2010) Nat Rev Neurol 6:373–382
120. Axtell RC, de Jong BA, Boniface K, van der Voort LF, Bhat R, De Sarno P, Naves R, Han M, Zhong F, Castellanos JG, Mair R, Christakos A, Kolkowitz I, Katz L, Killestein J, Polman CH, de Waal Malefyt R, Steinman L, Raman C (2010) Nat Med 16:406–412
121. Koch AE, Kunkel SE, Harlow SE, Mazarakis DD, Haines GK, Burdick MD, Pope RM, Strieter RM (1994) J Clin Invest 93:921–928
122. Macchioni P, Boiardi L, Meliconi R, Pulsatelli L, Maldini MC, Ruggeri R, Facchini A, Salvarani C (1998) J Rheumatol 25:320–325
123. Katschke KJ Jr, Rottman JB, Ruth JH, Qin S, Wu L, LaRosa G, Ponath P, Park CC, Pope RM, Koch AE (2001) Arthritis Rheum 44:1022–1032
124. Kasama T, Strieter RM, Lukacs NW, Lincoln PM, Burdick MD, Kunkel SL (1995) J Clin Investig 95:2868–2876
125. Szekanecz Z, Szucs G, Szanto S, Koch AE (2006) Curr Drug Targets 7:91–102
126. Barnes DA, Tse J, Kaufhold M, Owen M, Hesselgesser J, Strieter R, Horuk R, Perez HD (1998) J Clin Invest 101:2910–2919
127. Haas CS, Martinez RJ, Attia N, Haines GK 3rd, Campbell PL, Koch AE (2005) Arthritis Rheum 52:3718–3730
128. Haringman JJ, Smeets TJ, Reinders-Blankert P, Tak PP (2006) Ann Rheum Dis 65:294–300
129. Amat M, Benjamim CF, Williams LM, Prats N, Terricabras E, Beleta J, Kunkel SL, Godessart N (2006) Br J Pharmacol 149:666–675
130. Vallet S, Raje N, Ishitsuka K, Hideshima T, Podar K, Chhetri S, Pozzi S, Breitkreutz I, Kiziltepe T, Yasui H, Ocio EM, Shiraishi N, Jin J, Okawa Y, Ikeda H, Mukherjee S, Vaghela N, Cirstea D, Ladetto M, Boccadoro M, Anderson KC (2007) Blood 110:3744–3752
131. Vergunst CE, Gerlag DM, von Moltke L, Karol M, Wyant T, Chi X, Matzkin E, Leach T, Tak PP (2009) Arthritis Rheum 60:3572–3581
132. Gladue RP, Brown MF, Zwillich SH (2010) Curr Top Med Chem 10:1268–1277
133. Gladue RP, Cole SH, Roach ML, Tylaska LA, Nelson RT, Shepard RM, McNeish JD, Ogborne KT, Neote KS (2006) J Immunol 176:3141–3148
134. Dairaghi DJ, Zhang P, Wang Y, Seitz LC, Johnson DA, Miao S, Ertl LS, Zeng Y, Powers JP, Pennell AM, Bekker P, Schall TJ, Jaen JC (2011) Clin Pharmacol Ther 89:726–734
135. Zhang P, Pennell AMK, Wright JJK, Chen W, Leleti MR, Li Y, Xu Y (2009) US 7,524,845
136. Tak PP, Balanescu A, Tseluyko V, Bojin S, Drescher E, Dairaghi D, Miao S, Marchesin V, Jaen J, Bekker P, Schall TJ (2011) Safety and efficacy of oral chemokine receptor 1 antagonist CCX354-C in a phase 2 rheumatoid arthritis study. In: American College of Rheumatology, Chicago https://acr.confex.com/acr/2011/webprogram/Paper24548.html
137. Gladue RP, Brown MF, Zwillich SH (2010) Curr Top Med Chem 10(13):1268–1277
138. Cavallaro CL, Briceno S, Chen J, Cvijic ME, Davies P, Hynes J, Liu R-Q, Mandlekar S, Rose AV, Tebben AJ, Van Kirk K, Watson A, Wu H, Yang G, Carter PH (2012) J Med Chem 55:9643–9653
139. Gardner DS, Santella JB 3rd, Duncia JV, Carter PH, Dhar TG, Wu H, Guo W, Cavallaro C, Van Kirk K, Yarde M, Briceno SW, Grafstrom RR, Liu R, Patel SR, Tebben AJ, Camac D,

Khan J, Watson A, Yang G, Rose A, Foster WR, Cvijic ME, Davies P, Hynes J Jr (2013) Bioorg Med Chem Lett 23:3833–3840
140. (2011) Proof-of-concept study with BMS-817399 to treat moderate to severe rheumatoid arthritis (RA) http://clinicaltrials.gov/show/NCT01404585
141. Koch AE, Kunkel SL, Harlow LA, Johnson B, Evanoff HL, Haines GK, Burdick MD, Pope RM, Strieter RM (1992) J Clin Invest 90:772–779
142. Ogata H, Takeya M, Yoshimura T, Takagi K, Takahashi K (1997) J Pathol 182:106–114
143. Gong JH, Ratkay LG, Waterfield JD, Clark-Lewis I (1997) J Exp Med 186:131–137
144. Vergunst CE, Gerlag DM, Lopatinskaya L, Klareskog L, Smith MD, van den Bosch F, Dinant HJ, Lee Y, Wyant T, Jacobson EW, Baeten D, Tak PP (2008) Arthritis Rheum 58:1931–1939
145. Haringman JJ, Gerlag DM, Smeets TJ, Baeten D, van den Bosch F, Bresnihan B, Breedveld FC, Dinant HJ, Legay F, Gram H, Loetscher P, Schmouder R, Woodworth T, Tak PP (2006) Arthritis Rheum 54:2387–2392
146. Xue C, Feng H, Cao G, Huang T, Glenn J, Anand R, Meloni D, Zhang K, Kong L, Wang A, Zhang Y, Zheng C, Xia M, Chen L, Tanaka T, Han Q, Robinson DJ, Modi D, Storace L, Shao L, Sharief V, Li M, Galya LG, Covington M, Scherle P, Diamond S, Emm T, Yeleswaram S, Contel N, Vaddi K, Newton R, Hollis G, Friedman S, Metcalf B (2011) Med Chem Lett 2:450–454
147. United State Securities and Exchange Commission, Washington, D.C. 20549 Form 8-k Date of report: November 1, 2005 Filed by Incyte corporation
148. Carter PH, Cherney RJ, Mangion IK (2007) Ann Rep Med Chem 42:211–227
149. Braddock M (2007) Expert Opin Investig Drugs 16:909–917
150. Gerlag DM, Hollis S, Layton M, Vencovsky J, Szekanecz Z, Braddock M, Tak PP (2010) Arthritis Rheum 62:3154–3160
151. Fleishaker DL, Garcia Meijide JA, Petrov A, Kohen MD, Wang X, Menon S, Stock TC, Mebus CA, Goodrich JM, Mayer HB, Zeiher BG (2012) Arthritis Res Ther 14:R11
152. van Kuijk AW, Vergunst CE, Gerlag DM, Bresnihan B, Gomez-Reino JJ, Rouzier R, Verschueren PC, van de Leij C, Maas M, Kraan MC, Tak PP (2010) Ann Rheum Dis 69:2013–2016
153. Plater-Zyberk C, Hoogewerf AJ, Proudfoot AE, Power CA, Wells TN (1997) Immunol Lett 57:117–120
154. Vierboom MP, Zavodny PJ, Chou CC, Tagat JR, Pugliese-Sivo C, Strizki J, Steensma RW, McCombie SW, Celebi-Paul L, Remarque E, Jonker M, Narula SK, Hart B (2005) Arthritis Rheum 52:627–636
155. Yang YF, Mukai T, Gao P, Yamaguchi N, Ono S, Iwaki H, Obika S, Imanishi T, Tsujimura T, Hamaoka T, Fujiwara H (2002) Eur J Immunol 32:2124–2132
156. Pokorny V, McQueen F, Yeoman S, Merriman M, Merriman T, Harrison A, Highton J, McLean L (2004) Ann Rheum Dis 64:4897–4490
157. Schall TJ, Proudfoot AE (2011) Nat Rev Immunol 11:355–363
158. Murphy PM (2002) Pharmacol Rev 54:227–229
159. Pease JE, Murphy PM (1998) Sem Immunol 10:169–178
160. Cocchi F, DeVico AL, Garzino-Demo A, Arya SK, Gallo RC, Lusso P (1996) Science 270:1811–1815
161. Combadiere C, Ahuja SK, Tiffany HL, Murphy PM (1996) J Leukoc Biol 60:147–152
162. Raport CJ, Gosling J, Schweickart VL, Gray PW, Charo IF (1996) J Biol Chem 271:17161–17166
163. Samson M, Labbe O, Mollereau C, Vassart G, Parmentier M (1996) Biochemistry 35:3362–3367
164. Feng Y, Broder CC, Kennedy PE, Berger EA (1996) Science 272:872–877
165. Bleul CC, Farzan M, Choe H, Parolin C, Clark-Lewis I, Sodroski J, Springer TA (1996) Nature 382:829–833

166. Oberlin E, Amara A, Bachelerie F, Bessia C, Virelizier JL, Arenzana-Seisdedos F, Schwartz O, Heard JM, Clark-Lewis I, Legler DF, Loetscher M, Baggiolini M, Moser B (1996) Nature 382:833–835
167. Alkhatib G, Combadiere C, Broder CC, Feng Y, Kennedy PE, Murphy PM, Berger EA (1996) Science 272:1955–1958
168. Choe H, Farzan M, Sun Y, Sullivan N, Rollins B, Ponath PD, Wu L, Mackay CR, LaRosa G, Newman W, Gerard N, Gerard C, Sodroski J (1996) Cell 85:1135–1148
169. Deng H, Liu R, Ellmeier W, Choe S, Unutmaz D, Burkhart M, Di Marzio P, Marmon S, Sutton RE, Hill CM, Davis CB, Peiper SC, Schall TJ, Littman DR, Landau NR (1996) Nature 381:661–666
170. Dragic T, Litwin V, Allaway GP, Martin SR, Huang Y, Nagashima KA, Cayanan C, Maddon PJ, Koup RA, Moore JP, Paxton WA (1996) Nature 381:667–673
171. Alkhatib G, Locati M, Kennedy PE, Murphy PM, Berger EA (1997) Virology 234:340–348
172. Liu R, Paxton WA, Choe S, Ceradini D, Martin SR, Horuk R, MacDonald ME, Stuhlmann H, Koup RA, Landau NR (1996) Cell 86:367–377
173. Samson M, Libert F, Doranz BJ, Rucker J, Liesnard C, Farber CM, Saragosti S, Lapoumeroulie C, Cognaux J, Forceille C, Muyldermans G, Verhofstede C, Burtonboy G, Georges M, Imai T, Rana S, Yi Y, Smyth RJ, Collman RG, Doms RW, Vassart G, Parmentier M (1996) Nature 382:722–725
174. Huang Y, Paxton WA, Wolinsky SM, Neumann AU, Zhang L, He T, Kang S, Ceradini D, Jin Z, Yazdanbakhsh K, Kunstman K, Erickson D, Dragon E, Landau NR, Phair J, Ho DD, Koup RA (1996) Nat Med 2:1240–1243
175. Zimmerman PA, Buckler-White A, Alkhatib G, Spalding T, Kubofcik J, Combadiere C, Weissman D, Cohen O, Rubbert A, Lam G, Vaccarezza M, Kennedy PE, Kumaraswami V, Giorgi JV, Detels R, Hunter J, Chopek M, Berger EA, Fauci AS, Nutman TB, Murphy PM (1997) Mol Med 3:23–36
176. He J, Chen Y, Farzan M, Choe H, Ohagen A, Gartner S, Busciglio J, Yang X, Hofmann W, Newman W, Mackay CR, Sodroski J, Gabuzda D (1997) Nature 385:645–649
177. Liao F, Alkhatib G, Peden KW, Sharma G, Berger EA, Farber JM (1997) J Exp Med 185:2015–2023
178. Pleskoff O, Treboute C, Brelot A, Heveker N, Seman M, Alizon M (1997) Science 276:1874–1878
179. Biti R, Ffrench R, Young J, Bennetts B, Stewart G, Liang T (1997) Nat Med 3:252–253
180. O'Brien TR, Winkler C, Dean M, Nelson JA, Carrington M, Michael NL, White 2nd GC (1997) Lancet 349:1219
181. Theodorou I, Meyer L, Magierowska M, Katlama C, Rouzioux C (1997) Lancet 349:1219–1220
182. Simmons G, Clapham PR, Picard L, Offord RE, Rosenkilde MM, Schwartz TW, Buser R, Wells TNC, Proudfoot AEI (1997) Science 276:276–279
183. Townson JR, Graham GJ, Landau NR, Rasala B, Nibbs RJ (2000) J Biol Chem 275:39254–39261
184. Mack M, Luckow B, Nelson PJ, Cihak J, Simmons G, Clapham PR, Signoret N, Marsh M, Stangassinger M, Borlat F, Wells TN, Schlondorff D, Proudfoot AE (1998) J Exp Med 187:1215–1224
185. Amara A, Gall SL, Schwartz O, Salamero J, Montes M, Loetscher P, Baggiolini M, Virelizier JL, Arenzana-Seisdedos F (1997) J Exp Med 186:139–146
186. Baba M, Nishimura O, Kanzaki N, Okamoto M, Sawada H, Iizawa Y, Shiraishi M, Aramaki Y, Okonogi K, Ogawa Y, Meguro K, Fujino M (1999) Proc Natl Acad Sci U S A 96:5698–5703
187. Dragic T, Trkola A, Thompson DA, Cormier EG, Kajumo FA, Maxwell E, Lin SW, Ying W, Smith SO, Sakmar TP, Moore JP (2000) Proc Natl Acad Sci U S A 97:5639–5644
188. Imamura S, Ichikawa T, Nishikawa Y, Kanzaki N, Takashima K, Niwa S, Iizawa Y, Baba M, Sugihara Y (2006) J Med Chem 49:2784–2793

189. Seto M, Aikawa K, Miyamoto N, Aramaki Y, Kanzaki N, Takashima K, Kuze Y, Iizawa Y, Baba M, Shiraishi M (2006) J Med Chem 49:2037–2048
190. Klibanov OM, Williams SH, Iler CA (2010) Curr Opin Investig Drugs 11:940–950
191. Thoma G, Nuninger F, Schaefer M, Akyel KG, Albert R, Beerli C, Bruns C, Francotte E, Luyten M, MacKenzie D, Oberer L, Streiff MB, Wagner T, Walter H, Weckbecker G, Zerwes HG (2004) J Med Chem 47:1939–1955
192. Tagat JR, McCombie SW, Steensma RW, Lin S, Nazareno DV, Baroudy B, Vantuno N, Xu S, Liu J (2001) Bioorg Med Chem Lett 11:2143–2146
193. Tagat JR, Steensma RW, McCombie SW, Nazareno DV, Lin SI, Neustadt BR, Cox K, Xu S, Wojcik L, Murray MG, Vantuno N, Baroudy BM, Strizki JM (2001) J Med Chem 44:3343–3346
194. Este JA (2002) Curr Opin Investig Drugs 3:379–383
195. Tagat JR, McCombie SW, Nazareno D, Labroli MA, Xiao Y, Steensma RW, Strizki JM, Baroudy BM, Cox K, Lachowicz J, Varty G, Watkins R (2004) J Med Chem 47:2405–2408
196. Gulick RM, Su Z, Flexner C, Hughes MD, Skolnik PR, Wilkin TJ, Gross R, Krambrink A, Coakley E, Greaves WL, Zolopa A, Reichman R, Godfrey C, Hirsch M, Kuritzkes DR (2007) J Infect Dis 196:304–312
197. Watson C, Jenkinson S, Kazmierski W, Kenakin T (2005) Mol Pharmacol 67:1268–1282
198. Armour D, de Groot MJ, Edwards M, Perros M, Price DA, Stammen BL, Wood A (2006) ChemMedChem 1:706–709
199. Ryan CT (2005) AIDS Clin Care 17:107–108
200. Lieberman-Blum SS, Fung HB, Bandres JC (2008) Clin Therapeut 30:1228–1250
201. Cipriani S, Francisci D, Mencarelli A, Renga B, Schiaroli E, D'Amore C, Baldelli F, Fiorucci S (2013) Circulation 127:2114–2124
202. Tan Q, Zhu Y, Li J, Chen Z, Han GW, Kufareva I, Li T, Ma L, Fenalti G, Zhang W, Xie X, Yang H, Jiang H, Cherezov V, Liu H, Stevens RC, Zhao Q, Wu B (2013) Science 341:1387–1390
203. Garcia-Perez J, Rueda P, Alcami J, Rognan D, Arenzana-Seisdedos F, Lagane B, Kellenberger E (2011) J Biol Chem 286:33409–33421
204. Donzella GA, Schols D, Lin SW, Este JA, Nagashima KA, Maddon PJ, Allaway GP, Sakmar TP, Henson G, De Clercq E, Moore JP (1998) Nat Med 4:72–77
205. De Clercq E, Yamamoto N, Pauwels R, Baba M, Schols D, Nakashima H, Balzarini J, Debyser Z, Murrer BA, Schwartz D et al (1992) Proc Natl Acad Sci U S A 89:5286–5290
206. Rosenkilde MM, Gerlach LO, Jakobsen JS, Skerlj RT, Bridger GJ, Schwartz TW (2004) J Biol Chem 279:3033–3041
207. Hatse S, Princen K, Gerlach LO, Bridger G, Henson G, De Clercq E, Schwartz TW, Schols D (2001) Mol Pharmacol 60:164–173
208. Zhan W, Liang Z, Zhu A, Kurtkaya S, Shim H, Snyder JP, Liotta DC (2007) J Med Chem 50:5655–5664
209. Hendrix CW, Collier AC, Lederman MM, Schols D, Pollard RB, Brown S, Jackson JB, Coombs RW, Glesby MJ, Flexner CW, Bridger GJ, Badel K, MacFarland RT, Henson GW, Calandra G (2004) J Acquir Immune Defic Syndr 37:1253–1262
210. Hernandez PA, Gorlin RJ, Lukens JN, Taniuchi S, Bohinjec J, Francois F, Klotman ME, Diaz GA (2003) Nat Genet 34:70–74
211. McDermott DH, Lopez J, Deng F, Liu Q, Ojode T, Chen H et al (2011) J Cell Mol Med 15 (10):2071–2081
212. Dale DC, Bolyard AA, Kelley ML, Westrup EC, Makaryan V, Aprikyan A, Wood B, Hsu FJ (2011) Blood 118:4963–4966
213. McDermott DH, Liu Q, Ulrick J, Kwatemaa N, Anaya-O'Brien S, Penzak SR et al (2011) Blood 118(18):4957–4962
214. Murakami T, Nakajima T, Koyanagi Y, Tachibana K, Fujii N, Tamamura H, Yoshida N, Waki M, Matsumoto A, Yoshie O, Kishimoto T, Yamamoto N, Nagasawa T (1997) J Exp Med 186:1389–1393

215. Nakashima H, Masuda M, Murakami T, Koyanagi Y, Matsumoto A, Fujii N, Yamamoto N (1992) Antimicrob Agents Chemother 36:1249–1255
216. Wu B, Chien EY, Mol CD, Fenalti G, Liu W, Katritch V, Abagyan R, Brooun A, Wells P, Bi FC, Hamel DJ, Kuhn P, Handel TM, Cherezov V, Stevens RC (2010) Science 330:1066–1071
217. Lebre MC, Vergunst CE, Choi IY, Aarrass S, Oliveira AS, Wyant T, Horuk R, Reedquist KA, Tak PP (2011) PLoS One 6:e21772
218. Takeda A, Baffi JZ, Kleinman ME, Cho WG, Nozaki M, Yamada K, Kaneko H, Albuquerque RJ, Dridi S, Saito K, Raisler BJ, Budd SJ, Geisen P, Munitz A, Ambati BK, Green MG, Ishibashi T, Wright JD, Humbles AA, Gerard CJ, Ogura Y, Pan Y, Smith JR, Grisanti S, Hartnett ME, Rothenberg ME, Ambati J (2009) Nature 460:225–230
219. Wang H, Wittchen ES, Jiang Y, Ambati B, Grossniklaus HE, Hartnett ME (2011) Invest Ophthalmol Vis Sci 52:8271–8277
220. Ahmad I, Balasubramanian S, Del Debbio CB, Parameswaran S, Katz AR, Toris C, Fariss RN (2011) Invest Ophthalmol Vis Sci 52:2868–2878
221. Mizutani T, Ashikari M, Tokoro M, Nozaki M, Ogura Y (2013) Invest Ophthalmol Vis Sci 54:1564–1572
222. Li Y, Huang D, Xia X, Wang Z, Luo L, Wen R (2011) PLoS One 6:e17106
223. Villeda SA, Luo J, Mosher KI, Zou B, Britschgi M, Bieri G, Stan TM, Fainberg N, Ding Z, Eggel A, Lucin KM, Czirr E, Park JS, Couillard-Despres S, Aigner L, Li G, Peskind ER, Kaye JA, Quinn JF, Galasko DR, Xie XS, Rando TA, Wyss-Coray T (2011) Nature 477:90–94
224. Fernandez-Egea E, Scoriels L, Theegala S, Giro M, Ozanne SE, Burling K, Jones PB (2013) Prog Neuropsychopharmacol Biol Psychiatry 46:25–28

Top Med Chem (2015) 14: 41–86
DOI: 10.1007/7355_2014_77

Published online: 26 October 2014

# Role of 3D Structures in Understanding, Predicting, and Designing Molecular Interactions in the Chemokine Receptor Family

**Irina Kufareva, Ruben Abagyan, and Tracy M. Handel**

**Abstract** The recently solved crystallographic structures of two chemokine receptors, CXCR4 and CCR5, provided valuable insights into the molecular mechanisms of chemokine receptor function and interaction with various ligands. However, they did not answer all of the questions. It remains an important role of the computational community to complement and expand the structural insights into areas where experimental structure determination efforts have not yet succeeded, such as studying receptor functional states or their complexes with small molecule and protein ligands of different classes. In this chapter, we provide an overview of pre- and post-structure efforts in understanding, predicting, and designing chemokine receptor interactions with small molecules and peptides, chemokines, and HIV gp120 proteins, as well as structure-guided insights regarding chemokine receptor dimerization and the impact of structures on rational molecular design initiatives. As an inherently challenging family of GPCRs, chemokine receptors may only reveal their secrets when tackled by the efficient symbiosis of computational approaches with experimental structure determination.

**Keywords** Chemokine binding, Chemokine receptor, Docking, G-protein-coupled receptor, HIV entry, Homology modeling, Molecular design, Virtual screening

## Contents



---

I. Kufareva (✉), R. Abagyan, and T.M. Handel
UCSD Skaggs School of Pharmacy and Pharmaceutical Sciences, La Jolla, CA, USA
e-mail: ikufareva@ucsd.edu

## Abbreviations

| | |
|---|---|
| 3D | Three-dimensional |
| AUC | Area under curve |
| BiFC | Biomolecular fluorescence complementation |
| BRET | Bioluminescence resonance energy transfer |
| CCR5 | CC chemokine receptor 5 |
| CXCR4 | CXC chemokine receptor 4 |
| ECL | Extracellular loop |
| FRET | Fluorescence resonance energy transfer |
| GPCR | G-protein-coupled receptor |
| HIV | Human immunodeficiency virus |
| NMR | Nuclear magnetic resonance |
| PDB | Protein data bank |
| ROC | Receiver operating characteristic |
| SDF-1 | Stromal cell derived factor 1 |
| TM | Transmembrane |
| VLS | Virtual ligand screening |
| WT | Wild type |

## 1 Introduction

The need for high-resolution structures of proteins, especially as therapeutically important as G-protein-coupled receptors (GPCRs), is appreciated by most scientists and rarely questioned. It has been proved by numerous examples that an accurate structure of a protein is invaluable for understanding the principles of its interaction with various molecular partners and may ultimately lead to discovery of new ligands, often with better affinity, selectivity, or safety profiles [1, 2]. Such advances have been made possible by joint efforts of biologists, biochemists, biophysicists, structural biologists, and computational chemists, and as of 2014, these advances continue to be on a steep rise, with additional structures of GPCRs solved every year and more structure-based drug discovery initiatives undertaken [3]. "Give us more structures like this," – computational chemists plead – "and we will overturn drug discovery." "Give us appropriate funding," – the structural biology community echoes – "and we will saturate the field with the best structures of the most important and challenging proteins." But is it really so straightforward? Do emerging structures really answer all of the questions? Is it always true that structures are more valuable than 3D models built by homology? Finally, how do the structure determination and molecular modeling efforts change the field? In this chapter, we attempt to answer these questions in the context of the chemokine receptors.

Chemokine receptors are a subfamily of GPCRs that play a major role in controlling cell migration in the context of immune system function and development [4] and in the pathology of many diseases including asthma, rheumatoid arthritis, multiple sclerosis, heart disease, and cancer [5–7]. Several chemokine receptors (most famously CCR5 and CXCR4) are also known for their ability to serve as portals for HIV entry into host immune cells [8–10]. Similarly, the most common malarial parasite, *Plasmodium vivax*, uses the chemokine receptor DARC to enter red blood cells [11].

Chemokine receptors are challenging targets for structural studies as well as computational modeling. First, as all GPCRs, they are integral membrane proteins that are highly unstable and prone to aggregation in detergent solutions – properties that make crystallization extremely challenging. Second, they have evolved to optimally interact with chemokines, which are approximately 70–120 amino acid secreted proteins. As many other protein-protein interaction interfaces, the receptor:chemokine interface is large, polar, and difficult to target with small molecules (i.e., poorly *druggable*). Many of the available small-molecule modulators, while strong binders, are larger than an average orally available drug, very flexible, and either highly lipophilic or excessively polar, being unfavorable both in experimental settings and in modeling applications. Despite these difficulties, and thanks to the breakthrough advances in GPCR crystallography [12, 13], the first structure of a chemokine receptor, that of CXCR4, was solved in 2010 [14] and the structure of CCR5 was solved more recently in 2013 [15]. These advances boosted structure-based insights into chemokine receptor interactions with their cognate, pathological, and synthetic ligands as well as mechanisms of receptor activation and dimerization, as reviewed below.

## 2 Chemokine Receptor Structures and Complexes to Date

The first structure of a chemokine receptor was solved by Wu and coworkers [14] in 2010 as a part of the GPCR Network initiative [16]. Their target receptor, CXCR4, was crystallized in several symmetry groups and in complexes with two distinct antagonists: a small molecule of isothiourea class, IT1t [17], and a cyclic 16-residue peptide, CVX15 [18]. These structures had several unusual features that distinguished them from other GPCR structures solved by that time. First, a β-hairpin conformation of the extracellular loop 2 (ECL2) did not resemble any other non-rhodopsin ECL2 structures observed thus far and was different from the similarly shaped rhodopsin ECL2 by much higher placement with respect to the transmembrane (TM) bundle, thus allowing access to a wide and open polar binding pocket. Since that time, β-hairpins were demonstrated to be a preferred ECL2 conformation for a number of peptide-binding GPCRs such as opioid receptors [19–23], neurotensin receptor 1 [24], and protease-activated receptor 1 [25]. Second, all CXCR4 structures have a 13-residue long stretch of residues in their TM4, spanning residues Leu150$^{4.39}$ through I162$^{4.51}$ that adopts a loose π-helical conformation instead of a canonical α-helix, resulting in a helical bend that brings the intracellular end of TM4 closer to that of TM3. Quite remarkably, the TM4 residue that is most conserved in class A GPCRs, Trp161$^{4.50}$, is a part of this π-helix, with its backbone amide forming a hydrogen bond with the carbonyl oxygen of Val156$^{4.45}$ instead of the more traditional residue 4.46 observed in other class A receptors. This is potentially due to a unique Pro residue at Ballesteros position 4.52 (Pro163$^{4.52}$) that breaks TM4 helicity in CXCR4, the functional consequences of which are yet to be understood. Third, in all CXCR4 structures, residues following TM7 are either disordered or unfolded, whereas in other GPCRs, they form the so-called helix 8 running parallel to the membrane plane. Several unusual residues in CXCR4 may provide a potential explanation for this observation, including Arg77$^{2.43}$, Ala313$^{8.54}$, and Ala316$^{8.57}$, all three of which are found as medium to large hydrophobic amino acids in 94%, 87%, and 76% of class A GPCRs, respectively. Finally, in all five CXCR4 structures by Wu et al., the receptors formed consistent parallel dimers with twofold rotational symmetry stabilized by an interface spanning TM helices 5 and 6. This dimer geometry was somewhat similar to the previously observed crystallographic dimers of squid rhodopsin [26] and was (and still is) among the few X-ray dimer geometries that pass minimal filters of potentially being biologically relevant.

More recently, Tan et al. [15] presented the first X-ray structure of the chemokine receptor CCR5. The receptor was crystallized in the presence of the small-molecule antagonist, maraviroc, which is an approved anti-HIV therapeutic [27, 28]. While highly similar to CXCR4 in its CVX15 peptide bound state, including the conformation of ECL2, the CCR5 structure has the canonical conformation of TM4 and helix 8. Although parallel, its crystallographic dimer has only translational (i.e., not twofold rotational) symmetry and is therefore an artifact of crystallization rather than a biologically significant phenomenon. This is despite the fact that CCR5 has been shown to form dimers in live cells under physiological conditions [29–31].

Both chemokine receptors crystallized thus far represent inactive state structures, consistent with the antagonist nature of their co-crystallized ligands. The inactive state of CCR5 was additionally stabilized by a rationally designed point mutation that improved its stability and homogeneity in detergent, but also severely compromised its functional response to chemokine stimulation, as determined from intracellular $Ca^{2+}$ mobilization experiments [15].

# 3 Structural Basis of Receptor Interactions with Small Molecules

## *3.1 CXCR4 Interactions with Small Molecules and Peptides*

From the CXCR4:ligand complex structures, it appears that the binding site is a large, solvent-exposed cavity lined by multiple polar residues. Its volume (~2,000 $Å^3$) far exceeds the size of typical druggable pockets in other proteins (300–1,000 $Å^3$). The pocket involves residues from all seven TM helices and resembles a wide cognac glass (the analogy by R.C. Stevens, personal communication) whose outer edges are formed by the polar stretch of the N-terminus on one side and the no less polar ECL2 on the other side (Fig. 1a,b). The co-crystallized CVX15 peptide (16 residues, 150 non-hydrogen atoms, total molecular volume of 2,200 $Å^3$) occupies a major part of this cavity and makes extensive van der Waals and polar contacts with the surrounding receptor residues. On the contrary, the co-crystallized small molecule, IT1t (27 non-hydrogen atoms, molecular volume of 430 $Å^3$), occupies only a small fraction of the available volume and loses as little as 450 $Å^2$ of its solvent-accessible area in complex as compared to the unbound state. A large part of its binding energy is contributed by electrostatic interactions with the acidic side chains of $Glu288^{7.39}$ and $Asp97^{2.63}$. However, the anatomy of the pocket is such that in the absence of ligand, these side chains are favorably exposed to solvent, thus diminishing the net contribution of these interactions to the calculated binding energies. As such, the CXCR4:IT1t complex represents an extremely difficult case for docking and screening programs that are trained to recognize shape complementarity and favorable buried polar interactions. It is thus not surprising that even when docking IT1t to its cognate structure, most docking programs analyzed by Planesas et al. [35] had difficulties discriminating the near-native pose from incorrect conformations. On the other hand, the CXCR4:CVX15 complex is not an easy target either, because of the large size and flexibility of the peptide. Peptide pose prediction frequently fails [35] not only because of pose scoring and ranking errors but also because of insufficient conformational sampling.

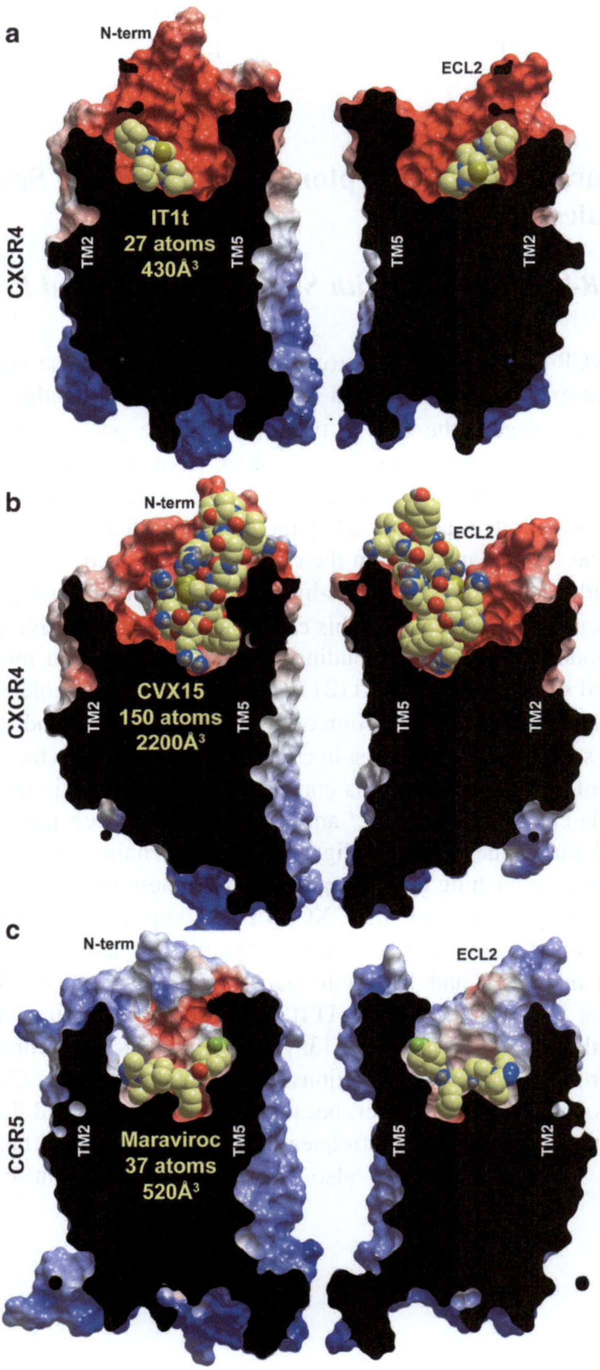


**Fig. 1** Cut-and-open view of CXCR4 in complex with IT1t (**a**), CVX15 peptide (**b**), and CCR5 in complex with maraviroc (**c**). Molecular surfaces are colored by the electrostatic potential (*red* – negative, *blue* – positive) calculated in the ICM molecular modeling program [32, 33]. The number of non-hydrogen atoms and the approximate molecular volume are given for each ligand.

### 3.2 CCR5 Interaction with Maraviroc and Structural Basis for Allostery

The overall architecture of CCR5 in complex with maraviroc is highly similar to CXCR4, especially in its CVX15-bound form. As for CXCR4, CCR5 has a widely open polar binding pocket formed by the TM helices and bounded by the β-hairpin-shaped ECL2 and part of the N-terminus between the conserved cysteine and TM1 (Fig. 1c). Yet on closer inspection, a number of important conformational differences between CCR5 and CXCR4 are observed. At this time, the maraviroc-bound structure is the only structure of CCR5; therefore, one can only hypothesize which differences are due to the inherent properties of CCR5 as opposed to conformational peculiarities induced by maraviroc binding; however, these differences reconcile well the experimentally observed complex allosteric behavior of maraviroc [36].

First, TM4 in CCR5 is shifted by about 1.5 Å towards the center of the helical bundle as compared to its respective conformation in CXCR4. This may be an inherent property of CCR5 and possibly other CC chemokine receptors, which do not have the unique proline-containing π-helix observed in CXCR4. It may also be an artifact of the rationally designed stabilizing mutation Gly163Asn in TM4 of CCR5. Second, the tip of the ECL2 β-hairpin in CCR5 is pointing up and out (towards the extracellular space) as compared to down and in (towards the binding pocket) in CXCR4. This region is highly flexible, as evidenced by comparison of the multiple CXCR4 structures, and the conformation observed in CCR5:maraviroc complex is probably one of multiple possibilities, additionally stabilized by the specific amino acids in ECL2. Third, CCR5 TM1 is moved outwards by 2.2 Å, allowing space for the favorable placement of the maraviroc substituted triazole group and its interaction with $Tyr37^{1.39}$. However, the movement of the extracellular end of TM1 is restricted by its connection through the N-terminus and a disulfide bond to ECL3; therefore, the observed outward movement of TM1 in CCR5:maraviroc complex is coupled to a strong kink in the first helical turn of TM1, a shift of ECL3 and the extracellular ends of TM6 and 7, as if they were pulled towards TM1, and a kink in the last (extracellular) helical turn of TM6 (Fig. 2a). These coupled conformational changes translate the binding of maraviroc, which occurs deep in the TM bundle, into rather dramatic conformational rearrangements in the extracellular parts of the receptor, probably making them incompatible with the binding of selected HIV gp120 variants and chemokines. Additionally, the movement of $Tyr108^{3.32}$ and $Glu283^{7.39}$, as well as a rotamer flip in $Trp86^{2.60}$, result in opening of two subpockets at the bottom of the CCR5 TM pocket that are not present in CXCR4 and consequent binding of

**Fig. 1** (continued) Molecular volume was calculated as the volume of the Connolly surface of the ligand [34], i.e., a smooth envelope touching the van der Waals surface of atoms as the solvent probe of radius 1.4 Å size rolls over the molecule

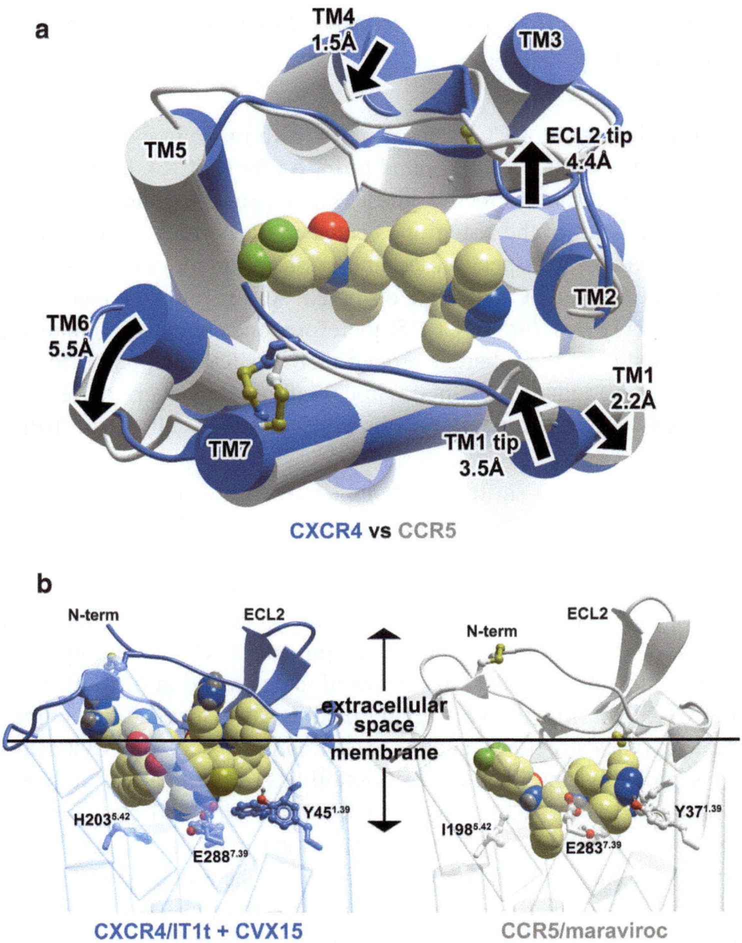


**Fig. 2** Crystallographic conformations of CXCR4 and CCR5 and possible structural basis for allosteric inhibition of CCR5 by Maraviroc. (**a**) View from the extracellular side across the membrane plane. TM movements observed in Maraviroc-bound CCR5 but not in CXCR4 are shown by *arrows*. (**b**) Lateral view along the plane of the membrane, with Maraviroc binding deeper in the TM pocket without engaging the extracellular domains

maraviroc deep in the pocket without engaging the extracellular loops (Fig. 2b). The latter observation may explain the noncompetitive nature of inhibition of chemokine signaling by maraviroc and related molecules [36]. The lack of interaction of maraviroc with the ECLs is also corroborated by biochemical studies which show that substituting the extracellular loops in CCR5 with those of the homologous receptor, CCR2, produces a chimera that is perfectly capable of recognizing maraviroc but not the CCR5-specific chemokines [37].

Along with biochemical evidence, these structural observations hint at the remarkable plasticity of CCR5 and provide insights into its possible conformational heterogeneity. While enabling CCR5 to bind a variety of molecular partners and to mediate multiple signaling pathways, such heterogeneity also provides a route for HIV to escape inhibition by CCR5-targeting therapeutics, which in turn may dramatically impact the future of CCR5 as HIV therapeutic target [38].

### 3.3 *Before the CXCR4 Structure: Computational Elucidation of the Structural Basis of CXCR4 Antagonist Binding*

CXCR4 presents an attractive drug target not only for HIV entry inhibitors but also in tumor metastasis and autoimmune and inflammatory diseases [39]. Numerous chemotypes of small-molecule CXCR4 antagonist have been developed by structure-guided as well as structure-independent approaches [18, 40, 41]; however, only one of them, AMD3100 (also known under its generic name of *plerixafor* or the trade name of *mozobil*), has reached the clinic so far and only for the non-chronic indication of stem cell mobilization in hematopoietic stem cell transplant patients [42, 43]. Many other CXCR4-targeting candidate compounds failed in clinical trials owing to their toxicity or poor pharmacokinetic profiles.

A common feature of all CXCR4 antagonists known to date is the excessive number of positive charges and rotatable bonds. The best binding affinities were achieved with highly basic peptides obtained by SAR optimization and size reduction of polyphemusin, an antimicrobial peptide from horseshoe crab, for example, T140 (H-Arg-Arg-Nal-Cys-Tyr-Arg-Lys-dLys-Pro-Tyr-Arg-Cit-Cys-Arg-OH) [18], CVX15 (H-Arg-Arg-Nal-Cys-Tyr-Gln-Lys-dPro-Pro-Tyr-Arg-Cit-Cys-Arg-Gly-dPro-OH), and FC131 (cyclo-Arg-Arg-Nal-Gly-Tyr) [44].

Numerous studies used homology modeling and molecular docking to elucidate the structural basis of the interaction of diverse antagonists with CXCR4. Extensive site-directed mutagenesis provided insights into the roles that the individual CXCR4 pocket residues played in the interaction with mono- and bicyclams [45, 46], non-cyclam antagonists [47, 48], or peptides [47]. In 2006, using the mutagenesis data and conformational optimization of a series of FC131 analogs, Våbenø et al. came up with a 3D pharmacophore that efficiently explained the existing SAR [49] and generated predictions for potential binding poses of several cyclopentapeptides by docking them into a bovine rhodopsin (bRho)-based CXCR4 model [50]. In 2011, Kawatkar et al. also used a bRho-based homology model to analyze the structural mechanisms of CXCR4 inhibition by small-molecule antagonists [51] and came up with antagonist pose hypotheses that were also consistent with site-directed mutagenesis.

Unfortunately, it is rarely possible to unambiguously interpret the results of a mutagenesis study as the observed effects of a mutation may be due to the mutation-induced allosteric effects or destabilization of receptor structure rather than alterations of direct interactions with the ligand. While one can often distinguish these causes in the context of soluble proteins by appropriate biophysical characterization, it is more difficult with membrane proteins. Additionally, with bRho being the closest available homology modeling template until 2007 (and with bRho, $\beta_2$AR, and $AA_{2A}R$ being the only available and rather distant templates until 2010), the accuracy of modeling the TM bundle and especially the extracellular loops was rather low. For example, the binding location or pose of T140 predicted by MD simulations in conjunction with alanine scanning mutagenesis [47] have very little in common with the crystallographically observed pose of the structurally related CVX15. A marginally better accuracy was achieved by Boulais et al. who studied T140-CXCR4 interaction using molecular modeling and photoaffinity labeling [52]; while correctly identifying TM4 as a part of the T140 interaction site, they docked the peptide upside down. Similarly, the poses predicted by docking IT1t in $\beta_2$AR-based models of CXCR4 [53] are rather distant from the crystallographic pose of IT1t in the crystal structure [14].

In order to objectively evaluate the extent to which homology modeling and compound docking may replace or complement the experimental structure determination, the GPCR Network [16] established a series of community-wide modeling and docking assessments called GPCR Dock. In 2010, prior to the release of the experimental coordinates of CXCR4 complexes, 35 groups used their best practices to generate receptor-ligand complex structure predictions that were later compared to the experimental structures [54]. This evaluation was conducted side by side with the dopamine D3 complex structure prediction assessment, which allowed the comparison of the prediction challenges between the aminergic and chemokine receptor families. The assessment results indicated that using best practices, relatively accurate predictions could be made for the dopamine D3 receptor complex with an orthosteric antagonist; however, the CXCR4 predictions were below the acceptable accuracy level. The two best predictions for the CXCR4:IT1t complex featured (1) a 4.88 Å RMSD for the ligand pose with 36% of correctly predicted ligand-receptor contacts and (2) a 2.47 Å RMSD with only 18% of correctly predicted contacts. A ligand RMSD of 8.88 Å and contact recall of 6% was the best result achieved for the CVX15 peptide complex [54]. These results reflect the challenges associated with modeling by distant homology (the closest modeling template available at the time of the assessment was only 25% identical to CXCR4 in the TM domain) and the inherent poor druggability of the chemokine receptors.

### *3.4 Drug Discovery Efforts by Virtual Ligand Screening Against Models and Structures of CXCR4*

Structure-based drug discovery efforts against CXCR4 had been undertaken long before the X-ray structure was solved in 2010 [55–57]. Virtual screening was performed against homology models of CXCR4 that, with rare exceptions, were built using the bRho structure as a template. Comparison of these models to the experimental CXCR4 structures [54] showed that these models were largely

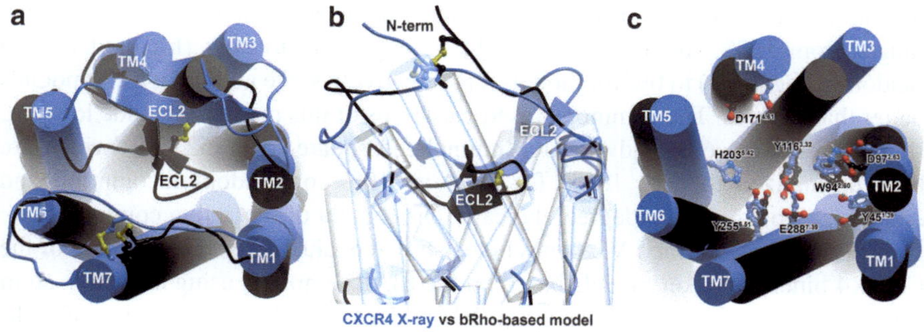


**Fig. 3** Bovine rhodopsin (bRho) as a template for chemokine receptor modeling. (**a**) The major distinctions between bRho and CXCR4 are observed in the lateral shift of TM3 and in position of the ECL2. (**b**) Lateral view along the plane of the membrane. (**c**) After ECL2 is removed, a bRho-based model provides a reasonable approximation for the main TM pocket residues involved in ligand binding. The model was built in the ICM molecular modeling package [32, 33] using the X-ray structure of bRho in trigonal crystal form [58] as a template. CXCR4 sequence was threaded through the template coordinates in all regions except N-terminus and ECL3; the latter two were modeled ab initio by extending TM6 and TM7 helices as far towards the extracellular space as possible and by introducing the disulfide bond between the N-terminus and ECL3. Following exhaustive sampling of residue side chain conformation, the model was subjected to full-atom gradient minimization in internal coordinates with the objective function including steric, electrostatic, and hydrogen binding terms as well as intra- and inter-helical distance restraints for preservation of the model topology

structurally inaccurate – partly because the arrangement of TM helices in CXCR4 is substantially different from that in bRho and partly because bRho ECL2 is placed so much deeper in the binding pocket (Fig. 3a, b). However, with an exception of the ECL2-related issues, these models position the important ligand-interacting residues in reasonable proximity to their crystallographically observed positions (Fig. 3c) and, after proper refinement, can be instrumental in CXCR4 antagonist virtual screening campaigns. Part of their success can be summarized by the statement that "a structure or a model is only good to the extent that it agrees with experimental data" (C. de Graaf, personal communication). Specifically, the ability of a model to recognize known high-affinity modulators among inactives or decoys of similar size and atomic composition in virtual screening, also called *VLS* (virtual ligand screening) *enrichment*, has been recognized as a robust criterion of the quality and accuracy of atomic-level predictions of ligand-receptor interactions [59–65]. It represents a way of evaluating model compliance with existing experimental data in the form of small-molecule chemical activity against the modeled protein.

Model selection by VLS enrichment has led to the most accurate predictions in all three GPCR Dock assessments [54, 66, 67]. For quantitative evaluation of VLS enrichment, the docked active and inactive/decoy compounds are ranked by their predicted binding scores, and a receiver operating characteristic (ROC) curve is built by plotting the fraction of true positives vs. false positives in the top scoring part of the ranked list for each value of the score cutoff. The VLS performance of the model is reported in terms of the area under the ROC curve (ROC AUC), which ranges from 100% (for models with ideal recognition) to 50% (for models with random recognition) to 0% for models that selectively score all inactive compounds better than actives. The commonly used variations of this measure include logAUC (logarithmic, [68, 69]) and nsAUC (normalized square root, [62, 70]) which are both normalized to give 100% for ideal and 0% for random recognition and emphasize, to a different degree, the early enrichment of the active compounds.

Neves et al. performed VLS enrichment-based evaluation and refinement of the CXCR4 binding pocket models based on the $\beta_2$AR template, using all heavy-atom elastic network normal mode analysis (EN-NMA [71–75]) for generation of additional pocket conformations. Their best model recognized 55 known actives of 6 diverse chemotypes among 1,000 random CXCR4-inert compounds from the GLIDA database [76] with ROC AUC of 88.3% (compared to 50% for random and 100% for ideal). Moreover, the authors showed that early enrichment of the actives can be improved when using an ensemble of pocket models, each tailored for recognition of its individual CXCR4 antagonist chemotype [53]. Although the predicted binding poses for isothiourea compounds were later proved incorrect, this study hints at the great plasticity of the CXCR4 binding pocket that may play role in its ability to recognize diverse small-molecule peptide and protein ligands.

In 2008, Perez-Nueno et al. performed a comprehensive analysis of VLS enrichment performance by bRho-based models of CXCR4 and CCR5 antagonist activity [77]. They compared pocket docking-based and shape matching-based approaches and benchmarked the models against a library of 602 known chemically diverse CXCR4 and CCR5 antagonists and ~4,700 decoys. The composition of the benchmarking set was such that CXCR4 actives could be discriminated from decoys by trivial ligand-based functions such as total positive charge (almost ideal discrimination) and logP (nonrandom discrimination with high initial enrichment) but not by other functions [35]. When evaluated using this set, the performance of different docking programs and scoring functions varied greatly. Among conventional approaches, the best CCR5 retrospective screening performance was achieved by AutoDock [78, 79], which retrieved ~12%, 42%, and 58% of the known actives in the top 1%, 5%, and 10% of the hit list, respectively. However, the same approach retrieved only 4%, 18%, and 29% of actives in the top 1%, 5%, and 10% of the hit list for CXCR4. The OpenEye FRED [80] docking program was more robust achieving 7%, 25%, and 40% active retrieval for CCR5 and 7%, 30%, 45% for CXCR4. Although far from perfect, this level of initial enrichment makes homology models useful for prospective screening. Surprisingly, a consensus scoring approach incorporating several FRED functions was able to retrieve all actives for both receptors in the top 10% of their respective hit lists, although it did poorly in the initial (1% and 5%) recognition. Finally, for both targets, shape matching of the library chemicals against the docked pose of a single ligand for each receptor far outperformed the tested pocket docking algorithms, supporting the hypothesis of only limited contribution of the model pocket features (as compared to the shape) to the recognition performance. In a follow-up study [81], using their previously developed models and a set of highly conservative consensus selection criteria, the authors identified 5 antagonists inhibiting CXCR4-mediated HIV entry with $EC_{50}$ values of 11.5 μM, 1.55 μM, 1 μM, 46 nM, and 125 nM. Unfortunately, higher affinity in these hits was achieved at the cost of chemical diversity; for example, the two most active compounds represent a hybrid of the monocyclam-amine (such as AMD3465) and amine-amine compounds used in model construction.

When the five CXCR4 structures were solved, the authors repeated their study [35] using the same compound set as in the 2008 publication [77]. Due to changes in either the docking protocol (possibly switching to newer versions of the software), the bRho-based model described in [77] performed dramatically better in 2012 than in the earlier 2008 publication, e.g., retrieving 15.5%, 45%, and 57% of the actives in the top 1%, 5%, and 10% of the hit list ordered by GoldScore [82] (compared with the 2008 result of 2.5%, 11.5%, and 23%). Using a different scoring function (LigandFit Jain), results improved to the point where 19.3%, 48.5%, and 66% of the actives were retrieved in the corresponding fractions of the hit list. This level of screening performance puts the bRho-based homology model on par with the crystal structures. Among the crystal structures, 18.9%, 62%, and 76% actives were retrieved in the top 1%, 5%, and 10% of the hit list by Schrodinger Glide [83] and 19.3% 77.5%, and 94% by Discovery Studio LigandFit DockScore [84], both using PDB 3oe0. LigandFit LigScore2 which was specifically designed for better screening discrimination between different compounds [84] performed slightly worse, retrieving 15.5%, 57.5%, and 80% of the actives when screening against the highest resolution structure, PDB 3odu. Of note, because of the aforementioned unusual nature of the binding pocket, the LigandFit docking module (which is based primarily on shape matching) was not able to find any poses at all in the default setup for the majority of the known CXCR4 antagonists; to alleviate the problem, the binding site had to be separated into three distinct sub-sites. Finally, all docking approaches were surpassed by a simple 5-point 3D pharmacophore model derived from the co-crystallized pose of IT1t in PDB 3oe6 [85]. Despite being derived from a single compound, this pharmacophore successfully retrieved all antagonist scaffolds starting with amine-amines, phenanthrolines, and tetrahydroquinolines, followed by bicyclams and macrocycles and then finally by isothiourea series compounds between 2% and 5% of the hit list. Again, the benchmarking set used in this study is not entirely unbiased as the logP and charge distribution differ significantly between actives and decoys [35]. Nevertheless it is clear that for CXCR4, ligand-based 3D models outperform ligand-independent pocket-based models, possibly due to the challenging nature of its binding pocket. The utility of the described pharmacophore for prospective screening remains to be studied, as does its ability to discriminate between heavily charged active and inactive compounds.

Prior to the release of the experimental CXCR4 structure coordinates, and as a part of their targeted modeling effort for the GPCR Dock 2010 assessment [54], Mysinger et al. analyzed VLS enrichment in more than 2000 models of CXCR4 based on the templates of $\beta_1$AR, $\beta_2$AR, $AA_{2A}$R, and bRho and enriched by conformers generated by elastic normal mode analysis [65]. They used their best performing model (which had normalized logAUC of 21%, compared to 0% for a random and a 100% for an ideal model) for prospective screening of a virtual library containing 3.3 million compounds. 1,800 top-ranking molecules, which represented 0.05% of the screened library, were manually inspected, and of them, 24 molecules were selected for experimental testing. Of this subset, the compound that ranked 1,725th in the initial hit list was found to be a weak inhibitor of SDF-1-induced $Ca^{2+}$ mobilization ($IC_{50}$ of 107 μM). Once the experimental CXCR4 coordinates were released, the authors repeated this exercise: the IT1t-bound structure was shown to have the normalized logAUC of 28% on the retrospective screening set (compared to 21% of the best screening model) and was used for prospective screening of the updated version of the same chemical library. Similarly, 23 molecules were manually selected from the top ~1,500 for testing. Of them, four compounds inhibited SDF-1-induced $Ca^{2+}$ mobilization with $IC_{50}$ values ranging from 55 μM to 77 μM. Moreover, two compounds disrupted SDF-1 binding to CXCR4 with $IC_{50}$ values of 306 nM and 14 μM, respectively.

To the best of our knowledge, and as of March 2014, this study by Mysinger et al. [65] is the only report of a partially successful structure-based lead discovery effort using CXCR4 structures. It also highlights weaknesses of the structure-based chemical library screening approach and the critical obstacles in its application to CXCR4 and possibly other chemokine receptors. First, binding site preparation, docking, and solution scoring algorithms had to be customized for CXCR4 in order to guide the procedure towards sampling more orientations at the bottom of the widely open polar active site. Second, the subsequent compound selection was also performed with heavy human intervention and over a far larger part of the automatically ranked compound list than is typically done in the field [65]. The human expert input was used to eliminate potential false positives resulting from incorrect 3D conformations or ionization states of the docked compounds, to avoid bias towards large hydrophobic (and thus insoluble) compounds that are abundant in the chemical databases and typically score best in screening, and to ensure engagement of proper expected interactions in the receptor binding site (namely, compound shape complementarity to the binding site and formation of the key salt bridge to $Glu288^{7.39}$ and at least one other anionic residue of CXCR4). Finally, the identified antagonists (even those obtained by screening against the X-ray structure) are rather weak, with functional $IC_{50}$ values exceeding 50 μM [65]. In a similar initiative of our own, all 40 compounds selected from the virtual screening hit list were found inactive at a single concentration of 5 μM and were not pursued any further (unpublished data). The CXCR4 hits by Mysinger et al. are far weaker than the best hits obtained by the same and other groups in their virtual screening campaigns against the structures of $\beta_2$ adrenergic (9 nM [86], 150–500 nM [87]), adenosine $A_{2A}$ (150 nM [88], 32 nM [89]), dopamine D3 (200–300 nM [59], 7 nM [90]), or muscarinic M2 and M3 (150–300 nM [91]) receptors. In other words, it appears that the CXCR4 structure has not yet led to therapeutic breakthroughs through structure-based virtual screening approaches, at least not to the level comparable with the structures of other GPCRs, either in terms of hit rates or the affinity of the identified modulators.

This issue raises the question of whether computational scientists can do better in terms of structure-based discovery of diverse, high-affinity chemokine receptor ligands with favorable pharmacokinetic profiles. And if not, what are the barriers? The answer seems to involve several issues:

- Deficiencies of the available compound libraries (which are biased towards poor pharmacokinetic profiles, large size, low solubility)
- Deficiencies of the 2D to 3D structural conversion for the chemicals and conformational sampling algorithms that result in unrealistic conformations of the docked ligands, especially for nontrivial fused ring systems and chiral compounds
- The approximate nature and incompleteness of the compound scoring functions
- The need for target-specific adjustment of the docking algorithms and compound scoring functions so that they give greater weight to expert-prioritized interactions.

The latter aspect is probably the most important as it provides a way to alleviate the inherent difficulties associated with the poorly druggable pockets in the chemokine receptors. As our analysis shows, models and scoring functions can be trained to recognize and predict binders to even challenging pockets. This can be achieved by assigning weight to important interactions, either manually or by incorporating them in the scoring function, and by using the consensus compound scoring approaches. When designing such customized scoring schemes, there is always the danger of overtraining; therefore, using diverse and objectively challenging compound benchmarks becomes extremely important.

Four years after publication of the CXCR4 structure, numerous VLS campaigns continue to be based on models rather than the experimental structures. For example, in 2013, Vinader et al. reported the results of their virtual screening against a bRho-based model of CXCR4 which led to identification of a weak (3.8 μM $IC_{50}$ for inhibition of SDF-1-induced $Ca^{2+}$ mobilization) antagonist of CXCR4; subsequent model-guided SAR and optimization yielded a 120 nM derivative, albeit with lower solubility [92]. Similarly, a 2012 publication by Kim et al. reports the discovery of the potential of an antimalarial drug in inhibiting CXCR4 by virtual screening of an antimalarial library against an ab initio CXCR4 model [93]. Although one of the top predictions in the GPCR Dock 2010 assessment [54], the model bears only limited resemblance to the X-ray structure, especially in the ligand binding region; however, the success of the aforementioned VLS study [93] proves its utility for drug discovery.

### 3.5 Molecular Design and Optimization Studies Assisted by the Structures of CXCR4

All the above taken into account, lead discovery efforts against CXCR4 still benefit from the 2010 X-ray structures. By means of molecular docking and structure-guided optimization, they help rationalize molecular design in retrospective and prospective manner. For example, by chemical synthesis and SAR studies on known scaffolds, a group from Emory University identified a potent CXCR4 p-xylyl-enediamine-based antagonist, MSX-122 [94, 95], that did not block the binding of radiolabeled SDF-1 to the receptor but efficiently inhibited SDF-1-induced cAMP response as well as SDF-1-mediated cell migration and invasion. Docking of the antagonist to the structure helped explain its noncompetitive inhibition mechanism, as the compound docked at the bottom of the transmembrane pocket of CXCR4 without engaging the extracellular loops, similar to maraviroc in the structure of CCR5 [15]. A different tetrahydroisoquinoline-based antagonist developed by the same group [96] was also studied from the ligand-receptor docking perspective and shown to engage the same residues as IT1t, thus resulting in competitive inhibition.

Cyclopentapeptide antagonists of CXCR4 have also been studied by the structure-based approaches. Because the CVX15 peptide co-crystallized with CXCR4 bears substantial similarity with FC131 and its active analogs, the CXCR4:CVX15 complex structure represented an excellent starting point for 3D modeling of cyclopentapeptide binding to CXCR4. The models helped rationalize the existing SAR [97] and guided the design of isosteric substitutions to the cyclopeptide scaffold [98, 99], as well as conformationally constrained [100] and dimeric [101] cyclopentapeptides variants.

Finally, an interesting structure-based approach was undertaken by Aboye et al. who developed a novel potent CXCR4 antagonist ($IC_{50}$ of 20 nM and 2 nM in SDF-1 activation and HIV-1 entry assays, respectively) by grafting a CVX15-based sequence onto the scaffold of a cyclotide, a small globular protein with a cyclized backbone and three intramolecular disulfide bonds [102]. This represents an example of a protein design project directly inspired by the 2010 CXCR4: CVX15 peptide complex structure.

### 3.6 Computational Prediction of Small-Molecule Antagonist Binding to CCR5

Before the structure of CCR5 in complex with Maraviroc was solved, multiple efforts were undertaken to build this complex, as well as other CCR5:antagonist complexes, by computational methods. As in the case of CXCR4, some of the earlier studies used bRho-based models of the receptor. More recent attempts were based on the CXCR4 structure or, in some cases, ab initio modeling of the receptor TM bundle. The consensus and variable parts of the binding site for several small-molecule CCR5 antagonists were mapped by site-directed mutagenesis [103–106]. These studies correctly identified Trp86$^{2.60}$, Tyr108$^{3.32}$, Tyr251$^{6.51}$, and Glu283$^{7.39}$ as making direct contacts with the antagonists in the binding site, but they also proposed other residues that, in light of the now available crystal structure, are quite distant from the ligand. The observed effect of mutations at these other positions does not originate from the loss of direct contacts with the ligand but is allosteric in nature and involves destabilization of distant parts of the receptor.

Labrecque et al. used molecular modeling informed by site-directed mutagenesis, binding, and functional studies [104] to elucidate the molecular basis of CCR5 interactions with aplaviroc, maraviroc, vicriviroc, TAK-779, SCH-C, and a benzyloxycarbonyl-aminopiperidin-1-yl-butane derivative 3bb [107]. The receptor model was built using bRho as the template. Although the interaction fingerprint derived from the docked poses of the ligands was consistent with mutagenesis studies, none of the complex models successfully reproduced the maraviroc binding pose as revealed later by crystallography. The uncertainty in compound placement originated from the fact that while the salt bridge between the primary amine of the compound and Glu283$^{7.39}$ of the receptor is established and unambiguous, the compound may be rotated in at least two ways with respect to this pivotal point, each way providing favorable interactions for the remaining polar groups of the compound.

Berro et al. studied conformational heterogeneity of CCR5 and its effects on binding to the anti-HIV antagonist vicriviroc [108]. They demonstrated that G$\alpha_i$ binding on the intracellular side of CCR5 decreases the potency of the compound against vicriviroc-sensitive viruses and is favorable for the entry of vicriviroc-resistant strains. As a part of this study, the authors performed docking of maraviroc and other CCR5 antagonists into an ab initio model of CCR5 obtained by the GEnSeMBLE method [109]. While capturing the general orientation and the mutagenesis-validated interaction fingerprint of maraviroc, their prediction placed it entirely in the major pocket formed by TM helices 3, 4, 5, 6, and 7 and rotated it around the major principal axis of the compound by approximately 120°, as compared to its experimental pose [15].

More recent maraviroc docking attempts made use of the CXCR4 structure as the closest available homology template for CCR5 modeling. For example, Kothandan et al. built models of CCR2 and CCR5 based on CXCR4 and used them for studying maraviroc binding using docking and molecular dynamics [110]. Their model correctly predicted interaction of maraviroc with Tyr37$^{1.39}$ but missed Glu283$^{7.39}$. Garcia-Perez et al. [61, 111] modeled the CCR5:maraviroc complex using a series of homology models that were first evaluated for their VLS enrichment properties. The model based on the peptide-bound CXCR4 performed best and was used for docking. The predicted pose of maraviroc correctly captures its overall orientation and interactions with Glu283$^{7.39}$, Trp86$^{2.60}$, and Tyr251$^{6.51}$ but misses the rotamer flip in Trp86$^{2.60}$ and the functionally important interaction of the ligand with the side chains of Tyr37$^{1.39}$. Nevertheless, it is among the most accurate pre-structure predictions of the CCR5-maraviroc complex. This achievement stresses the importance of a closely related experimentally determined structure (in this case, CXCR4) for generation of an accurate and predictive homology model, as well as the role of the conformational variations in the template (in this case, CVX15-bound CXCR4).

As these examples show, even in view of the available structure of the close homolog, CXCR4, modeling of CCR5:antagonist complexes remains a challenging task. Both chemokine receptors exhibit remarkable conformational plasticity that allows for accommodation of diverse ligands not only through pocket side chain rearrangements but also through lateral, axial, and rotational movements of the helices, all of which are virtually impossible to predict exclusively by computational methods. It is therefore expected to be no less challenging to predict CCR5 complex structures with small-molecule antagonists from other chemical classes, such as TAK-779, SCH-C [112], SCH-D/vicriviroc, or aplaviroc. All of these antagonists inhibit CCR5-mediated $Ca^{2+}$ mobilization responses to all tested chemokines [113] in a dose-dependent and *insurmountable* fashion, i.e., in the presence of antagonist, the maximum response cannot be restored, even at very high chemokine concentrations. Most small-molecule CCR5 antagonists also insurmountably inhibit CCL3/MIP-1α and CCL4/MIP-1β binding [113]. On the contrary, increasing concentrations of the *antagonists* can fully displace CCL3/MIP-1α and CCL4/MIP-1β at multiple chemokine concentrations. TAK779 also fully displaces CCL5/RANTES, whereas maraviroc and Sch-D/vicriviroc allow for some residual CCL5/RANTES binding and aplaviroc causes almost no CCL5/RANTES displacement [113]. Taken together, this behavior is interpreted as an evidence of mixed competitive-noncompetitive and/or allosteric inhibition of chemokine binding by the antagonists, possibly based on the conformational heterogeneity of the receptor population with different preexisting conformations preferentially targeted by each of the ligands. Furthermore, these observations demonstrate that the binding determinants of CCL5/RANTES to CCR5 are different from those for CCL3/MIP1α and CCL4/MIP1β and also that binding of the small molecules is driven by different hot spots inside the CCR5 binding pocket.

Multiple studies indicate, for example, that in contrast to other small-molecule antagonists of CCR5, the binding of aplaviroc to CCR5 may directly involve ECL2. Aplaviroc inhibits binding of a monoclonal antibody directed against the C-terminal end of ECL2 that faces the binding pocket, whereas TAK-779 and SCH-C cause only partial inhibition [114]. Mutations in ECL2 and its interface with TM4 (Gly163) or TM5 (Lys191) have a substantial impact on binding of aplaviroc but not TAK-779 or SCH-C. The disruption of the ECL2-TM3 disulfide bridge by a C178A mutation abrogates binding of aplaviroc but not TAK-779 or SCH-C [105]. Finally, a chimeric receptor built from the TM domain of CCR5 and the extracellular loops of CCR2 cannot bind aplaviroc [37] although it efficiently interacts with maraviroc and related compounds; however, aplaviroc binding is restored in a chimera that, in addition to the TM domain, possesses the native sequence for the C-terminal part of the ECL2.

In summary, elucidation of the structural basis of conformational heterogeneity of CCR5 (and, in fact, many other chemokine receptors) and its role in ligand recognition remains an unsolved problem that is extremely important to address by both the computational and experimental methods. The emerging homologous X-ray structures in their multiple functional states may stimulate computational method development and enable more accurate and reliable predictions.

### 3.7 Drug Discovery Efforts by Virtual Ligand Screening Against Models and Structures of CCR5

At the time when the present chapter was written, we were not aware of any virtual screening campaigns against the crystallographic structure of CCR5 [15], probably due to the short time that has passed since its publication in 2013. Moreover, to the best of our knowledge, no prospective VLS studies have yet been published against CXCR4-based models of CCR5. As it stands, the existing structures of the two receptors may provide modeling templates for each other in distinct biologically relevant conformations and hopefully enable and inspire such screening efforts in the future.

Nonetheless, as in case of CXCR4, CCR5 VLS campaigns have been conducted using the bRho-based models. For example, Kellenberger et al. used in silico compound screening against a bRho-based CCR5 model as a part of a pipeline for identification of novel small-molecule modulators of CCR5 [115]. The models were first evaluated for VLS enrichment using 7 known CCR5 antagonists and 993 random drug-like decoys and shown to retrieve about half of the antagonists in the top 5% of the hit lists. The prospective structure-based screening of the chemical library was preceded by a filtering step on the basis of drug-likeness and agreement with a simple 2D pharmacophore shared by all known CCR5 antagonists and followed up by clustering and manual selection of the experimental candidates from the top-ranking hits. Two compounds with 2–3 μM $IC_{50}$ in the CCL4/MIP1β competition-binding assay were identified in this screen; an additional 1 μM compound was found by a subsequent chemical similarity search. Upon experimental evaluation, all compounds appeared to be agonists of CCR5. This result may seem counterintuitive given the inactive state of the receptor model used for compound identification; however, in reality, it is expected that the approximate nature of the model does not allow reliable discrimination of agonists from antagonists.

VLS enrichment properties of multiple CCR5 models built from different templates were studied by Garcia-Perez et al. [61, 111]. In their hands, opsin-, β2AR-, and $AA_{A2}R$-based homology models of CCR5 were unsatisfactory in retrospective screening of 26 known CCR5 antagonists against a library of 154 inactive and decoy compounds (less than 10% recall at less than 20% precision). Interestingly, the model based on the IT1t-complexed CXCR4 also performed poorly. A bRho-based model performed only slightly better (35% recall at 33% precision). The best recognition of actives was achieved by a hybrid model including ab initio built TM2 and TM3 in the context of the bRho-based model (73%

recall at 34% precision) and a model built using the CVX15 peptide-complexed form of CXCR4 (46% recall at 71% precision). This result is understandable from the point of view of the now possible comparison of the CCR5 structure to the previously available templates. CCR5 indeed appeared to be most similar to CXCR4 in its CVX15-bound form. However, as described above, even with this closely homologous template, a number of conformational features of CCR5 could not be captured by homology modeling. Crystallographic elucidation of each receptor in complexes with multiple ligands and in multiple conformational states is therefore a worthy (but hard-to-achieve) goal.

### 3.8 Model-Assisted Molecular Design and Optimization for CCR5

Despite their relative structural inaccuracy, CCR5 proved instrumental in rationalizing existing SAR or prospective optimization of small-molecule antagonist candidates. Numerous 3D QSAR and fragment assembly studies were informed by compound poses predicted by docking to bRho-based CCR5 models [116–119]. More recently, by compound docking to a CXCR4-based CCR5 model, Gadhe et al. built and evaluated a comprehensive 3D QSAR model which provided good agreement with the experimental compound affinity data [120]. An example of successful prospective molecular design is given by the work of Metz et al. who docked previously discovered high-affinity CCR5 antagonists into the bRho-based model developed earlier by Labrecque et al. [104] as well as a newly generated CXCR4-based model [121] and used the predicted poses for prospective molecular optimization. Because the authors could not resolve the uncertainty in the compound placement in either of the models, they came up with several equally plausible hypotheses using Glu283$^{7.39}$ as a single anchor and a pivotal point for the ligand interactions. Despite this uncertainty, and possibly due to their use of multiple complex models, they successfully identified compound analogs with improved potency, specificity, and reduced hERG binding.

## 4 Structure-Guided Understanding of Receptor Interactions with Chemokines

Chemokine receptors mediate cell migration in response to binding and activation by their endogenous chemokine ligands. Chemokines are small (~70–120 amino acids) secreted proteins that share a conserved topology consisting of an unstructured N-terminus and a folded globular core. The latter contains a three-strand β-sheet and a C-terminal α-helix. The N-terminus is stapled to the folded globular core by two disulfide bonds, one to the loop connecting $\beta_1$ and $\beta_2$ strands and another to the $\beta_3$ strand. The sequence segment between the N-terminal disulfides and the $\beta_1$ strand is referred to as the *N-loop*. Although many chemokines oligomerize, it is well established that monomers are responsible for activating cell migration [122, 123].

The prevalent model of receptor-chemokine interaction involves two distinct sites [124–127], one formed by binding of the receptor N-terminus to the globular core of the chemokine (the so-called chemokine recognition site 1 or CRS1 [9]) and another by binding of the chemokine N-terminus in the TM domain pocket of the receptor (chemokine recognition site 2 or CRS2). The CRS1 interaction is believed to mainly provide the binding affinity, whereas the CRS2 interaction is important for both affinity and receptor activation. Consistent with this model, for multiple chemokines, as little as a single residue truncation or modification in the chemokine N-terminus (CRS2) can have a dramatic effect on the chemokine pharmacology, such as changing the chemokine from an agonist to an antagonist [128–132]. On the other hand, mutations in the receptor N-terminal residues, including the Tyr residues whose sulfation is important for chemokine recognition and is assumed to promote CRS1 interactions [127, 133–136], have only a mild impact on the signaling capacity of the chemokine, affecting its $EC_{50}$ but not the maximal response [137, 138]. A guided molecular dynamics simulation of the CXCR4 N-terminus in the context of the entire receptor [127] suggested that the sulfate groups promote an extended conformation of the N-terminus making it more accessible for chemokine binding, which may partially explain the experimentally observed affinity changes of the sulfotyrosine mutants; however, the role of the sulfotyrosines in direct interaction with the chemokines has been established as well [135, 139].

Attempts to detach the N-terminus and the associated CRS2 functionality from the rest of the chemokine typically result in extremely weak (often undetectable) binders. However, this can be partially rescued by a longer peptide including the chemokine N-terminus and a part of or a complete N-loop which is universally recognized as one on the determinants of receptor interaction [128, 135, 140–144]. For example, a 17-residue N-terminal peptide of SDF-1 has been shown to bind CXCR4 with the $K_d$ of ~850 nM and to activate CXCR4-mediated chemotaxis with an $EC_{50}$ of ~2.2 μM [145]. Dimerization of an N-terminal 9-residue peptide of SDF-1α produces a more potent agonist with a binding $K_d$ of ~730 nM and chemotactic $EC_{50}$ of ~500 nM [145]. Although ineffective against CCR5, an isolated 21-residue N-terminal peptide of the viral chemokine, vMIP-II, inhibited CXCR4 binding to SDF-1α and the 12G5 antibody with an $IC_{50}$ of 190 nM and 640 nM, respectively [146, 147]. CTCE-9908, a covalently dimerized N-terminal peptide (residues 1–8) of the SDF-1(P2G) variant by Chemokine Therapeutics, has been described as a potent antagonist of CXCR4 [148]. Finally, Lefrançois et al. were able to develop a high-affinity partial agonist by linking the SDF-1 N-terminus to a T140 antagonist scaffold [149]. Along with chimeric chemokine studies [128], such experiments support the two-site model and the independent signaling role of the chemokine N-terminus.

Apart from the N-loop, the role of specific regions within the chemokine globular core is unclear. It was demonstrated that the SDF-1 N-terminus/N-loop (residues 1–14) connected by a 4-Gly linker to the C-terminal helix (residues 55–67) retains most of the binding affinity ($IC_{50}$ of 225–250 nM compared to 39 nM for WT SDF-1 in the same assay) as well as the biological activity of WT SDF-1, as long as the helical conformation in the C-terminal residues is stabilized by a lactam ring [150]. This study thus suggests the relatively minor role of the three-strand β-sheet of SDF-1 that has been postulated important in other studies [135, 140, 141, 151]. The β-sheet was also shown important in other chemokines [142, 144]. Similarly, the role of receptor extracellular domains is not clear and is not consistent between different receptors. For example, CXCR4 ECL2 is believed to be important for HIV gp120 binding but not for SDF-1 binding [152], whereas in CCR1, ECL2 was described as the point of interaction with the N-terminus of the MIP-1α chemokine [153].

Based on these and other biochemical insights, attempts to elucidate the structure of the receptor-chemokine complex using molecular modeling were undertaken as early as 2001 [138, 154]. The pivotal publication of the CXCR4 structure in 2010 [14], unfortunately, did not provide a direct and immediate answer for the structural basis of chemokine recognition by the receptor, partly because the entire N-terminus of the receptor (residues 1–26) is disordered in the crystallographic density and partly because of the notorious plasticity of the TM domain as well as the extracellular loops that may be present in SDF-1-incompatible conformations in these antagonist-bound structures.

However, in conjunction with the 2008 NMR structure of the CXCR4 N-terminus (residues 1–38) with a disulfide-locked SDF-1 dimer [135], the CXCR4 X-ray structure informed a series of molecular modeling studies aimed at elucidating the geometry of the complex. Unfortunately, the hybrid model resulting from spatial superposition of the highly rigid part of the receptor present in both structures (CXCR4 residues 27–29) is inconsistent with the two-site interaction hypothesis described above and supported by substantial biochemical evidence (Fig. 4b). Among the possible reasons for this discrepancy is the small 3-residue overlap between the two structures as well as the fact that the NMR structure was that of an SDF-1 dimer. Furthermore, at this time, the stoichiometry of the receptor:chemokine interaction is not clear [14, 127] apart from the fact that chemokine monomers are involved [122, 123]. Chemokine receptors are known to homo- and heterodimerize, and simultaneous interaction of a single monomeric chemokine with two protomers within the receptor dimer, with possibly decoupled CRS1 and CRS2, remains a plausible hypothesis; moreover, unlike the one-to-one interaction model, such two-to-one interaction models can reconcile the available pieces of structural information from the NMR and the X-ray structures (Fig. 4a). The two-to-one receptor:chemokine stoichiometry provides a feasible explanation for the experimentally observed negative chemokine binding cooperativity and transinhibition across chemokine receptor heterodimers [155, 156] and the proposed triggering effect of chemokine on receptor dimer formation [135, 157].

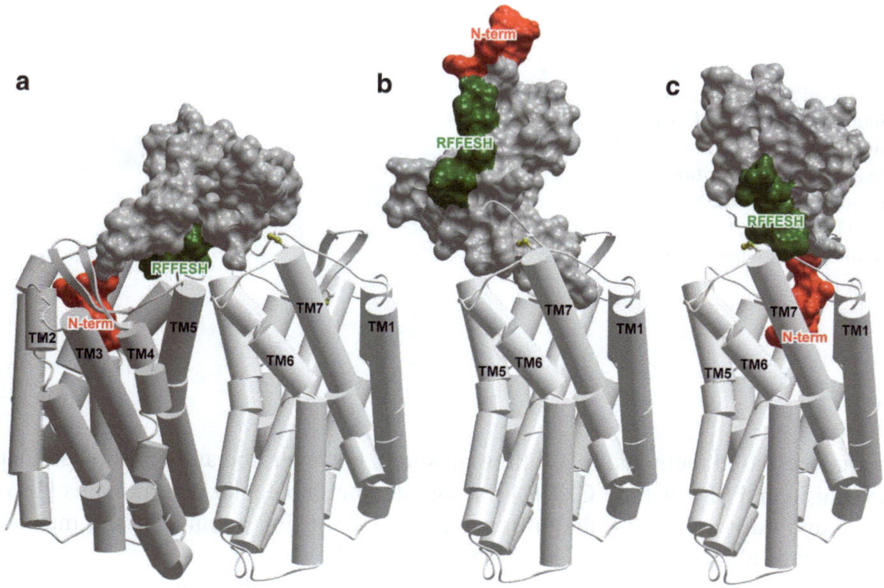


**Fig. 4** Chemokine binding geometry and stoichiometry hypotheses. (**a**) A 1:2 model, with a single chemokine molecule simultaneously interacting with the N-terminus of one receptor molecule and the TM pocket of its dimer partner, reconciles both the mutagenesis-derived insights and the relative orientation of the CXCR4 N-term to SDF-1 observed in the NMR complex structures. (**b–c**) A 1:1 model contradicts either mutagenesis data (**b**) or the receptor-chemokine contacts observed in the NMR [135] (**c**). The models were built in the ICM molecular modeling package [32, 33] by docking of the full-atom chemokine molecule into the binding pocket of a monomer or a dimer of the receptor represented as potential grid maps. The backbone variables of the chemokine N-terminus (residue 1–10) and all side chain variables were kept fully flexible through the simulation. Interactions observed in the NMR structure [135] between the chemokine and the receptor residues 25–30 were imposed as restraints in docking for (**a**) and (**b**) but not (**c**)

Other studies remain faithful to the one-to-one stoichiometry hypothesis and thus have to disregard some of the constraints of the NMR structure of SDF-1 and the CXCR4 N-terminal peptide (Fig. 4c). For example, Xu et al. performed docking and a molecular dynamics simulation of SDF-1 in a hybrid model of CXCR4 built from the 2010 X-ray structure of the TM domain [14] with the N-terminus extracted from the 2008 NMR structure [135]. The resulting one-to-one complex [158] reconciled much of the mutagenesis data but not the residue chemical shifts and NMR distance restraints from the NMR structure [135]. On the contrary, a series of 1:1 models of interaction of CXCL8 chemokine with CXCR1, obtained via molecular docking and molecular dynamics simulations [159], never places the chemokine N-terminus in the receptor TM binding pocket.

In 2010, Saini et al. demonstrated that in addition to SDF-1, CXCR4 is a receptor for extracellular ubiquitin [160]. Chemokines and ubiquitin share distant fold similarity (Figure 5) although the distinct connectivity of their secondary structure elements does not allow detection or quantification of their similarity by

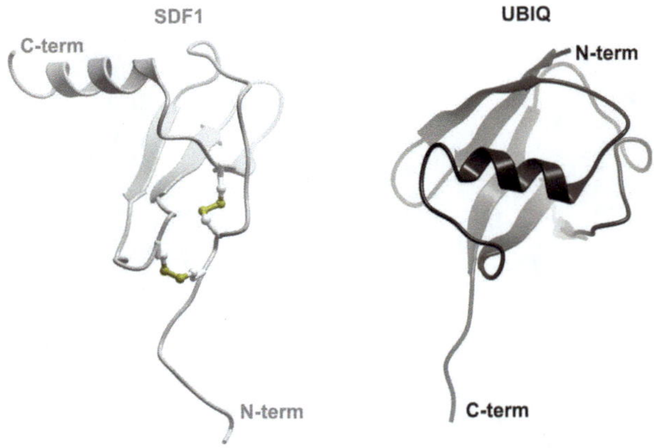


**Fig. 5** Distant fold similarity between SDF-1 chemokine and ubiquitin may explain their ability to interact with CXCR4. This figure was obtained by manual superimposition of ubiquitin (PDB 4m0w [161]) onto the structure of SDF-1 chemokine (PDB 2j7z [162])

sequence-based methods. In a follow-up study, however, the authors showed that ubiquitin and SDF-1 bind CXCR4 through distinct molecular determinants [163, 164] and built a CXCR4:ubiquitin complex model using automated and manual docking of ubiquitin to the CXCR4 structure [163].

Despite these advances, a complete understanding of the structural principles underlying the interaction of chemokine receptors with chemokines is still lacking and most likely will not be reached until the structure of a complex is solved. The incompleteness (lacking N-terminus) and the inactive conformation of the receptor structures available to date make it impossible to predict receptor interactions with chemokines with certainty. Moreover, some chemokine antagonist variants have been shown to bind CXCR4 without inhibiting SDF-1α-induced signaling [165], a seemingly paradoxical observation that hints at conformational heterogeneity of the receptor with distinct subpopulations preferentially binding SDF-1α versus the antagonist variants. Elucidation of such heterogeneity by means of X-ray crystallography will enable more reliable computational studies of receptor-chemokine complexes in different functional states.

## 5 Structure-Guided Understanding of Receptor Interactions with HIV gp120

Both CCR5 and CXCR4 are known for their role as co-receptors for HIV entry into host immune cells. Viral entry is a complicated molecular process initiated by binding of the HIV envelope glycoprotein (gp120) trimer to the CD4 glycoprotein on the surface of the host cells, which induces a conformational change in gp120 and exposes the previously buried variable loop 3 (V3). Using this loop and probably other surface residues, the CD4-bound gp120 then binds to one of the co-receptors (CCR5, CXCR4, or other chemokine receptors) which promotes more stable attachment, further conformational changes, and ultimately penetration of

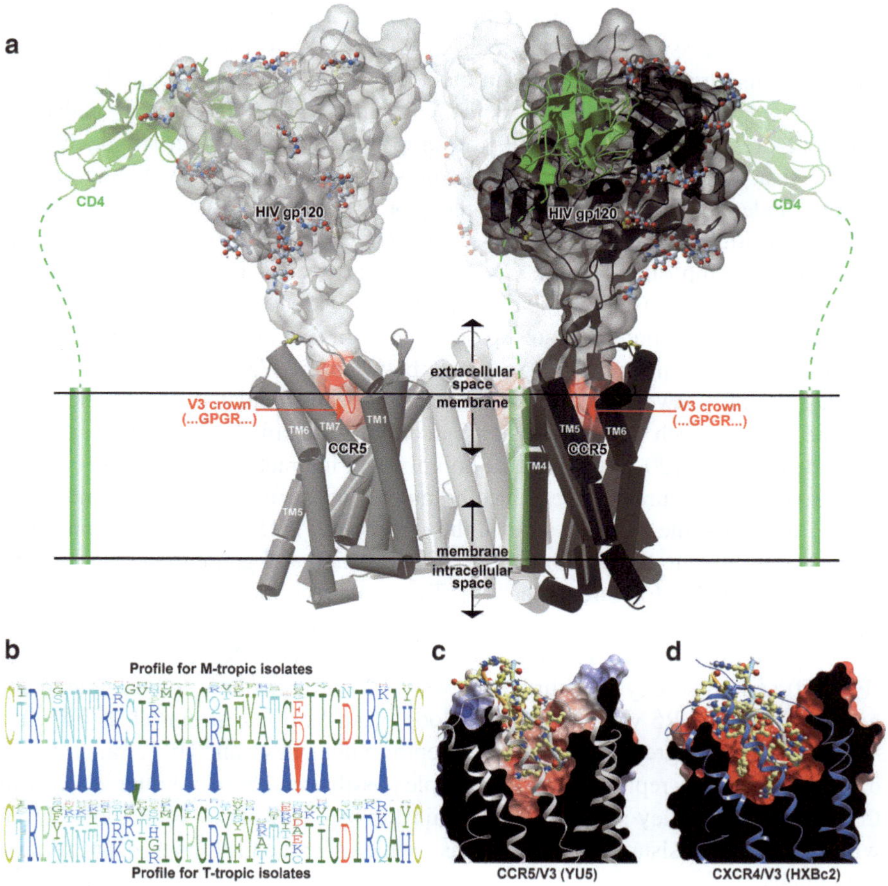


**Fig. 6** Chemokine receptor interaction with HIV gp120 protein. (**a**) A hypothetical depiction of the overall architecture of molecular complexes involved in the initial stages of HIV entry. This approximate model was built in the ICM molecular modeling package [32, 33] by first superimposing the globular core of gp120 from the crystallographic structure of gp120/CD4 complex (PDB 2qad [133]) onto the electron microscopy (EM) structure of a CD4-bound gp120 trimer (PDB 3dno [166]) and then by superimposing the crown of the V3 loop from our model of V3 binding to the crystal structure of CCR5 (PDB 4mbs [15]) onto the corresponding region of gp120. The protomers in the gp120 trimer are colored *white*, *grey*, and *black*. (**b**) Comparison of sequence profiles of gp120 V3 loops from M-tropic and T-tropic isolates indicates that coreceptor switching from CCR5 to CXCR4 involves accumulation of basic residues at multiple positions of the V3 loop. To generate this image, alignments were built separately for 1307 unique M-tropic and 495 unique T-tropic or dual-tropic V3 sequences from Los Alamos HIV database (http://www.hiv.lanl.gov/). (**c–d**) Structural models of V3 loop interaction with CCR5 (**c**) and CXCR4 (**d**) receptors

the target cell membrane by the N-terminal fusion peptide gp41 (Fig. 6a). For entry, the primary HIV-1 isolates almost exclusively use CCR5, which is expressed on the surface of macrophages and activated and memory T-cells. These isolates, referred to as R5, appear critical in establishing the persistent infection [167]. In the course of disease progression, the virus evolves to use CXCR4 as a co-receptor, thus expanding its target cell repertoire to include resting and naive T-cells [168]. Isolates using exclusively CXCR4 or both CCR5 and CXCR4 for entry are called X4 and R5X4, respectively, and typically develop at the advanced stages of the disease.

The structural determinants of HIV tropism and co-receptor switching are not entirely understood. It is clear that co-receptor switching involves accumulation of the basic residues in the stem of the V3 loop [169] (Fig. 6b), in accordance with the more acidic composition of the CXCR4 binding pocket as compared to that of CCR5. Knowledge-based and statistical potential-based methods [169–172] have been developed that predict co-receptor usage from sequence or structural models. However, the specific molecular contacts between the V3 loop residues and the binding site of the two co-receptors have not yet been elucidated. Moreover, it also may be true that such specific contacts can only be identified in the context of each individual HIV gp120 strain and that no universal interaction determinants exist.

Due to the structural and sequence homology between the CVX15 peptide and the V3 loop of some T-tropic strains, the CXCR4:CVX15 complex structure may provide insights into the possible mode of interaction and guide molecular modeling. In their CCR5 structure paper [15], Tan et al. used molecular docking guided by the CVX15-derived chemical fields [173] to build the model of the HIV gp120 V3 loop (T-tropic strain HXBc2) interacting with the CXCR4 binding pocket (Fig. 6d) and homology modeling followed by local refinement to build an analogous model of CCR5 with a gp120 V3 loop of an M-tropic strain YU2 (Fig. 6c). While the models agreed well with the receptor mutagenesis and HIV strain tropism data, they may only represent one of multiple possible modes of V3 interaction with the co-receptors. They also did not explain the mechanisms of CCR5 interaction with maraviroc-resistant gp120 strains, which frequently emerge in treated individuals.

The emergence of drug resistance raised serious concerns about the future of co-receptor antagonists as HIV therapeutics [38]. Two mechanisms have been proposed for this. The first mechanism involves the remarkable conformational heterogeneity of CCR5. While the small-molecule antagonists, chemokines, and most anti-CCR5 antibodies seem to explore selected conformations rather than the entire conformational spectrum, the HIV gp120 protein strains have been shown to not only efficiently utilize multiple co-receptor conformations [174] but also evolve towards conformations incompatible with either chemokine or small-molecule antagonist binding [175]. An alternative resistance mechanism involves development of the ability to enter cells via small-molecule-occupied CCR5, thus forming ternary complexes [176]. The direction of resistant strain evolution is dependent on the antagonist; for example, maraviroc and vicriviroc resistance has been mapped to different amino acid changes in the gp120 V3 loop [177]. Resistance to CCR5 inhibition by chemokines can be acquired by switching to alternative (i.e., chemokine-incompatible) conformations or by co-receptor switching [175]. The phenomenon of noncompetitive resistance of T-tropic isolates to CXCR4 antagonists has also been described [178], indicating that the resistance-forming mechanisms may be common across the co-receptors.

The importance of conformational ensembles in GPCRs is being increasingly recognized by the modelers and structural chemists. For example, Berro et al. used ab initio prediction of the preferred functional conformers of CCR5 and its signaling dead or constitutively active mutants in complexes with maraviroc or vicriviroc to derive plausible hypotheses about TM helix movements associated with these mutations and/or binding events and possibly relevant to the HIV strain resistance mechanisms [108]. However, our understanding of chemokine receptor conformational equilibrium, its role in ligand and HIV interactions, and the computational tools needed for elucidating receptor conformational heterogeneity are still in their infancy. Additional successes in the determination of chemokine receptor structures in different functional and conformational states will be helpful in promoting such understanding.

## 6 Structure-Guided Understanding of Receptor Dimers

The persistently dimeric behavior of chemokine receptors [179–181] has been characterized by multiple experimental methods including co-immunoprecipitation, fluorescence and bioluminescence resonance energy transfer (FRET and BRET, respectively), and biomolecular fluorescence complementation (BiFC) [179]. The functional consequences of dimers have yet to be fully understood but are already known to include changes in trafficking, G-protein coupling, and positive or negative cooperativity in ligand binding and activation (reviewed in [179, 181–183]).

However, there is no consensus about the structural basis for dimerization. Crystallographic studies of the detergent-solubilized G-protein-coupled receptors provide only limited information about their dimerization geometry in native membranes. For most GPCR crystal structures, the inter-receptor packing interfaces can be immediately discarded for reasons of insufficient size, lack of specific interactions, or geometric incompatibility with the parallel orientation and twofold rotational symmetry of the expected biological dimers. For other GPCR structures, the observed crystallographic dimers pass these filters but are still lacking biochemical evidence required for their unambiguous designation as biological dimers. Among them, there is the dimer interface involving helices 5 and 6 and consistently observed in all five crystal structures of CXCR4, despite the structures being solved in different space groups and in different complex compositions [14]. A similar interface is also observed in the structure of the μ-opioid receptor [21] and in several structures of squid rhodopsin [26]. Because the GPCR activation process involves a rather dramatic conformational change in these TM segments [184], it is not clear whether such geometry is compatible with activation. Other GPCR dimer interfaces observed in crystallography involve TM segments 1 and 2 along with the C-terminal helix 8 [23, 185] or TM helices 4 and 5 [183, 184].

Several attempts have been made to reveal the biological dimerization interfaces by designing point mutations that would disrupt the dimer formation and render the receptor monomeric [29, 186, 187]. Importantly, the monomeric character of the mutants has to be probed in live cells and preferably at the physiological levels of receptor expression. However, the dimers proved to be exceptionally stable, possibly due to scaffolding effects from other cellular proteins (e.g., G-proteins [188]) or due to the existence of higher-order oligomers which make separation of dimer subunits problematic [189]. The dimers also appear important for receptor folding and trafficking, frequently resulting in modified expression profiles of the mutants and effectively making it impossible to design and characterize a fully monomeric but functional GPCR in live cells. Moreover, live cell studies using BRET, FRET, and other proximity-based techniques invariably run into the caveat of altered mutant expression and distribution between the cellular compartments thus not being able to quantitatively distinguish strongly and weakly dimerizing constructs [187]. Even experiments with fluorescent labeling of receptors exclusively at the cell surface [29, 186] lack sufficient resolution for characterization of subtle changes in dimerization brought about by single point mutations.

For example, Hernanz-Falcon et al. reported the identification of I52$^{1.54}$V and V150$^{4.47}$A as residues directly involved in CCR5 dimerization using FRET [29]; however, this finding could not be reproduced with co-immunoprecipitation or BRET [190]. Similarly, Kufareva et al. probed the relevance of the crystallographic dimer of CXCR4 by introducing computationally designed mutations at the interface and characterizing their effects with BRET [187]. Although changes were observed in the donor-acceptor saturation BRET experiments with mutants compared to wild-type CXCR4, they could not be unambiguously interpreted as changes in the dimerization affinity due to limitations in the quantification of the BRET data when comparing different transfectants.

An alternative approach for identifying dimer interfaces involves competition with small exogenously introduced peptides mimicking the selected transmembrane domains [191]. Using FRET between surface-labeled CCR5, Hernanz-Falcon et al. discovered that 7-residue peptides mimicking TM domains 2 and 4 of the wild-type CCR5 disrupted the preformed dimer, while those containing mutations did not [29]. For CXCR4, peptides mimicking TM domains 2 and 4 had no effect on preformed dimers but inhibited agonist-induced changes in the dimer geometry, as shown by BRET [192]. A similar effect was observed with peptides mimicking TM6 and 7, albeit to a lower extent. It is hard to say, however, whether such peptides inhibit dimerization by directly binding at the dimer interface or by an allosteric mechanism that disrupts TM bundle packing at a site distant from the interface.

In some GPCRs (but to the best of our knowledge, not in chemokine receptors), the dimerization interfaces have also been studied using computational approaches. By rigid-body docking of homology models of neurotensin receptors, Casciari et al. concluded that the dimer interface involves helices TM1, 2, and 4 [193]. Using coarse-grained biased molecular dynamics simulations in POPC lipid bilayers [194], Periole et al. identified TM1/H8 and TM4/3 as alternative dimerization interfaces in bovine rhodopsin dimers [195]. With a similar simulation, Johnston and colleagues evaluated the relative stability of the TM1/H8 and TM4/3 dimers in β1- and β2-adrenergic receptors [196], concluding that the TM1/H8 interface is more stable. Finally, by exploring the concept of residual hydrophobic mismatch between the GPCR TM helices and the surrounding lipids in the context of a coarse-grain molecular dynamics simulation, Mondal et al. were able to rationalize the distinct oligomerization behaviors of β1- and β2-adrenergic receptors [197]. These and similar computational studies may provide interesting dimerization geometry hypotheses, although in view of the above experimental challenges, their validation may not be straightforward.

## 7 Structure-Based Design of Crystallographic Constructs for Chemokine Receptors

G-protein-coupled receptors represent challenging targets for structural and biophysical studies because of their notorious instability outside of their native membrane environment. Except for bovine rhodopsin, all crystallized GPCRs to date have been artificially stabilized to improve their behavior in detergent. The three stabilization approaches that proved successful thus far are point mutations, N-terminal or internal fusions with soluble proteins, and stabilizing antibodies. In particular, the multiple structures of $\beta_1$-adrenergic [198–201] and adenosine $A_{2A}$ [202–205] receptors were obtained using engineered constructs containing, in some cases, as many as nine simultaneous point mutations. Molecular dynamics simulations provided a rational explanation for the observed improvements in the stability of these crystallized mutant receptors [206, 207]. It is worth noting, however, that the stabilizing mutations were identified by comprehensive alanine scanning mutagenesis, which is an expensive and time-consuming process. In view of this, there is a considerable interest in rational methods that can help design crystallizable GPCR constructs with minimal experimental trial and error.

Methods for predicting stabilizing substitutions have been successfully developed in the context of soluble proteins and include both force field-based and knowledge-based approaches [208, 209]. However, they are not ideally suited for prediction of GPCR stability in detergents. To address this challenge, we developed a method for rational structure- or model-guided design of stabilizing mutations in GPCRs and benchmarked it on the available GPCR stabilizing mutations [66]. The method was applied to the design of stabilizing mutations in CCR5 using an ensemble of CXCR4-based and bRho-based homology models of the receptor. Of the 24 proposed mutations, 4 have been shown to improve the stability and homogeneity of detergent-solubilized CCR5. In combination with a rubredoxin fusion in ICL3, these mutations enabled the crystallization of the receptor in complex with maraviroc [15]. Retrospective analysis of the mutations in the context of the structure showed that three of them (C58$^{1.60}$Y, A233$^{6.33}$D, and K303$^{8.49}$E, all

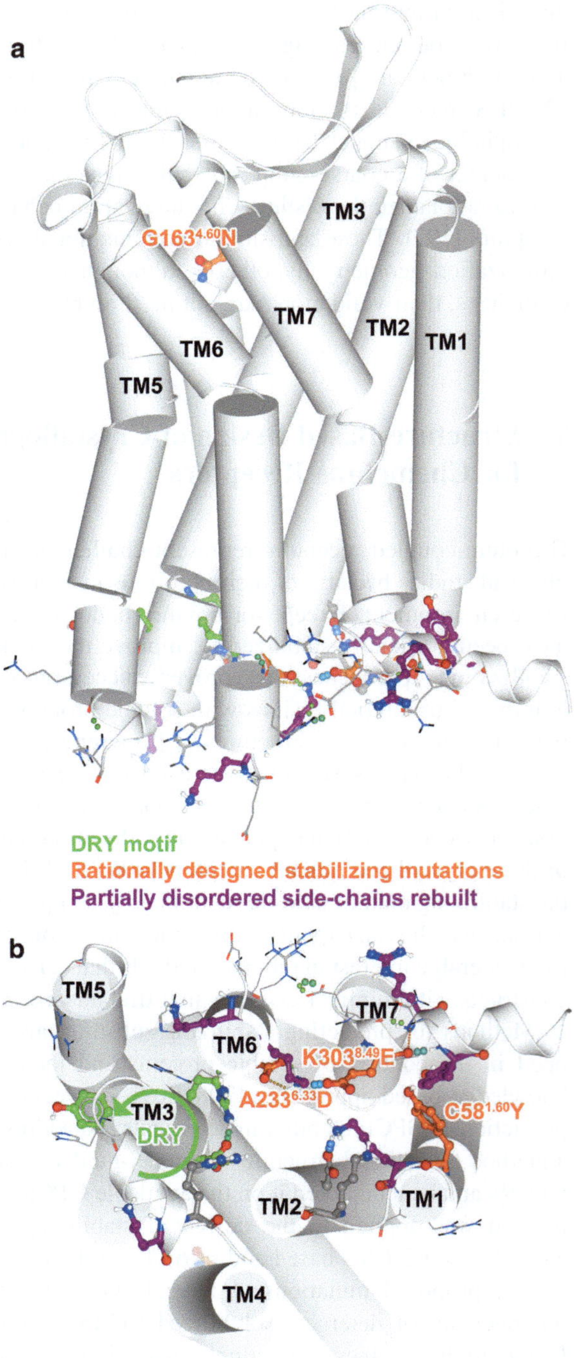


**Fig. 7** Rationally designed point mutations that stabilized and enabled crystallization of CCR5. (**a**) Lateral view along the plane of the membrane. (**b**) View across the membrane from the intracellular side. The mutations are likely to stabilize the receptor through an intricate hydrogen bonding network of WT basic residues (some of which are disordered in the X-ray structure and were rebuilt and optimized to produce this image) and the artificially introduced acids on the intracellular side of the helical bundle (A233$^{6.33}$D and K303$^{8.49}$E)

located at the intracellular termini of TM helices) stabilize the receptor by forming a ring of alternating acids and bases and thus a favorable network of hydrogen bonds (Fig. 7). Of note, the A233$^{6.33}$D mutation, but not other mutations, selectively stabilized the inactive state and rendered the receptor signaling dead. The G153$^{4.60}$ N mutation in the extracellular half of TM4 improves the helix rigidity and residue packing in this region.

Similarly, by a combination of bioinformatics and structural analysis of the $\beta_1$-adrenergic receptor, Chen et al. identified regions of local instability in its TM domain, namely, six nonconserved polar residues that were not significantly stabilized by hydrogen bonds. These nonconserved unsatisfied polar residue positions, as well as poorly packed hydrophobic regions, were used to prospectively design stabilizing residue substitutions using the Rosetta Membrane software. After stringent filtering of solutions, the authors arrived at nine single- and multiple-point mutants, all of which were found to have higher melting temperatures (improvements ranging from 0.3°C to 8.7°C) than the crystallized construct that was used to design them. Moreover, the designs could be combined to achieve as much as 11°C in stability increase [210].

These successes illustrate that structure-based design of stabilizing mutations is an attainable goal, although the success rates heavily depend on the accuracy of the structure or, in the case of CCR5, of the model used for mutant design. Because the main application of such methods is for receptors for which structures do not exist, higher success rates are expected for close homologs of already crystallized receptors. Moreover, selective stabilization of different functional states requires prediction of such states in the model, which, at the present state of the art, is not achievable in the absence of the relevant templates.

## 8 Modeling of Other Chemokine Receptors by Homology with Existing Structures

GPCR Dock 2010 illustrated how challenging receptor modeling is in the absence of closely related homology templates [54, 211]. In 2010, the CXCR4 structures provided the first atomic-resolution glimpse of the organization of chemokine receptors, which was enriched in 2013 by the CCR5 structure. These advances now enable modeling of other chemokine receptors with previously unattainable degree of prediction reliability [212]. Figure 8 summarizes the levels of homology observed between the solved structures and other receptors in the chemokine family.

As such, the available chemokine receptor structures have already informed and empowered studies of CXCR2 [214], CXCR3 [215], CXCR7 [216], and other chemokine receptors. For example, De Kruijf et al. used a CXCR4-based model of CXCR2, comparative structure analysis, and in silico-guided mutagenesis studies to locate a novel allosteric binding site for imidazolylpyrimidine compounds between TM helices 3, 5, and 6 [214]. Similarly, using site-directed mutagenesis

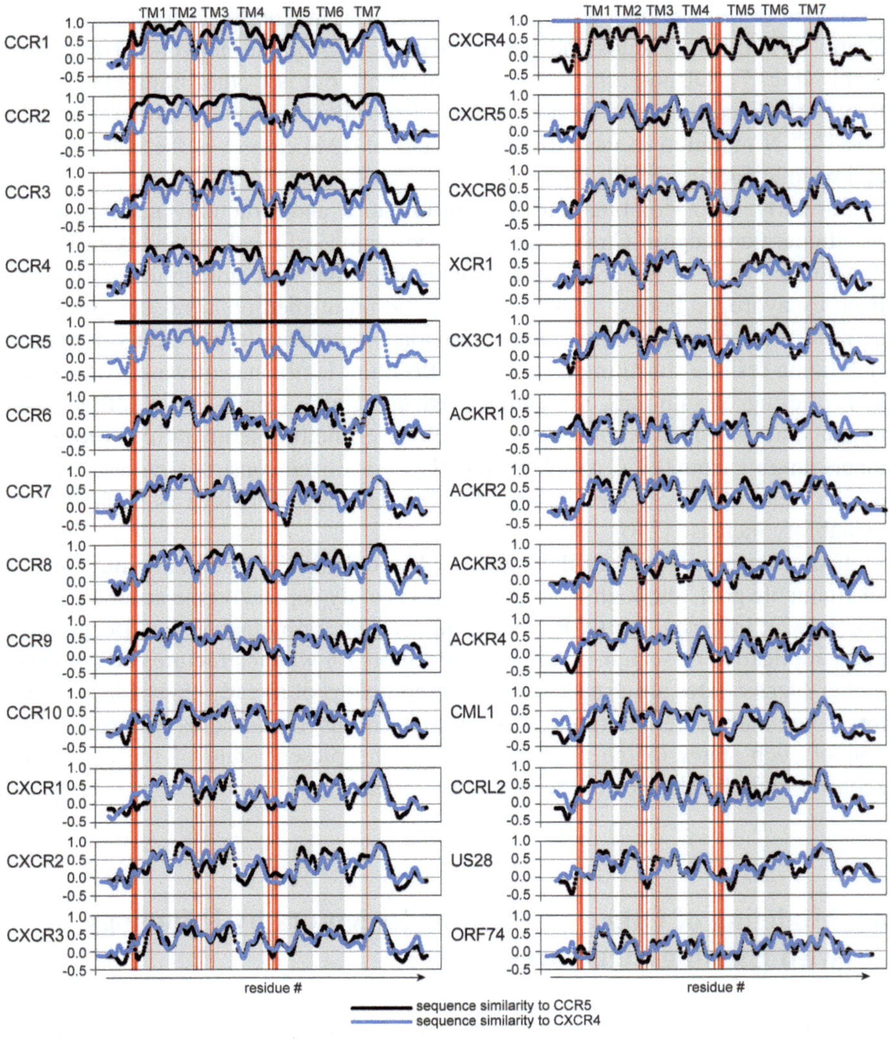


**Fig. 8** Per-residue sequence similarity of CCR5 (*black*) and CXCR4 (*blue*) with all chemokine and chemokine-like receptors was iteratively window-averaged along the sequence. Similarity was quantitated using normalized Gonnet matrix [213]. The location of TM domains is marked with grey stripes. Residues tentatively predicted to interact with chemokines in CXCR4 are projected through the sequence alignment and shown as *red bars*

and in silico CXCR4-based modeling of CXCR3, Scholten and colleagues elucidated the binding modes of distinct chemotypes of CXCR3 antagonists [215]. A virtual screening protocol using the consensus of pharmacophore-based and model-based approaches was proposed for the identification of CXCR3 antagonists, using the CXCR4 structure for the CXCR3 modeling [217]. Finally, using CXCR4-based conformational ensemble homology modeling, docking, and virtual screening,

Yoshikawa et al. were able to identify novel chemotypes of CXCR7 antagonists [216]. While the accuracy of compound binding pose predictions can only be confirmed (or disproved) by future X-ray crystallography efforts, these studies support the role of the structures in the chemokine receptor research and discovery.

**Conclusion**

The recently solved crystallographic structure of two chemokine receptors, CXCR4 and CCR5, provided valuable insights into the molecular mechanisms of chemokine receptor function and interaction with various ligands. However, they did not answer all the questions and have not led to molecular discovery breakthroughs to the extent comparable with the structures of other GPCRs. A possible reason for this shortfall is the inherently challenging nature of chemokine receptors that evolved to efficiently interact with chemokines via their conformationally variable, polar, and poorly druggable interfaces. Along with the experimentally determined structures, important molecular discoveries could be and were made using computational models of the receptors built from distantly homologous templates that were subsequently refined and trained to be predictive via application of modern computational refinement methodologies. Ligand- and pharmacophore-based techniques that are independent of receptor pocket models also proved productive.

In view of challenges posed by this family of receptors, only the efficient symbiosis of computational modeling with experimental determination of chemokine receptor structures with multiple ligands and in multiple functional states has the potential of answering all the questions.

**Acknowledgements** Authors thank Dr. Seva Katritch, The Scripps Research Institute, for valuable discussions and insights, and Eugene Raush, Molsoft LLC, for the help with molecular graphics. This work is partially funded by National Institutes of Health grants R01 GM071872, U01 GM094612, U54 GM094618.

## References

1. Salon JA, Lodowski DT, Palczewski K (2011) The significance of G protein-coupled receptor crystallography for drug discovery. Pharmacol Rev 63:901–937
2. Sliwoski G, Kothiwale S, Meiler J, Lowe EW (2014) Computational methods in drug discovery. Pharmacol Rev 66:334–395
3. Jacobson KA, Costanzi S (2012) New insights for drug design from the X-ray crystallographic structures of G-protein-coupled receptors. Mol Pharmacol 82:361–371
4. Allen SJ, Crown SE, Handel TM (2007) Chemokine: receptor structure, interactions, and antagonism. Annu Rev Immunol 25:787–820
5. Contento RL, Molon B, Boularan C, Pozzan T, Manes S, Marullo S, Viola A (2008) CXCR4-CCR5: a couple modulating T cell functions. Proc Natl Acad Sci U S A 105:10101–10106

6. Koelink PJ, Overbeek SA, Braber S, de Kruijf P, Folkerts G, Smit MJ, Kraneveld AD (2012) Targeting chemokine receptors in chronic inflammatory diseases: an extensive review. Pharmacol Ther 133:1–18
7. Viola A, Luster AD (2008) Chemokines and their receptors: drug targets in immunity and inflammation. Annu Rev Pharmacol Toxicol 48:171–197
8. Berger EA, Murphy PM, Farber JM (1999) Chemokine receptors as HIV-1 coreceptors: roles in viral entry, tropism, and disease. Annu Rev Immunol 17:657–700
9. Scholten DJ, Canals M, Maussang D, Roumen L, Smit MJ, Wijtmans M, de Graaf C, Vischer HF, Leurs R (2011) Pharmacological modulation of chemokine receptor function. Br J Pharmacol 165:1617–1643
10. Wu Y, Yoder A (2009) Chemokine coreceptor signaling in HIV-1 infection and pathogenesis. PLoS Pathog 5:e1000520
11. Horuk R, Chitnis CE, Darbonne WC, Colby TJ, Rybicki A, Hadley TJ, Miller LH (1993) A receptor for the malarial parasite *Plasmodium vivax*: the erythrocyte chemokine receptor. Science 261:1182–1184
12. Bill RM, Henderson PJF, Iwata S, Kunji ERS, Michel H, Neutze R, Newstead S, Poolman B, Tate CG, Vogel H (2011) Overcoming barriers to membrane protein structure determination. Nat Biotech 29:335–340
13. Cherezov V, Abola E, Stevens RC (2010) Recent progress in the structure determination of GPCRs, a membrane protein family with high potential as pharmaceutical targets. Methods Mol Biol 654:141–168
14. Wu B, Chien EYT, Mol CD, Fenalti G, Liu W, Katritch V, Abagyan R, Brooun A, Wells P, Bi FC et al (2010) Structures of the CXCR4 Chemokine GPCR with small-molecule and cyclic peptide antagonists. Science 330:1066–1071
15. Tan Q, Zhu Y, Li J, Chen Z, Han GW, Kufareva I, Li T, Ma L, Fenalti G, Li J et al (2013) Structure of the CCR5 chemokine receptor-HIV entry inhibitor maraviroc complex. Science 341:1387–1390
16. Stevens RC, Cherezov V, Katritch V, Abagyan R, Kuhn P, Rosen H, Wuthrich K (2013) The GPCR Network: a large-scale collaboration to determine human GPCR structure and function. Nat Rev Drug Discov 12:25–34
17. Thoma G, Streiff MB, Kovarik J, Glickman F, Wagner T, Beerli C, Zerwes H-G (2008) Orally bioavailable isothioureas block function of the chemokine receptor CXCR4 in vitro and in vivo. J Med Chem 51:7915–7920
18. Oishi S, Fujii N (2012) Peptide and peptidomimetic ligands for CXC chemokine receptor 4 (CXCR4). Org Biomol Chem 10:5720–5731
19. Fenalti G, Giguere PM, Katritch V, Huang X-P, Thompson AA, Cherezov V, Roth BL, Stevens RC (2014) Molecular control of delta-opioid receptor signalling. Nature 506:191–196
20. Granier S, Manglik A, Kruse AC, Kobilka TS, Thian FS, Weis WI, Kobilka BK (2012) Structure of the delta-opioid receptor bound to naltrindole. Nature 485:400–404
21. Manglik A, Kruse AC, Kobilka TS, Thian FS, Mathiesen JM, Sunahara RK, Pardo L, Weis WI, Kobilka BK, Granier S (2012) Crystal structure of the [mu]-opioid receptor bound to a morphinan antagonist. Nature 485:321–326
22. Thompson AA, Liu W, Chun E, Katritch V, Wu H, Vardy E, Huang X-P, Trapella C, Guerrini R, Calo G et al (2012) Structure of the nociceptin/orphanin FQ receptor in complex with a peptide mimetic. Nature 485:395–399
23. Wu H, Wacker D, Mileni M, Katritch V, Han GW, Vardy E, Liu W, Thompson AA, Huang X-P, Carroll FI et al (2012) Structure of the human k-opioid receptor in complex with JDTic. Nature 485:327–332
24. White JF, Noinaj N, Shibata Y, Love J, Kloss B, Xu F, Gvozdenovic-Jeremic J, Shah P, Shiloach J, Tate CG et al (2013) Structure of the agonist-bound neurotensin receptor. Nature 490:508–513

25. Zhang C, Srinivasan Y, Arlow DH, Fung JJ, Palmer D, Zheng Y, Green HF, Pandey A, Dror RO, Shaw DE et al (2012) High-resolution crystal structure of human protease-activated receptor 1. Nature 492:387–392
26. Murakami M, Kouyama T (2008) Crystal structure of squid rhodopsin. Nature 453:363–367
27. Dorr P, Westby M, Dobbs S, Griffin P, Irvine B, Macartney M, Mori J, Rickett G, Smith-Burchnell C, Napier C et al (2005) Maraviroc (UK-427,857), a potent, orally bioavailable, and selective small-molecule inhibitor of chemokine receptor CCR5 with broad-spectrum anti-human immunodeficiency virus type 1 activity. Antimicrob Agents Chemother 49:4721–4732
28. Wood A, Armour D (2005) The discovery of the CCR5 receptor antagonist, UK-427,857, a new agent for the treatment of HIV infection and AIDS. Prog Med Chem 43:239–271
29. Hernanz-Falcon P, Rodriguez-Frade JM, Serrano A, Juan D, del Sol A, Soriano SF, Roncal F, Gomez L, Valencia A, Martinez-A C et al (2004) Identification of amino acid residues crucial for chemokine receptor dimerization. Nat Immunol 5:216–223
30. Issafras H, Angers S, Bulenger S, Blanpain C, Parmentier M, Labbé-Jullié C, Bouvier M, Marullo S (2002) Constitutive agonist-independent CCR5 oligomerization and antibody-mediated clustering occurring at physiological levels of receptors. J Biol Chem 277:34666–34673
31. Springael J-Y, Le Minh PN, Urizar E, Costagliola S, Vassart G, Parmentier M (2006) Allosteric modulation of binding properties between units of chemokine receptor homo- and hetero-oligomers. Mol Pharmacol 69:1652–1661
32. Abagyan R, Totrov M (1994) Biased probability Monte Carlo conformational searches and electrostatic calculations for peptides and proteins. J Mol Biol 235:983–1002
33. Abagyan RA, Totrov MM, Kuznetsov DA (1994) Icm: a new method for protein modeling and design: applications to docking and structure prediction from the distorted native conformation. J Comp Chem 15:488–506
34. Connolly M (1983) Analytical molecular surface calculation. J Appl Crystallogr 16:548–558
35. Planesas JM, Perez-Nueno VI, Borrell JI, Teixido J (2012) Impact of the CXCR4 structure on docking-based virtual screening of HIV entry inhibitors. J Mol Graph Model 38:123–136
36. Muniz-Medina VM, Jones S, Maglich JM, Galardi C, Hollingsworth RE, Kazmierski WM, Ferris RG, Edelstein MP, Chiswell KE, Kenakin TP (2009) The relative activity of "Function Sparing" HIV-1 entry inhibitors on viral entry and CCR5 internalization: is allosteric functional selectivity a valuable therapeutic property? Mol Pharmacol 75:490–501
37. Thiele S, Steen A, Jensen PC, Mokrosinski J, Frimurer TM, Rosenkilde MM (2011) Allosteric and orthosteric sites in CC chemokine receptor (CCR5), a chimeric receptor approach. J Biol Chem 286:37543–37554
38. Horster S, Goebel FD (2006) Serious doubts on safety and efficacy of CCR5 antagonists. Infection 34:110–113
39. Choi W-T, Duggineni S, Xu Y, Huang Z, An J (2012) Drug discovery research targeting the CXC chemokine receptor 4 (CXCR4). J Med Chem 55:977–994
40. Debnath B, Xu S, Grande F, Garofalo A, Neamati N (2013) Small Molecule Inhibitors of CXCR4. Theranostics 3:47–75
41. Pease J, Horuk R (2012) Chemokine receptor antagonists. J Med Chem 55:9363–9392
42. De Clercq E (2010) Recent advances on the use of the CXCR4 antagonist plerixafor (AMD3100, Mozobil(TM)) and potential of other CXCR4 antagonists as stem cell mobilizers. Pharmacol Ther 128:509–518
43. Steinberg M, Silva M (2010) Plerixafor: a chemokine receptor-4 antagonist for mobilization of hematopoietic stem cells for transplantation after high-dose chemotherapy for non-hodgkin's lymphoma or multiple myeloma. Clin Ther 32:821–843
44. Fujii N, Oishi S, Hiramatsu K, Araki T, Ueda S, Tamamura H, Otaka A, Kusano S, Terakubo S, Nakashima H et al (2003) Molecular-size reduction of a potent CXCR4-chemokine antagonist using orthogonal combination of conformation- and sequence-based libraries. Angew Chem Int Ed 42:3251–3253

45. Gerlach LO, Skerlj RT, Bridger GJ, Schwartz TW (2001) Molecular interactions of cyclam and bicyclam non-peptide antagonists with the CXCR4 chemokine receptor. J Biol Chem 276:14153–14160
46. Rosenkilde MM, Gerlach L-O, Hatse S, Skerlj RT, Schols D, Bridger GJ, Schwartz TW (2007) Molecular mechanism of action of monocyclam versus bicyclam non-peptide antagonists in the CXCR4 chemokine receptor. J Biol Chem 282:27354–27365
47. Trent JO, Wang Z-X, Murray JL, Shao W, Tamamura H, Fujii N, Peiper SC (2003) Lipid bilayer simulations of CXCR4 with inverse agonists and weak partial agonists. J Biol Chem 278:47136–47144
48. Wong RSY, Bodart V, Metz M, Labrecque J, Bridger G, Fricker SP (2008) Comparison of the potential multiple binding modes of bicyclam, monocylam, and noncyclam small-molecule CXC chemokine receptor 4 inhibitors. Mol Pharmacol 74:1485–1495
49. Våbenø J, Nikiforovich GV, Marshall GR (2006) A minimalistic 3D pharmacophore model for cyclopentapeptide CXCR4 antagonists. Pept Sci 84:459–471
50. Våbenø J, Nikiforovich GV, Marshall GR (2006) Insight into the binding mode for cyclopentapeptide antagonists of the CXCR4 receptor. Chem Biol Drug Des 67:346–354
51. Kawatkar SP, Yan M, Gevariya H, Lim MY, Eisold S, Zhu X, Huang Z, An J (2011) Computational analysis of the structural mechanism of inhibition of chemokine receptor CXCR4 by small molecule antagonists. Exp Biol Med 236:844–850
52. Boulais PE, Dulude D, Cabana J, Heveker N, Escher E, Lavigne P, Leduc R (2009) Photolabeling identifies transmembrane domain 4 of CXCR4 as a T140 binding site. Biochem Pharmacol 78:1382–1390
53. Neves MAC, Simoes S, Sae Melo ML (2010) Ligand-guided optimization of CXCR4 homology models for virtual screening using a multiple chemotype approach. J Comput Aided Mol Des 24:1023–1033
54. Kufareva I, Rueda M, Katritch V, Stevens RC, Abagyan R (2011) Status of GPCR modeling and docking as reflected by community-wide GPCR Dock 2010 assessment. Structure 19:1108–1126
55. Debnath AK (2013) Rational design of HIV-1 entry inhibitors. In: Kortagere S (ed) In silico models for drug discovery. Humana Press, pp 185–204. doi:10.1007/978-1-62703-342-8_13
56. Kooistra AJ, Leurs R, Esch IJP, Graaf C (2014) From three-dimensional GPCR structure to rational ligand discovery. In: Filizola M (ed) G protein-coupled receptors – modeling and simulation. Springer, Netherlands, pp 129–157
57. Roumen L, Scholten DJ, de Kruijf P, de Esch IJP, Leurs R, de Graaf C (2012) C(X)CR in silico: computer-aided prediction of chemokine receptor-ligand interactions. Drug Discov Today Technol 9:e281–e291
58. Li J, Edwards PC, Burghammer M, Villa C, Schertler GFX (2004) Structure of bovine rhodopsin in a trigonal crystal form. J Mol Biol 343:1409–1438
59. Carlsson J, Coleman RG, Setola V, Irwin JJ, Fan H, Schlessinger A, Sali A, Roth BL, Shoichet BK (2011) Ligand discovery from a dopamine D3 receptor homology model and crystal structure. Nat Chem Biol 7:769–778
60. Fan H, Irwin JJ, Webb BM, Klebe G, Shoichet BK, Sali A (2009) Molecular docking screens using comparative models of proteins. J Chem Inf Model 49:2512–2527
61. Garcia-Perez J, Rueda P, Alcami J, Rognan D, Arenzana-Seisdedos F, Lagane B, Kellenberger E (2011) Allosteric model of maraviroc binding to CC chemokine receptor 5 (CCR5). J Biol Chem 286:33409–33421
62. Katritch V, Rueda M, Lam PC-H, Yeager M, Abagyan R (2010) GPCR 3D homology models for ligand screening: lessons learned from blind predictions of adenosine A2a receptor complex. Proteins 78:197–211
63. Kennedy DP, McRobb FM, Leonhardt SA, Purdy M, Figler H, Marshall MA, Chordia M, Figler RA, Linden J, Abagyan R et al (2013) The second extracellular loop of the adenosine A1 receptor mediates activity of allosteric enhancers. Mol Pharmacol 85(2):301–309

64. McRobb FM, Capuano B, Crosby IT, Chalmers DK, Yuriev E (2010) Homology modeling and docking evaluation of aminergic G protein-coupled receptors. J Chem Inf Model 50:626–637
65. Mysinger MM, Weiss DR, Ziarek JJ, Gravel S, Doak AK, Karpiak J, Heveker N, Shoichet BK, Volkman BF (2012) Structure-based ligand discovery for the protein-protein interface of chemokine receptor CXCR4. Proc Natl Acad Sci U S A 109:5517–5522
66. Kufareva I, Katritch V, Participants of GPCR Dock, Stevens RC, Abagyan R (2014) Advances in GPCR modeling evaluated by the GPCR Dock 2013 assessment: meeting new challenges. Structure 22(8):1120–1139
67. Michino M, Abola E, Participants GD, Brooks CL, Dixon JS, Moult J, Stevens RC (2009) Community-wide assessment of GPCR structure modelling and ligand docking: GPCR Dock 2008. Nat Rev Drug Discov 8:455–463
68. Clark R, Webster-Clark D (2008) Managing bias in ROC curves. J Comput Aided Mol Des 22:141–146
69. Jain A, Nicholls A (2008) Recommendations for evaluation of computational methods. J Comput Aided Mol Des 22:133–139
70. Katritch V, Rueda M, Abagyan R (2012) Ligand-guided receptor optimization. Methods Mol Biol 857:189–205
71. Brooks B, Karplus M (1983) Harmonic dynamics of proteins: normal modes and fluctuations in bovine pancreatic trypsin inhibitor. Proc Natl Acad Sci U S A 80:6571–6575
72. Dobbins SE, Lesk VI, Sternberg MJ (2008) Insights into protein flexibility: the relationship between normal modes and conformational change upon protein–protein docking. Proc Natl Acad Sci U S A 105:10390–10395
73. Hayward S, de Groot BL (2008) Normal modes and essential dynamics. Methods Mol Biol 443:89–106
74. Rueda M, Bottegoni G, Abagyan R (2009) Consistent improvement of cross-docking results using binding site ensembles generated with elastic network normal modes. J Chem Inf Model 49(3):716–725
75. Tama F, Miyashita O, Brooks CL 3rd (2004) Flexible multi-scale fitting of atomic structures into low-resolution electron density maps with elastic network normal mode analysis. J Mol Biol 337:985–999
76. Okuno Y, Tamon A, Yabuuchi H, Niijima S, Minowa Y, Tonomura K, Kunimoto R, Feng C (2008) GLIDA: GPCR–ligand database for chemical genomics drug discovery–database and tools update. Nucleic Acids Res 36:907–912
77. Perez-Nueno VI, Ritchie DW, Rabal O, Pascual R, Borrell JI, Teixido J (2008) Comparison of ligand-based and receptor-based virtual screening of HIV entry inhibitors for the CXCR4 and CCR5 receptors using 3D ligand shape matching and ligand-receptor docking. J Chem Inf Model 48:509–533
78. Morris GM, Huey R, Lindstrom W, Sanner MF, Belew RK, Goodsell DS, Olson AJ (2009) AutoDock4 and AutoDockTools4: automated docking with selective receptor flexibility. J Comput Chem 30:2785–2791
79. Trott O, Olson AJ (2010) AutoDock Vina: improving the speed and accuracy of docking with a new scoring function, efficient optimization, and multithreading. J Comput Chem 31:455–461
80. McGann M (2011) FRED pose prediction and virtual screening accuracy. J Chem Inf Model 51:578–596
81. Perez-Nueno VI, Pettersson S, Ritchie DW, Borrell JI, Teixido J (2009) Discovery of novel HIV entry inhibitors for the CXCR4 receptor by prospective virtual screening. J Chem Inf Model 49:810–823
82. Verdonk ML, Cole JC, Hartshorn MJ, Murray CW, Taylor RD (2003) Improved protein-ligand docking using GOLD. Proteins 52:609–623

83. Friesner RA, Banks JL, Murphy RB, Halgren TA, Klicic JJ, Mainz DT, Repasky MP, Knoll EH, Shelley M, Perry JK et al (2004) Glide: a new approach for rapid, accurate docking and scoring. 1. Method and assessment of docking accuracy. J Med Chem 47:1739–1749
84. Montes M, Miteva MA, Villoutreix BO (2007) Structure-based virtual ligand screening with LigandFit: pose prediction and enrichment of compound collections. Proteins Struct Funct Bioinformat 68:712–725
85. Karaboga AS, Planesas JM, Petronin F, Teixido J, Souchet M, Perez-Nueno VI (2013) Highly specific and sensitive pharmacophore model for identifying CXCR4 antagonists. Comparison with docking and shape-matching virtual screening performance. J Chem Inform Model 53:1043–1056
86. Kolb P, Rosenbaum DM, Irwin JJ, Fung JJ, Kobilka BK, Shoichet BK (2009) Structure-based discovery of b2-adrenergic receptor ligands. Proc Natl Acad Sci U S A 106:6843–6848
87. Weiss DR, Ahn S, Sassano MF, Kleist A, Zhu X, Strachan R, Roth BL, Lefkowitz RJ, Shoichet BK (2013) Conformation guides molecular efficacy in docking screens of activated b2 adrenergic G protein coupled receptor. ACS Chem Biol 8:1018–1026
88. Carlsson J, Yoo L, Gao Z-G, Irwin JJ, Shoichet BK, Jacobson KA (2010) Structure-based discovery of A2A adenosine receptor ligands. J Med Chem 53:3748–3755
89. Katritch V, Jaakola V-P, Lane JR, Lin J, Ijzerman AP, Yeager M, Kufareva I, Stevens RC, Abagyan R (2010) Structure-based discovery of novel chemotypes for adenosine A2A receptor antagonists. J Med Chem 53(4):1799–1809
90. Lane JR, Chubukov P, Liu W, Canals M, Cherezov V, Abagyan R, Stevens RC, Katritch V (2013) Structure-based ligand discovery targeting orthosteric and allosteric pockets of dopamine receptors. Mol Pharmacol 84:794–807
91. Kruse AC, Weiss DR, Rossi M, Hu J, Hu K, Eitel K, Gmeiner P, Wess J, Kobilka BK, Shoichet BK (2013) Muscarinic receptors as model targets and antitargets for structure-based ligand discovery. Mol Pharmacol 84:528–540
92. Vinader V, Ahmet DS, Ahmed MS, Patterson LH, Afarinkia K (2013) Discovery and computer aided potency optimization of a novel class of small molecule CXCR4 antagonists. PLoS One 8:e78744
93. Kim J, Yip MLR, Shen X, Li H, Hsin L-YC, Labarge S, Heinrich EL, Lee W, Lu J, Vaidehi N (2012) Identification of anti-malarial compounds as novel antagonists to chemokine receptor CXCR4 in pancreatic cancer cells. PLoS One 7:e31004
94. Liang Z, Zhan W, Zhu A, Yoon Y, Lin S, Sasaki M, Klapproth J-MA, Yang H, Grossniklaus HE, Xu J et al (2012) Development of a unique small molecule modulator of CXCR4. PLoS One 7:e34038
95. Zhan W, Liang Z, Zhu A, Kurtkaya S, Shim H, Snyder JP, Liotta DC (2007) Discovery of small molecule CXCR4 antagonists. J Med Chem 50:5655–5664
96. Truax VM, Zhao H, Katzman BM, Prosser AR, Alcaraz AA, Saindane MT, Howard RB, Culver D, Arrendale RF, Gruddanti PR et al (2013) Discovery of tetrahydroisoquinoline-based CXCR4 antagonists. ACS Med Chem Lett 4:1025–1030
97. Mungalpara J, Zachariassen ZG, Thiele S, Rosenkilde MM, Vabeno J (2013) Structure-activity relationship studies of the aromatic positions in cyclopentapeptide CXCR4 antagonists. Org Biomol Chem 11:8202–8208
98. Inokuchi E, Oishi S, Kubo T, Ohno H, Shimura K, Matsuoka M, Fujii N (2011) Potent CXCR4 antagonists containing amidine type peptide bond isosteres. ACS Med Chem Lett 2:477–480
99. Kobayashi K, Oishi S, Hayashi R, Tomita K, Kubo T, Tanahara N, Ohno H, Yoshikawa Y, Furuya T, Hoshino M et al (2012) Structure-activity relationship study of a CXC chemokine receptor type 4 antagonist, FC131, using a series of alkene dipeptide isosteres. J Med Chem 55:2746–2757
100. Mungalpara J, Thiele S, Eriksen O, Eksteen J, Rosenkilde MM, Vabeno J (2012) Rational design of conformationally constrained cyclopentapeptide antagonists for C-X-C chemokine receptor 4 (CXCR4). J Med Chem 55:10287–10291

101. Demmer O, Dijkgraaf I, Schumacher U, Marinelli L, Cosconati S, Gourni E, Wester H-J, Kessler H (2011) Design, synthesis, and functionalization of dimeric peptides targeting chemokine receptor CXCR4. J Med Chem 54:7648–7662
102. Aboye TL, Ha H, Majumder S, Christ F, Debyser Z, Shekhtman A, Neamati N, Camarero JA (2012) Design of a novel cyclotide-based CXCR4 antagonist with anti-human immunodeficiency virus (HIV)-1 activity. J Med Chem 55:10729–10734
103. Kazmierski W, Bifulco N, Yang H, Boone L, DeAnda F, Watson C, Kenakin T (2003) Recent progress in discovery of small-molecule CCR5 chemokine receptor ligands as HIV-1 inhibitors. Bioorg Med Chem 11:2663–2676
104. Labrecque J, Metz M, Lau G, Darkes MC, Wong RSY, Bogucki D, Carpenter B, Chen G, Li T, Nan S et al (2011) HIV-1 entry inhibition by small-molecule CCR5 antagonists: A combined molecular modeling and mutant study using a high-throughput assay. Virology 413:231–243
105. Maeda K, Das D, Ogata-Aoki H, Nakata H, Miyakawa T, Tojo Y, Norman R, Takaoka Y, Ding J, Arnold GF et al (2006) Structural and molecular interactions of CCR5 inhibitors with CCR5. J Biol Chem 281:12688–12698
106. Seibert C, Ying W, Gavrilov S, Tsamis F, Kuhmann SE, Palani A, Tagat JR, Clader JW, McCombie SW, Baroudy BM et al (2006) Interaction of small molecule inhibitors of HIV-1 entry with CCR5. Virology 349:41–54
107. Finke PE, Oates B, Mills SG, MacCoss M, Malkowitz L, Springer MS, Gould SL, DeMartino JA, Carella A, Carver G et al (2001) Antagonists of the human CCR5 receptor as anti-HIV-1 agents. Part 4: synthesis and structure-activity relationships for 1-[N-(Methyl)-N-(phenylsulfonyl)amino]-2-(phenyl)-4-(4-(N-(alkyl)-N-(benzyloxycarbonyl)amino)piperidin-1-yl)butanes. Bioorg Med Chem Lett 11:2475–2479
108. Berro R, Yasmeen A, Abrol R, Trzaskowski B, Abi-Habib S, Grunbeck A, Lascano D, Goddard WA, Klasse PJ, Sakmar TP et al (2013) Use of G-protein-coupled and -uncoupled CCR5 receptors by CCR5 inhibitor-resistant and -sensitive human immunodeficiency virus type 1 variants. J Virol 87:6569–6581
109. Abrol R, Griffith AR, Bray JK, Goddard WA III (2012) Structure prediction of G protein-coupled receptors and their ensemble of functionally important conformations. In: Vaidehi N, Klein-Seetharaman J (eds) Membrane protein structure and dynamics. Humana Press, pp 237–254. doi:10.1007/978-1-62703-023-6_14
110. Kothandan G, Gadhe CG, Cho SJ (2012) Structural Insights from binding poses of CCR2 and CCR5 with clinically important antagonists: a combined in silico study. PLoS One 7:e32864
111. Lagane B, Garcia-Perez J, Kellenberger E (2012) Modeling the allosteric modulation of CCR5 function by Maraviroc. Drug Discov Today Technol 10:e297–e305
112. Strizki JM, Xu S, Wagner NE, Wojcik L, Liu J, Hou Y, Endres M, Palani A, Shapiro S, Clader JW et al (2001) SCH-C (SCH 351125), an orally bioavailable, small molecule antagonist of the chemokine receptor CCR5, is a potent inhibitor of HIV-1 infection in vitro and in vivo. Proc Natl Acad Sci U S A 98:12718–12723
113. Watson C, Jenkinson S, Kazmierski W, Kenakin T (2005) The CCR5 receptor-based mechanism of action of 873140, a potent allosteric noncompetitive HIV entry inhibitor. Mol Pharmacol 67:1268–1282
114. Maeda K, Nakata H, Koh Y, Miyakawa T, Ogata H, Takaoka Y, Shibayama S, Sagawa K, Fukushima D, Moravek J et al (2004) Spirodiketopiperazine-Based CCR5 inhibitor which preserves CC-chemokine/CCR5 interactions and exerts potent activity against R5 human immunodeficiency virus type 1 in vitro. J Virol 78:8654–8662
115. Kellenberger E, Springael J-Y, Parmentier M, Hachet-Haas M, Galzi J-L, Rognan D (2007) Identification of nonpeptide CCR5 receptor agonists by structure-based virtual screening. J Med Chem 50:1294–1303
116. Liu Y, Zhou E, Yu K, Zhu J, Zhang Y, Xie X, Li J, Jiang H (2008) Discovery of a novel CCR5 antagonist lead compound through fragment assembly. Molecules 13:2426–2441

117. Song M, Breneman CM, Sukumar N (2004) Three-dimensional quantitative structure-activity relationship analyses of piperidine-based CCR5 receptor antagonists. Bioorg Med Chem 12:489–499
118. Xu Y, Liu H, Niu C, Luo C, Luo X, Shen J, Chen K, Jiang H (2004) Molecular docking and 3D QSAR studies on 1-amino-2-phenyl-4-(piperidin-1-yl)-butanes based on the structural modeling of human CCR5 receptor. Bioorg Med Chem 12:6193–6208
119. Zhuo Y, Kong R, Cong X-J, Chen W-Z, Wang C-X (2008). Three-dimensional QSAR analyses of 1,3,4-trisubstituted pyrrolidine-based CCR5 receptor inhibitors. Eur J Med Chem 43:2724–2734
120. Gadhe CG, Kothandan G, Cho SJ (2013) Binding site exploration of CCR5 using in silico methodologies: a 3D-QSAR approach. Arch Pharm Res 36:6–31
121. Metz M, Bourque E, Labrecque J, Danthi SJ, Langille J, Harwig C, Yang W, Darkes MC, Lau G, Santucci Z et al (2011) Prospective CCR5 small molecule antagonist compound design using a combined mutagenesis/modeling approach. J Am Chem Soc 133:16477–16485
122. Paavola CD, Hemmerich S, Grunberger D, Polsky I, Bloom A, Freedman R, Mulkins M, Bhakta S, McCarley D, Wiesent L et al (1998) Monomeric monocyte chemoattractant protein-1 (MCP-1) binds and activates the MCP-1 receptor CCR2B. J Biol Chem 273:33157–33165
123. Rajarathnam K, Sykes B, Kay C, Dewald B, Geiser T, Baggiolini M, Clark-Lewis I (1994) Neutrophil activation by monomeric interleukin-8. Science 264:90–92
124. Monteclaro FS, Charo IF (1996) The amino-terminal extracellular domain of the MCP-1 receptor, but not the RANTES/MIP-1a receptor, confers chemokine selectivity: evidence for a two-step mechanism fir MCP-1 receptor activation. J Biol Chem 271:19084–19092
125. Monteclaro FS, Charo IF (1997) The amino-terminal domain of CCR2 is both necessary and sufficient for high affinity binding of monocyte chemoattractant protein 1: receptor activation by a pseudo-tethered ligand. J Biol Chem 272:23186–23190
126. Rajagopalan L, Rajarathnam K (2006) Structural basis of chemokine receptor function – a model for binding affinity and ligand selectivity. Biosci Rep 26:325–339
127. Szpakowska M, Fievez V, Arumugan K, van Nuland N, Schmit J-C, Chevigne A (2012) Function, diversity and therapeutic potential of the N-terminal domain of human chemokine receptors. Biochem Pharmacol 84:1366–1380
128. Chevigne A, Fievez V, Schmit J-C, Deroo S (2011) Engineering and screening the N-terminus of chemokines for drug discovery. Biochem Pharmacol 82:1438–1456
129. Crump MP, Gong JH, Loetscher P, Rajarathnam K, Amara A, Arenzana-Seisdedos F, Virelizier JL, Baggiolini M, Sykes BD, Clark-Lewis I (1997) Solution structure and basis for functional activity of stromal cell-derived factor-1; dissociation of CXCR4 activation from binding and inhibition of HIV-1. Embo J 16:6996–7007
130. Dong C-Z, Tian S, Choi W-T, Kumar S, Liu D, Xu Y, Han X, Huang Z, An J (2012) Critical role in CXCR4 signaling and internalization of the polypeptide main chain in the amino terminus of SDF-1α probed by novel N-methylated synthetically and modularly modified chemokine analogues. Biochemistry 51:5951–5957
131. Gaertner H, Cerini F, Escola J-M, Kuenzi G, Melotti A, Offord R, Rossitto-Borlat IN, Nedellec R, Salkowitz J, Gorochov G et al (2008) Highly potent, fully recombinant anti-HIV chemokines: reengineering a low-cost microbicide. Proc Natl Acad Sci U S A 105:17706–17711
132. Loetscher P, Clark-Lewis I (2001) Agonistic and antagonistic activities of chemokines. J Leukoc Biol 69:881–884
133. Huang C-C, Lam SN, Acharya P, Tang M, Xiang S-H, Hussan SS-U, Stanfield RL, Robinson J, Sodroski J, Wilson IA, et al (2007). Structures of the CCR5 N terminus and of a tyrosine-sulfated antibody with HIV-1 gp120 and CD4. Science 317:1930–1934
134. Tan JHY, Ludeman JP, Wedderburn J, Canals M, Hall P, Butler SJ, Taleski D, Christopoulos A, Hickey MJ, Payne RJ et al (2013) Tyrosine sulfation of chemokine receptor

CCR2 enhances interactions with both monomeric and dimeric forms of the chemokine monocyte chemoattractant protein-1 (MCP-1). J Biol Chem 288:10024–10034
135. Veldkamp CT, Seibert C, Peterson FC, De la Cruz NB, Haugner JC III, Basnet H, Sakmar TP, Volkman BF (2008) Structural basis of CXCR4 sulfotyrosine recognition by the chemokine SDF-1/CXCL12. Sci Signal 1:ra4
136. Ziarek JJ, Getschman AE, Butler SJ, Taleski D, Stephens B, Kufareva I, Handel TM, Payne RJ, Volkman BF (2013) Sulfopeptide probes of the CXCR4/CXCL12 interface reveal oligomer-specific contacts and chemokine allostery. ACS Chem Biol 8:1955–1963
137. Brelot A, Heveker N, Montes M, Alizon M (2000) Identification of residues of CXCR4 critical for human immunodeficiency virus coreceptor and chemokine receptor activities. J Biol Chem 275:23736–23744
138. Zhou N, Luo Z, Luo J, Liu D, Hall JW, Pomerantz RJ, Huang Z (2001) Structural and functional characterization of human CXCR4 as a chemokine receptor and HIV-1 Co-receptor by mutagenesis and molecular modeling studies. J Biol Chem 276:42826–42833
139. Veldkamp CT, Ziarek JJ, Peterson FC, Chen Y, Volkman BF (2010) Targeting SDF-1/CXCL12 with a ligand that prevents activation of CXCR4 through structure-based drug design. J Am Chem Soc 132:7242–7243
140. Kofuku Y, Yoshiura C, Ueda T, Terasawa H, Hirai T, Tominaga S, Hirose M, Maeda Y, Takahashi H, Terashima Y et al (2009) Structural basis of the interaction between chemokine stromal cell-derived factor-1/CXCL12 and its g-protein-coupled receptor CXCR4. J Biol Chem 284:35240–35250
141. Murphy JW, Cho Y, Sachpatzidis A, Fan C, Hodsdon ME, Lolis E (2007) Structural and functional basis of CXCL12 (stromal cell-derived factor-1a) binding to heparin. J Biol Chem 282:10018–10027
142. Nardese V, Longhi R, Polo S, Sironi F, Arcelloni C, Paroni R, DeSantis C, Sarmientos P, Rizzi M, Bolognesi M et al (2001) Structural determinants of CCR5 recognition and HIV-1 blockade in RANTES. Nat Struct Mol Biol 8:611–615
143. Ohnishi Y, Senda T, Nandhagopal N, Sugimoto K, Shioda T, Nagal Y, Mitsui Y (2000) Crystal structure of recombinant native SDF-1alpha with additional mutagenesis studies: an attempt at a more comprehensive interpretation of accumulated structure-activity relationship data. J Interferon Cytokine Res 20:691–700
144. Vangelista L, Longhi R, Sironi F, Pavone V, Lusso P (2006) Critical role of the N-loop and b1-strand hydrophobic clusters of RANTES-derived peptides in anti-HIV activity. Biochem Biophys Res Commun 351:664–668
145. Loetscher P, Gong J-H, Dewald B, Baggiolini M, Clark-Lewis I (1998) N-terminal peptides of stromal cell-derived factor-1 with CXC chemokine receptor 4 agonist and antagonist activities. J Biol Chem 273:22279–22283
146. Luo Z, Fan X, Zhou N, Hiraoka M, Luo J, Kaji H, Huang Z (2000) Structure-function study and anti-HIV activity of synthetic peptide analogues derived from viral chemokine vMIP-II. Biochemistry 39:13545–13550
147. Zhou N, Luo Z, Luo J, Hall JW, Huang Z (2000) A novel peptide antagonist of CXCR4 derived from the N-terminus of viral chemokine vMIP-II. Biochemistry 39:3782–3787
148. Kim S, Lee C, Midura B, Yeung C, Mendoza A, Hong S, Ren L, Wong D, Korz W, Merzouk A et al (2008) Inhibition of the CXCR4/CXCL12 chemokine pathway reduces the development of murine pulmonary metastases. Clin Exp Metastasis 25:201–211
149. Lefrançois M, Lefebvre M-R, Saint-Onge G, Boulais PE, Lamothe S, Leduc R, Lavigne P, Heveker N, Escher E (2011) Agonists for the chemokine receptor CXCR4. ACS Med Chem Lett 2:597–602
150. Tudan C, Willick GE, Chahal S, Arab L, Law P, Salari H, Merzouk A (2002) C-Terminal cyclization of an SDF-1 small peptide analogue dramatically increases receptor affinity and activation of the CXCR4 receptor. J Med Chem 45:2024–2031

151. Gozansky EK, Louis JM, Caffrey M, Marius Clore G (2005) Mapping the binding of the N-terminal extracellular tail of the CXCR4 receptor to stromal cell-derived factor-1alpha. J Mol Biol 345:651–658
152. Tian S, Choi W-T, Liu D, Pesavento J, Wang Y, An J, Sodroski JG, Huang Z (2005) Distinct functional sites for human immunodeficiency virus type 1 and stromal cell-derived factor 1alpha on CXCR4 transmembrane helical domains. J Virol 79:12667–12673
153. Zoffmann S, Chollet A, Galzi J-L (2002) Identification of the extracellular loop 2 as the point of interaction between the N terminus of the chemokine MIP-1alpha and its CCR1 receptor. Mol Pharmacol 62:729–736
154. Huang X, Shen J, Cui M, Shen L, Luo X, Ling K, Pei G, Jiang H, Chen K (2003) Molecular dynamics simulations on SDF-1a binding with CXCR4 receptor. Biophys J 84:171–184
155. El-Asmar LL, Springael J-Y, Ballet SB, Andrieu EU, Vassart G, Parmentier M (2005) Evidence for negative binding cooperativity within CCR5-CCR2b heterodimers. Mol Pharmacol 67:460–469
156. Sohy D, Parmentier M, Springael J-Y (2007) Allosteric transinhibition by specific antagonists in CCR2/CXCR4 heterodimers. J Biol Chem 282:30062–30069
157. Vila-Coro AJ, Rodruguez-Frade JM, Martin De Ana A, Moreno-Ortiz MC, Martinez-A C, Mellado M (1999) The chemokine SDF-1a triggers CXCR4 receptor dimerization and activates the JAK/STAT pathway. FASEB J 13:1699–1710
158. Xu L, Li Y, Sun H, Li D, Hou T (2013) Structural basis of the interactions between CXCR4 and CXCL12/SDF-1 revealed by theoretical approaches. Mol BioSyst 9:2107–2117
159. Liou J-W, Chang F-T, Chung Y, Chen W-Y, Fischer WB, Hsu H-J (2014) In silico analysis reveals sequential interactions and protein conformational changes during the binding of chemokine CXCL-8 to its receptor CXCR1. PLoS One 9:e94178
160. Saini V, Marchese A, Majetschak M (2010) CXC chemokine receptor 4 is a cell surface receptor for extracellular ubiquitin. J Biol Chem 285:15566–15576
161. Chou C-Y, Lai H-Y, Chen H-Y, Cheng S-C, Cheng K-W, Chou Y-W (2014) Structural basis for catalysis and ubiquitin recognition by the severe acute respiratory syndrome coronavirus papain-like protease. Acta Crystallogr Sect D 70:572–581
162. Ryu EK, Kim TG, Kwon TH, Jung ID, Ryu D, Park Y-M, Kim J, Ahn KH, Ban C (2007) Crystal structure of recombinant human stromal cell-derived factor-1α. Proteins Struct Funct Bioinformat 67:1193–1197
163. Saini V, Marchese A, Tang W-J, Majetschak M (2011) Structural determinants of ubiquitin-CXC chemokine receptor 4 interaction. J Biol Chem 286:44145–44152
164. Saini V, Staren DM, Ziarek JJ, Nashaat ZN, Campbell EM, Volkman BF, Marchese A, Majetschak M (2011) The CXC chemokine receptor 4 ligands ubiquitin and stromal cell-derived factor-1alpha function through distinct receptor interactions. J Biol Chem 286:33466–33477
165. Choi W-T, Tian S, Dong C-Z, Kumar S, Liu D, Madani N, An J, Sodroski JG, Huang Z (2005) Unique ligand binding sites on CXCR4 probed by a chemical biology approach: implications for the design of selective human immunodeficiency virus type 1 inhibitors. J Virol 79:15398–15404
166. Liu J, Bartesaghi A, Borgnia MJ, Sapiro G, Subramaniam S (2008) Molecular architecture of native HIV-1 gp120 trimers. Nature 455:109–113
167. Unutmaz D, Littman DR (1997) Expression pattern of HIV-1 coreceptors on T cells: implications for viral transmission and lymphocyte homing. Proc Natl Acad Sci U S A 94:1615–1618
168. Bleul CC, Wu L, Hoxie JA, Springer TA, Mackay CR (1997) The HIV coreceptors CXCR4 and CCR5 are differentially expressed and regulated on human T lymphocytes. Proc Natl Acad Sci U S A 94:1925–1930
169. Lengauer T, Sander O, Sierra S, Thielen A, Kaiser R (2007) Bioinformatics prediction of HIV coreceptor usage. Nat Biotech 25:1407–1410

170. Bozek K, Lengauer T, Sierra S, Kaiser R, Domingues FS (2013) Analysis of physicochemical and structural properties determining HIV-1 coreceptor usage. PLoS Comput Biol 9: e1002977
171. Kumar R, Raghava GPS (2013) Hybrid approach for predicting coreceptor used by HIV-1 from Its V3 loop amino acid sequence. PLoS One 8:e61437
172. Masso M, Vaisman I (2010) Accurate and efficient gp120 V3 loop structure based models for the determination of HIV-1 co-receptor usage. BMC Bioinformat 11:494
173. Kufareva I, Chen Y-C, Ilatovskiy AV, Abagyan R (2012) Compound activity prediction using models of binding pockets or ligand properties in 3D. Curr Top Med Chem 12:1869–1882
174. Colin P, Benureau Y, Staropoli I, Wang Y, Gonzalez N, Alcami J, Hartley O, Brelot A, Arenzana-Seisdedos F, Lagane B (2013) HIV-1 exploits CCR5 conformational heterogeneity to escape inhibition by chemokines. Proc Natl Acad Sci U S A 110:9475–9480
175. Nedellec R, Coetzer M, Lederman MM, Offord RE, Hartley O, Mosier DE (2011) Resistance to the CCR5 Inhibitor 5P12-RANTES requires a difficult evolution from CCR5 to CXCR4 coreceptor use. PLoS One 6:e22020
176. Tilton JC, Amrine-Madsen H, Miamidian JL, Kitrinos KM, Pfaff J, Demarest JF, Ray N, Jeffrey JL, Labranche CC, Doms RW (2010) HIV type 1 from a patient with baseline resistance to CCR5 antagonists uses drug-bound receptor for entry. AIDS Res Hum Retroviruses 26:13–24
177. Berro R, Klasse PJ, Moore JP, Sanders RW (2012) V3 determinants of HIV-1 escape from the CCR5 inhibitors Maraviroc and Vicriviroc. Virology 427:158–165
178. Harrison JE, Lynch JB, Sierra L-J, Blackburn LA, Ray N, Collman RG, Doms RW (2008) Baseline resistance of primary human immunodeficiency virus type 1 strains to the CXCR4 inhibitor AMD3100. J Virol 82:11695–11704
179. Kramp BK, Sarabi A, Koenen RR, Weber C (2011) Heterophilic chemokine receptor interactions in chemokine signaling and biology. Exp Cell Res 317:655–663
180. Salanga CL, O'Hayre M, Handel T (2009) Modulation of chemokine receptor activity through dimerization and crosstalk. Cell Mol Life Sci 66:1370–1386
181. Stephens B, Handel TM (2013) Chemokine receptor oligomerization and allostery. In: Kenakin T (ed) Oligomerization and allosteric modulation in G-protein coupled receptors. Academic, pp 375–420. doi:10.1016/B978-0-12-394587-7.00009-9
182. Milligan G (2009) The role of dimerisation in the cellular trafficking of G-protein-coupled receptors. Curr Opin Pharmacol 10:23–29
183. Milligan G (2013) The prevalence, maintenance, and relevance of G protein-coupled receptor oligomerization. Mol Pharmacol 84:158–169
184. Katritch V, Cherezov V, Stevens RC (2013) Structure-function of the G protein-coupled receptor superfamily. Annu Rev Pharmacol Toxicol 53:531–556
185. Huang J, Chen S, Zhang JJ, Huang X-Y (2013) Crystal structure of oligomeric β1-adrenergic G protein–coupled receptors in ligand-free basal state. Nat Struct Mol Biol 20:419–425
186. Hern JA, Baig AH, Mashanov GI, Birdsall B, Corrie JET, Lazareno S, Molloy JE, Birdsall NJM (2010) Formation and dissociation of M1 muscarinic receptor dimers seen by total internal reflection fluorescence imaging of single molecules. Proc Natl Acad Sci U S A 107:2693–2698
187. Kufareva I, Stephens B, Gilliland CT, Wu B, Fenalti G, Hamel DJ, Stevens RC, Abagyan R, Handel TM (2013) A novel approach to quantify G-protein-coupled receptor dimerization equilibrium using bioluminescence resonance energy transfer. In: Cardona AE, Ubogu EE (eds) Chemokines: methods and protocols. New York, Springer, pp 93–127
188. Nobles M, Benians A, Tinker A (2005) Heterotrimeric G proteins precouple with G protein-coupled receptors in living cells. Proc Natl Acad Sci U S A 102:18706–18711
189. Sohy D, Yano H, de Nadai P, Urizar E, Guillabert A, Javitch JA, Parmentier M, Springael J-Y (2009) Hetero-oligomerization of CCR2, CCR5, and CXCR4 and the protean effects of "selective" antagonists. J Biol Chem 284:31270–31279

190. Lemay J, Marullo S, Jockers R, Alizon M, Brelot A (2005) On the dimerization of CCR5. Nat Immunol 6:535–535
191. Wang J, He L, Combs CA, Roderiquez G, Norcross MA (2006) Dimerization of CXCR4 in living malignant cells: control of cell migration by a synthetic peptide that reduces homologous CXCR4 interactions. Mol Cancer Ther 5:2474–2483
192. Percherancier Y, Berchiche YA, Slight I, Volkmer-Engert R, Tamamura H, Fujii N, Bouvier M, Heveker N (2005) Bioluminescence resonance energy transfer reveals ligand-induced conformational changes in CXCR4 homo- and heterodimers. J Biol Chem 280:9895–9903
193. Casciari D, Dell'Orco D, Fanelli F (2008) Homodimerization of neurotensin 1 receptor involves helices 1, 2, and 4: insights from quaternary structure predictions and dimerization free energy estimations. J Chem Inf Model 48:1669–1678
194. Johnston JM, Filizola M (2014) Beyond standard molecular dynamics: investigating the molecular mechanisms of G protein-coupled receptors with enhanced molecular dynamics methods. In: Filizola M (ed) G protein-coupled receptors – modeling and simulation. Springer, Netherlands, pp 95–125
195. Periole X, Knepp AM, Sakmar TP, Marrink SJ, Huber T (2012) Structural determinants of the supramolecular organization of G protein-coupled receptors in bilayers. J Am Chem Soc 134:10959–10965
196. Johnston JM, Wang H, Provasi D, Filizola M (2012) Assessing the relative stability of dimer interfaces in G protein-coupled receptors. PLoS Comput Biol 8:e1002649
197. Mondal S, Johnston JM, Wang H, Khelashvili G, Filizola M, Weinstein H (2013) Membrane driven spatial organization of GPCRs. Sci Rep 3
198. Christopher JA, Brown J, Dore AS, Errey JC, Koglin M, Marshall FH, Myszka DG, Rich RL, Tate CG, Tehan B et al (2013) Biophysical fragment screening of the beta1-adrenergic receptor: identification of high affinity arylpiperazine leads using structure-based drug design. J Med Chem 56:3446–3455
199. Moukhametzianov R, Warne T, Edwards PC, Serrano-Vega MJ, Leslie AGW, Tate CG, Schertler GFX (2011) Two distinct conformations of helix 6 observed in antagonist-bound structures of a b1-adrenergic receptor. Proc Natl Acad Sci U S A 108:8228–8232
200. Warne T, Edwards PC, Leslie AG, Tate CG (2012) Crystal structures of a stabilized b1-adrenoceptor bound to the biased agonists bucindolol and carvedilol. Structure 20:841–849
201. Warne T, Moukhametzianov R, Baker JG, Nehme R, Edwards PC, Leslie AGW, Schertler GFX, Tate CG (2011) The structural basis for agonist and partial agonist action on a b1-adrenergic receptor. Nature 469:241–244
202. Congreve M, Andrews SP, Dore AS, Hollenstein K, Hurrell E, Langmead CJ, Mason JS, Ng IW, Tehan B, Zhukov A et al (2012) Discovery of 1,2,4-triazine derivatives as adenosine A2A antagonists using structure based drug design. J Med Chem 55:1898–1903
203. Dore AS, Robertson N, Errey JC, Ng I, Hollenstein K, Tehan B, Hurrell E, Bennett K, Congreve M, Magnani F et al (2011) Structure of the adenosine A2A receptor in complex with ZM241385 and the xanthines XAC and caffeine. Structure 19:1283–1293
204. Hino T, Arakawa T, Iwanari H, Yurugi-Kobayashi T, Ikeda-Suno C, Nakada-Nakura Y, Kusano-Arai O, Weyand S, Shimamura T, Nomura N et al (2012) G-protein-coupled receptor inactivation by an allosteric inverse-agonist antibody. Nature 482:237–240
205. Lebon G, Warne T, Edwards PC, Bennett K, Langmead CJ, Leslie AGW, Tate CG (2011) Agonist-bound adenosine A2A receptor structures reveal common features of GPCR activation. Nature 474:521–525
206. Lee S, Bhattacharya S, Grisshammer R, Tate C, Vaidehi N (2014) Dynamic behavior of the active and inactive states of the adenosine A2A receptor. J Phys Chem B 118:3355–3365
207. Niesen MJM, Bhattacharya S, Grisshammer R, Tate CG, Vaidehi N (2013) Thermostabilization of the b1-adrenergic receptor correlates with increased entropy of the inactive state. J Phys Chem B 117:7283–7291

208. Bordner AJ, Abagyan RA (2004) Large-scale prediction of protein geometry and stability changes for arbitrary single point mutations. Proteins 57:400–413
209. Masso M, Vaisman II (2008) Accurate prediction of stability changes in protein mutants by combining machine learning with structure based computational mutagenesis. Bioinformatics 24:2002–2009
210. Chen K-YM, Zhou F, Fryszczyn BG, Barth P (2012) Naturally evolved G protein-coupled receptors adopt metastable conformations. Proc Natl Acad Sci U S A 109:13284–13289
211. Bhattacharya S, Lam AR, Li H, Balaraman G, Niesen MJM, Vaidehi N (2013) Critical analysis of the successes and failures of homology models of G protein-coupled receptors. Proteins Struct Funct Bioinformat 81:729–739
212. Zhu L, Zhao Q, Wu B (2013) Structure-based studies of chemokine receptors. Curr Opin Struct Biol 23:539–546
213. Gonnet G, Cohen M, Benner S (1992) Exhaustive matching of the entire protein sequence database. Science 256:1443–1445
214. de Kruijf P, Lim HD, Roumen L, Renjaan VA, Zhao J, Webb ML, Auld DS, Wijkmans JCHM, Zaman GJR, Smit MJ et al (2011) Identification of a novel allosteric binding site in the CXCR2 chemokine receptor. Mol Pharmacol 80:1108–1118
215. Scholten DJ, Roumen L, Wijtmans M, Verkade-Vreeker MCA, Custers H, Lai M, de Hooge D, Canals M, de Esch IJP, Smit MJ et al (2014) Identification of overlapping but differential binding sites for the high-affinity CXCR3 antagonists NBI-74330 and VUF11211. Mol Pharmacol 85:116–126
216. Yoshikawa Y, Oishi S, Kubo T, Tanahara N, Fujii N, Furuya T (2013) Optimized method of G-protein-coupled receptor homology modeling: its application to the discovery of novel CXCR7 ligands. J Med Chem 56:4236–4251
217. Huang D, Gu Q, Ge H, Ye J, Salam NK, Hagler A, Chen H, Xu J (2012) On the value of homology models for virtual screening: discovering hCXCR3 antagonists by pharmacophore-based and structure-based approaches. J Chem Inf Model 52:1356–1366

Top Med Chem (2015) 14: 87–118
DOI: 10.1007/7355_2014_82

Published online: 23 November 2014

# Allosteric Modulation of Chemokine Receptors

**Nuska Tschammer, Arthur Christopoulos, and Terry Kenakin**

**Abstract** A central role of chemokines and their receptors in inflammatory processes has spurred numerous screening campaigns dedicated to the search for chemokine-receptor antagonists, which largely failed to deliver drugs for the treatment of inflammatory diseases. The quest for effective chemokine-receptor drug candidates thus continues, and the concept of allosteric targeting of the receptors may be the way forward. In this review, the complex allosteric mechanisms by which the functions of chemokines and their receptors are fine-tuned will be discussed and their impact on preclinical drug discovery presented. The opportunities and challenges of bench-to-clinic approaches are elucidated. We propose that while allosteric modulation of chemokine receptors adds a level of complexity to analyses and approaches to drug discovery, it also introduces a tremendous capacity for pharmacologic control of this physiological system for therapeutic advantage.

## Contents



N. Tschammer (✉)
Department of Chemistry and Pharmacy, Medicinal Chemistry, Emil Fischer Center, Friedrich Alexander University, Schuhstraße 19, 91052 Erlangen, Germany
e-mail: nuska.tschammer@fau.de

A. Christopoulos
Department of Pharmacology, Drug Discovery Biology, Monash Institute of Pharmaceutical Sciences, Monash University, 399 Royal Parade, Parkville, VIC 3052, Australia

T. Kenakin
Department of Pharmacology, University of North Carolina School of Medicine, Chapel Hill, NC, USA

# 1 Introduction

The chemokine-receptor system is a delicately woven network of more than 50 chemokines, which exert their action by interactions with over 20 classical and atypical chemokine receptors, and orchestrates functions of various immune cells in health and disease. Their central role in the inflammatory processes has led to the initiation of programs for the development of chemokine-receptor antagonists by several major pharmaceutical companies. The development of orthosteric ligands as antagonists and agonists of chemokine receptors has proven to be a challenge because of a fairly large natural ligands chemokines (soluble proteins with the size up to 10 kDa) [1]. However, allosteric ligands, known to produce global changes in receptor protein conformation [2], fill the void, and a number of allosteric chemokine agonists and antagonists have been reported [3]. Approximately 40 candidates have progressed into human clinical trials, but these have largely failed to deliver drugs for the treatment of inflammatory and/or autoimmune diseases [1, 4, 5]. The only two drugs that inhibit chemokine receptors are approved for the noninflammatory indications: the treatment of HIV infection (maraviroc, allosteric inhibitor of the CCR5 receptor) and stem cell mobilization (plerixafor (AMD3100), inhibitor of the CXCR4 receptor). Inappropriate target selection, ineffective dosing, off-target effects, and poor drug-like properties of the small molecule antagonists were suggested as the most likely culprit for the lack of efficacy in clinical trials.

The quest for effective chemokine-receptor drug candidates continues. Most efforts are focused on the development of allosteric ligands, which can have three basic phenotypic effects on the receptor: (1) modulate or reduce endogenous agonist activity; (2) potentiate endogenous agonist activity; and/or (3) directly activate receptors (in this latter category, depression of constitutive activity to produce inverse agonism is also included). The complex properties of allosteric ligands, which include probe dependence and biased signaling, can present both challenges and opportunities for preclinical drug discovery [6]. In this review we discuss the mechanism of chemokine-receptor interactions and their role in health and disease. The mechanisms of allostery are also discussed in detail and their therapeutic benefits illustrated. Finally the opportunities and challenges in the development of therapeutics targeting chemokine receptors are presented.

## 2 Mechanism of Chemokine-Receptor Interactions

Despite many years of research, the precise mechanisms underlying the interaction of chemokines with their receptors and the subsequent activation processes remain incompletely understood. It is known that both CC and CXC family chemokines have the potential to interact with the receptors as monomers or dimers [7]. However, only monomeric CC chemokines have the ability to activate their cognate receptors, whereas monomeric or dimeric CXC family chemokines can activate their receptors [8–11]. A general model that has been proposed to describe the chemokine-receptor interaction is the so-called "two-site" or "two-step" model [3, 7]. As illustrated schematically in Fig. 1, the first step of the interaction involves the binding of the globular core region of the chemokine to the receptor's N terminus and extracellular loops, often referred to as "site 1." This is governed largely by ionic interactions and includes an important role of sulfated tyrosines in the receptor's N-terminal region, not only in terms of affecting chemokine affinity but also selectivity [12–14]. The second step of the interaction involves a conformational rearrangement of the N-terminal portion of the chemokine to engage "site 2," which is comprised of transmembrane and extracellular loop domains (Fig. 1). The chemokine N terminus is considered the "triggering domain," since it is this second step that results in receptor activation [3].

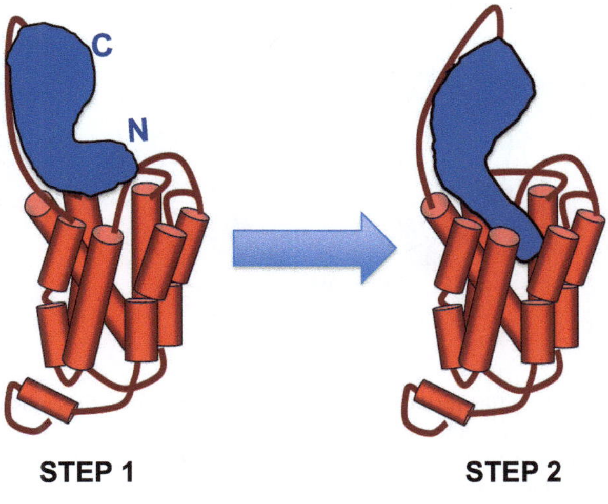


**Fig. 1** The "two-step" binding mechanism proposed for chemokine-receptor interactions. Step 1 involves interaction between the globular core domain of the chemokine with the receptor N terminus and extracellular regions prior to conformational rearrangement that occurs in step 2, leading to engagement by the chemokine N-terminal region of receptor transmembrane regions and subsequent activation

Although there is consensus for the two-step model as a general descriptor of the chemokine-receptor engagement process, the specific details underlying the relevant interactions can vary between receptors, as well as between different chemokines interacting at the same receptor. This has important implications given that chemokine receptors are highly promiscuous with regard to both their cognate ligands and associated intracellular signaling pathways and suggests that even subtly different networks of interactions at the level of the receptor can play a substantial role in governing functional outcomes.

## 3 Promiscuity of Chemokines and Their Receptors in Health and Disease

A common misinterpretation of the crossover activity of endogenous chemokines and their receptors is that the promiscuity of the chemokine system indicates redundancy (several chemokines or chemokine receptor can carry the same function in vivo [4]). This notion also promotes the concept that it will be impossible to derive therapeutic benefit by inhibiting a single chemokine receptor, as a separate chemokine receptor (or even many other chemokine receptors) will merely compensate for the role of the inhibited receptor, and the pathology will remain. However, a closer analysis shows that discrete chemokines are under differential temporal (Fig. 2a) and spatial (Fig. 2b) control in vivo and that the multiple binding events, for example, the binding of different chemokines to a given chemokine receptor, do not necessarily result in the same biological response [4, 5, 7, 15,

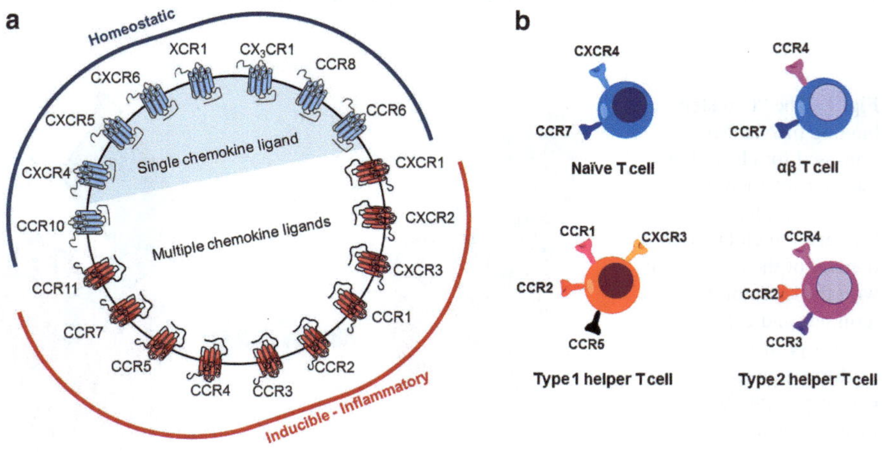


**Fig. 2** Differential temporal and spatial distribution of chemokines. (**a**) Chemokine receptors can be grouped (with some exceptions) as homeostatic or constitutive (developmentally regulated) or inducible (inflammatory). (**b**) T-cell polarization results in selective expression of chemokine receptors [15, 16]

16]. The chemokine receptor CXCR3 is one of the examples of this phenomenon. CXCR3 is rapidly induced on naïve T cells following activation and preferentially remains highly expressed on type1 helper (Th1)-type $CD4^+$ T cells, effector $CD8^+$ T cells, and innate-type lymphocytes, such as natural killer (NK) and natural killer T cells (NKT) [17]. CXCR3 is activated by three chemokine ligands: CXCL9, CXCL10, and CXCL11. Differential regulation of the three ligands at specific times in defined anatomically restricted locations *in vivo* likely participates in the fine control of T-cell trafficking over the course of an immune response. For example, CXCL10 is induced by a variety of innate stimuli that induce IFN-α/IFN-β as well as the adaptive immune cell cytokine IFN-γ [18, 19], whereas CXCL9 induction is restricted to IFN-γ [18, 20], the induction of CXCL11 is limited to IFN-γ and IFN-β but not by IFN-α [21]. Also there is a hierarchy of affinity for CXCR3 with CXCL11 having the highest affinity followed by CXCL10 with moderate affinity and with CXCL9 having the lowest affinity [22–25]. Also, while CXCL11 is reported to completely displace CXCL10 from CXCR3, a substantial amount of bound CXCL11 is not displaced by CXCL10 [23, 26]. As alluded to in the preceding section, the receptor regions required for activation upon the chemokine binding can differ between chemokine ligands. For example, the proximal 16 amino acid residues of the N terminus are required for CXCL10 and CXCL11 binding and activation of CXCR3, but not for activation by CXCL9 [27]. Also CXCR3 internalization is governed by two different domains distinctly used by its ligands. The C-terminus is predominantly required for CXCL9- and CXCL10-mediated internalization, whereas the third intracellular loop is required by CXCL11 [28]. The differences in the interactions between the three CXCR3 ligands with the receptor result ultimately in dissimilar functional responses. The chemokines CXCL9, CXCL10, and CXCL11 differentially stimulate $G_{\alpha i}$-independent signaling and actin response in human intestinal myofibroblasts [29]. The three CXCR3 ligands share the ability to activate PI3K and MAPK and actin reorganization; however, CXCL11 is unique in its ability to elevate intracellular calcium [29]. In the case of lymphocytic choriomeningitis virus-induced meningitis in mice, the results indicated a central role for CXCL10 in regulating the accumulation of effector T cells at sites of CNS inflammation, with no apparent compensatory effect of other CXCR3 ligands [30]. In a mouse tumor model, it has been shown that CXCR3-targeting human CXC chemokines CXCL9 and CXCL10 have potent antitumor activity through inhibition of angiogenesis, but not CXCL11 [31].

The CXCR3 chemokines do not regulate only the Th1 lymphocytes, but have a striking role also in the regulation of Th2 lymphocytes that express CCR3. The chemokines CXCL9, CXCL10, and CXCL11 act as antagonists for CCR3 and compete for the binding of CCL11- to CCR3-bearing cells and inhibit their migration. These results suggest that chemokines that attract Th1 cells via CXCR3 can concomitantly block the migration of Th2 cells in response to CCR3 ligands, thus enhancing the polarization of T-cell recruitment [32]. A further example of regulation of different functions by the same chemokine ligand targeting different receptors is the chemokine ligand CCL5. CCL5 is expressed and secreted by many cell types including activated T lymphocytes, fibroblasts, and renal and pulmonary epithelia [33, 34]. This CC chemokine is a chemoattractant for monocytes, T lymphocytes, NK cells, basophils, eosinophils, and dendritic cells [35, 36]. CCL5 binds to four different chemokine receptors: CCR1, CCR3, CCR4, and CCR5. Not surprisingly it was shown that distinct but overlapping epitopes of the N terminus of CCL5 are responsible for the interactions and activation of CCR1, CCR3, CCR4, and CCR5 [36, 37]. The activation of the abovementioned receptors by CCL5 triggers chemotaxis toward this ligand and also receptor-specific non-chemotactic responses [37, 38]. Monocytes and Th1-like T-cell clones arrest under the CCL5-mediated activation of CCR1; on the other hand the CCL5-mediated activation of CCR5 contributes to the spreading of cells in shear flow [38]. Probe-dependent allostery may serve to allow fine tuning of chemokine response in the seemingly redundant arena of multiple chemokine agonists for receptors. Thus, while CCL19 and CCL21 are both agonists for the CCR7 receptor, their signaling effects are quite different. Specifically, both activate G proteins but only CCL19 causes agonist-dependent phosphorylation of the receptor and recruitment of β-arrestin to terminate response [39].

The differential activation and antagonism of these multiple chemokine systems through conventional orthosteric mechanisms (interaction with the endogenous chemokine binding site) cannot be achieved since the occupancy of the receptor by a synthetic ligand will preclude binding of all other chemokines. However, allosteric mechanisms can produce a very much more diverse effect by selectively altering responses to some chemokines and not others.

## 4 Allostery and Chemokine Receptors

There are two essential characteristics of protein allostery that are relevant to chemokine-receptor signaling. The first relates to the relative geography of binding of the allosteric and endogenous ligand. The separation of allosteric and chemokine binding sites has been shown structurally through mutation studies for many chemokine receptors. For example, HIV-1 allosteric entry inhibitors have been shown to bind to a site distant from the chemokine binding site for CCR5 [40–42]. Similarly, a growing list of allosteric ligands has been shown to access chemokine receptors from the intracellular space [43–46]. The separation of binding sites has been suggested to be advantageous for finding selective ligands. Specifically, the prominent cross-reactivity of chemokine receptors for different chemokines ([4] – see Fig. 3) predicts difficulty in the design of orthosteric ligands selective for a given chemokine receptor. However, as has been suggested for some receptors, allosteric sites may be more diverse between subtypes of receptor [47–51] thereby allowing the development of selective chemokine-receptor ligands through allosteric means. On the other hand, there is no strict a priori reason that a given allosteric site could not be found in more than one chemokine receptor, and allosteric ligands for more than a single chemokine receptor are known (i.e., Sch527123 blocks both CXCR1 and CXCR2 receptors [52]). This issue is discussed in detail in the subchapter *Challenges and opportunities faced in the design of*

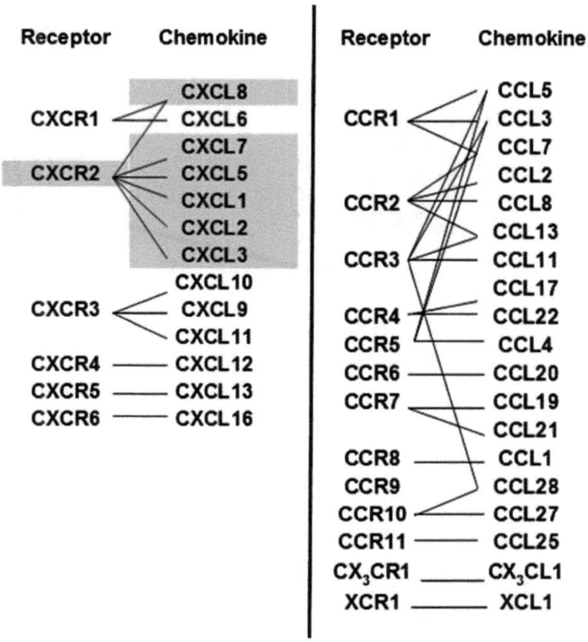


**Fig. 3** Cross-reactivity of natural chemokines for chemokine receptors. Redrawn from Proudfoot (2002) [4]

*allosteric ligands for chemokine receptors*. Interestingly, as will be seen with other allosteric modulators, there is no reliable way to predict the effects of modulation.

Since allosteric molecules bind to a site separate from that utilized by the endogenous agonist, there are no rules for the relationship between the effects of the allosteric modulator and the endogenous agonist. This is not true for orthosteric effects where the ligand binds to the same site as the endogenous agonist. Under these circumstances, there is an obligatory relationship between occupancy of the receptor by the orthosteric ligand and the endogenous agonist. For example, if the ligand does not have intrinsic efficacy for the receptor, then it will produce surmountable (or in cases of temporal disequilibrium) insurmountable antagonism with a strict relationship existing between the affinity of the ligand and its concentration in the receptor compartment [53]. If the orthosteric ligand is a partial agonist, then there will be a strict relationship between the direct agonism produced by the ligand and the alteration of response to the endogenous ligand, i.e., the effects to the endogenous ligand will be blocked. In contrast, allosteric effects are *permissive* [54] in that the endogenous agonist may be free to interact with the receptor even in the presence of the allosteric modulator. For example, reparixin was found to inhibit different signaling pathways of CXCR1 and CXCR2 activated by CXCL8 but had no effect on CXCL8 binding to the receptors [55]. Also, it was also possible to gain selectivity between CCR1 and CCR8 vs CCR5 and to modulate the allosteric properties of this class of ligands from being ago-allosteric modulators to being agonists without allosteric properties in CCR1 and simultaneously to being allosteric modulators with no intrinsic activity in

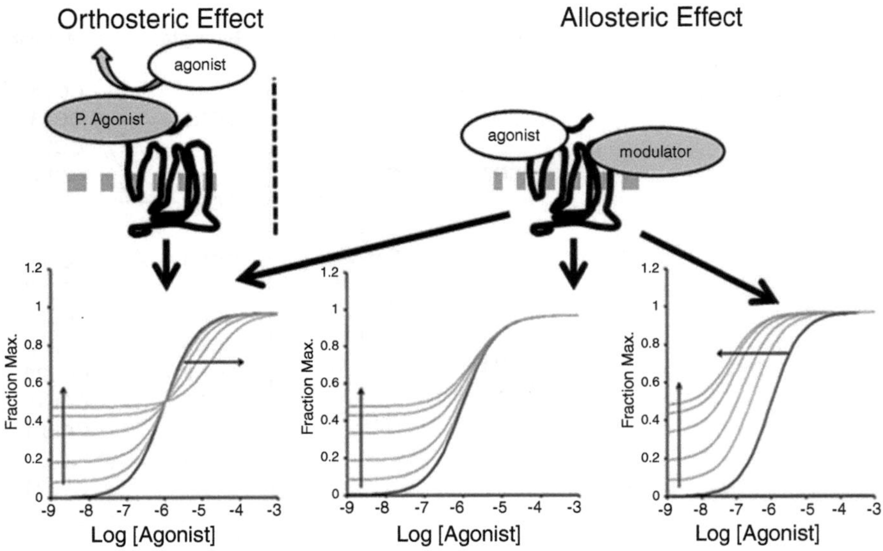


**Fig. 4** Two distinct pharmacological mechanisms for partial agonism. Orthosteric partial agonism links agonist effect with antagonism of more efficacious agonists in a compulsory manner. In contrast, allosteric partial agonists do not. Specifically, depending on the values of $\alpha$ and $\beta$ [see Eq. (1)], allosteric partial agonists may block, have effect on, or potentiate more efficacious agonists

CCR5 [56]. As will be seen from the current models of allosteric function, modulators can separately alter the affinity and/or efficacy of the endogenous agonist and thus produce a plethora of effects. Figure 4 shows the obligatory range of effects of a partial agonist on endogenous agonism and the considerably different range of effects of an allosteric partial agonist on endogenous agonism. It is this variety that constitutes the therapeutic potential of allosteric modulators.

The fact that allosteric molecules bind to their own site on the receptor protein also leads to the property of *saturation of effect*, i.e., whatever allosteric effect is operative on the receptor and endogenous signaling, there will be a maximal asymptote to that effect when the allosteric binding sites are fully occupied. This, coupled with the permissive quality of allosterism, can lead to unique therapeutic profiles whereby the receptor signaling system may be only modified, not completely inhibited. For example, the allosteric modulator UCB35625 produces only a 1.5-fold reduction in the affinity of the CCR1 receptor for CCL3 [57] – see Fig. 5. This permissive quality opens the therapeutic pharmacological toolbox to ligands that can have a range of activity from slight modulation of response to potentiation of weak endogenous signaling (through the action of positive allosteric modulators, PAMs).

The other unique feature of allosteric molecules is *probe dependence*. At this point it is useful to define allosteric effects on chemokine receptors in terms of changes in the activity of receptor "probes"; these are defined as the endogenous chemokine or other synthetic agonists that co-bind to the receptor with the allosteric

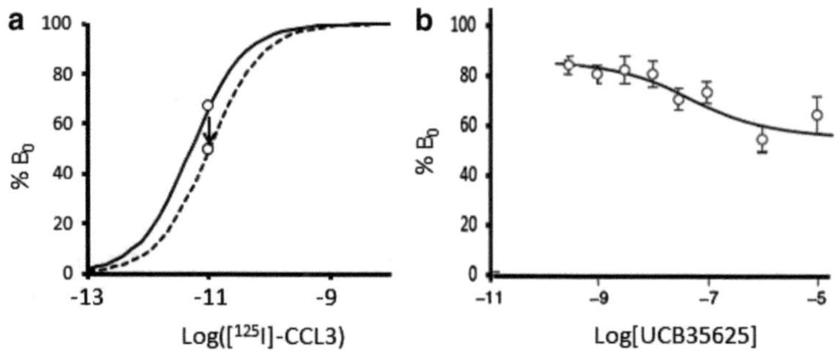


**Fig. 5** Effect of the allosteric modulator UCB35625 on the binding of [$^{125}$I]-CCL3. Translation of the partial displacement curve shown in *panel B* into the effect on a saturation binding curve for [$^{125}$I]-CCL3 (*panel A*) shows a very mild modulation of affinity for the chemokine. Redrawn from Sabroe et al. [57]

modulator. Allosteric molecules affect receptor signaling from other (endogenous) molecules through changes in the conformation of the receptor, and there is no reason that the change in conformation induced by a given allosteric molecule will have identical effects on two receptor probes. In fact, it is well known that allosteric modulators can have radically different effects on the interaction of two probes with the same receptor. For example, the allosteric modulator alcuronium produces an inhibition of the binding of the M2 receptor radioligand antagonist [$^3$H]-methyl-QNB but a *potentiation* of the binding of the M2 receptor antagonist radioligand [$^3$H]-atropine [58]. An example of a similar effect with chemokine receptors is found in the allosteric CXCR4 receptor modulator which AMD-3100 blocks CXCL12 binding but enhances the binding of CXCL12 to its other natural receptor CXCR7[59]. It will be seen that probe dependence can lead to major therapeutic advantages in the pharmacological manipulation of chemokine systems. The various effects of allosteric ligands on chemokine receptors will be described in terms of these general characteristics of saturation of effect and probe dependence.

## 5 The Allosteric Vector

Allosteric effects can be described in terms of the cooperative effects of two bodies with the receptor; the ternary complex defined by the two co-binding ligands and receptor is defined as the *allosteric vector* [2]. The allosteric vector is comprised of a *modulator* binding to a *conduit* (receptor) affecting the further interaction of this complex with a *guest* – see Fig. 6. It should be noted that the definition of the terms modulator and guest are essentially interchangeable in that the allosteric energy generated by the complete interaction is bidirectional in flow, i.e., while a defined modulator can be seen to affect the interaction of the receptor with a guest, the guest will impart the same effect on the interaction of the receptor with the modulator. Thus, the allosteric vector is relevant only to descriptions of overall outcome of the

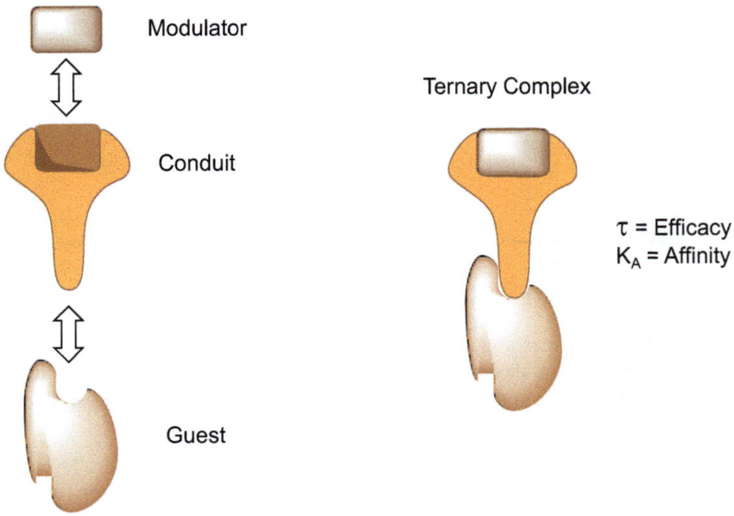


**Fig. 6** All activities of seven transmembrane receptors are allosteric and can be described in terms of a modulator that interacts with a conduit (the receptor) to cause a reciprocal effect on the interaction of the modulator-conduit unit with a guest (which can be another ligand, receptor, protein, signaling protein, etc). Thus, a complete description of the function of the receptor must include information about the nature and concentration of the modulator and guest

interaction. Chemokine receptors, like all 7TMRs, are defined by their interactants (modulators and guests). Thus, it is the ternary complex of modulator (i.e., agonist), receptor (conduit), and guest (particular signaling protein, i.e., β-arrestin) that defines the drug parameters efficacy (as defined by the Black and Leff's operational model [60]) as the term $\tau$ and affinity (as defined by the reciprocal of the equilibrium dissociation constant of the modulator (agonist)-receptor complex). It will be seen that some of the confusion surrounding chemokine function and chemokine-receptor-targeted drugs is associated with making drug parameter measurements in systems where the appropriate ternary complex is not present.

All activities of chemokine receptors are allosteric in nature, i.e., chemokine receptors interact with chemokines to change their conformation and the resulting complex interacts uniquely with another body (guest) to induce pharmacological response. These interactions can be classified in terms of three types of allosteric vector (see Fig. 7):

1. Allosterism along the plane of the cell membrane (oligomerization)
2. Guest allosterism – cooperativity between species binding to the receptor
3. Cell cytosol directed allosterism: (biased) agonism

The first to be considered is an allosteric vector oriented along the plane of the cell membrane – see Fig. 7.

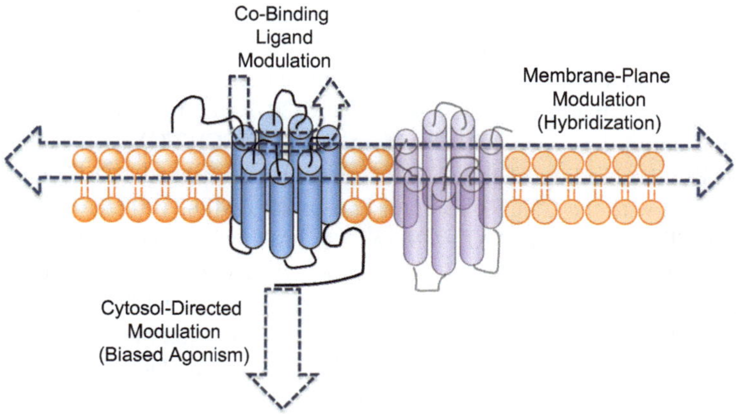


**Fig. 7** All allosteric functions can be described in terms of an allosteric vector which may have different directions pharmacologically. The allosterism may be directed along the plane of the cell membrane to mediate receptor oligomerization or interaction with membrane components such as RAMPS, of through the receptor between ligands (guest allostery). Thus, allosteric modulators (NAMs, PAMs) can affect the reactivity of the receptor toward natural chemokine ligands through this mechanism. Finally, the vector may be directed toward the cytosol since all agonism is allosteric in nature. Probe dependence at this level can lead to biased agonism

## 5.1 *Allosterism Along the Plane of the Membrane*

There is considerable evidence that chemokine receptors form functional dimers [61–66]. Dimerization can be thought of in allosteric terms as having two functional consequences. In one, the dimerizing receptor can be the allosteric modulator as is seen with co-expression of the CXCR7 receptor in cells containing the CXCR4 receptor [67]. Specifically, while the chemokine CXCL12 produces a calcium response through the CXCR4 receptor in HEK 293T cells, this ligand does not produce a calcium response through the CXCR7 receptor in the same cells. However, co-expression of the CXCR7 receptor in CXCR4-containing cells causes the concentration-response curve to CXCL12 to shift to the right by a factor of 9, i.e., the CXCR7 receptor functions as a negative allosteric modulator for CXCL12 acting on CXCR4 receptors [67]. Another example of membrane level allosteric modulation is with membrane-bound single transmembrane receptor activity modifying proteins (RAMPs) [68, 69]. For example, the human calcitonin receptor is less sensitive to the peptide amylin than it is to human calcitonin until the receptor is co-expressed with RAMP3; the new complex forms a new phenotype whereby amylin is more potent than human calcitonin and the potency of antagonists selectively increases to amylin response (over human calcitonin response) [68]. Receptor heterodimerization also can cause alteration of chemokine signaling. For example, CCR2/CCR5 heterodimerization causes recruitment of dissimilar signaling pathways, namely, $G_{q/11}$ association to cause a pertussis-insensitive calcium response leading to cell adhesion instead of chemotaxis [70].

In another scenario, receptor dimerization can form a new allosteric conduit through which antagonists can cross-react, i.e., a previously inactive antagonist for a receptor can attain activity at that receptor once heterodimerization has occurred. For example, in binding studies with the CCR2 and CXCR4 receptor, cross antagonist activity is seen with the formation of CCR2/CXCR4 heterodimers. Specifically, the binding of the CXCR4 receptor radioligand $^{125}$I-CXCL12/SDF-1α is blocked with the CXCR4 receptor antagonist AMD-3100 ($K_i = 0.81$ nM) but not the CCR2 antagonist TAK-779 ($K_i > 1$ μM); when the CCR2 receptor is co-expressed in the cell, then TAK-779 becomes a potent antagonist of $^{125}$I-CXCL12 binding ($K_i = 0.08$ nM) [61]. Such cross-receptor interaction can occur with chemokine and non-chemokine heteromers. For example, CXCR2 receptor antagonists can enhance the function of δ-opioid receptor (DOP) agonists through the formation of a CXCR2/DOP heterodimer [71].

In some cases, this dimerization can have therapeutic consequences as in the mediation of HIV-1 entry in AIDS. For example, it is known that homozygous CCR5Δ32 individuals are resistant to AIDS by virtue of the fact that the CCR5 receptor cannot reach the cell surface to mediate HIV-1 infection. However, it also is known that heterozygous CCR5Δ32 individuals (who still possess native CCR5 receptors capable of mediating HIV-1 entry) progress to AIDS more slowly than usual when infected [42, 72]; this effect is attributed to the formation of CCR5/CCR5Δ32 heterodimers in the endoplasmic reticulum leading to a reduced cell surface level of CCR5 receptors [73].

## 5.2 Guest Allosterism

The most extensively studied allosteric vector describes the interaction of an allosteric modulator on the effects of chemokines on chemokine receptors; this will be referred to as "guest allostery" – see Fig. 5. An allosterically modulated receptor should be considered a new receptor with possibly completely new properties of interaction with the endogenous agonist. Under these circumstances, both the affinity and efficacy of the endogenous agonist may change [54, 74, 75]. The factor quantifying the effect of the modulator on affinity is denoted $\alpha$, and it is defined as the ratio of the affinity of the endogenous agonist in the presence of the modulator to the affinity in the absence of the modulator. Therefore, an $\alpha$-value of 10 means that the modulator increased the affinity of the receptor for the endogenous agonist by a factor of 10. An independent property of the endogenous agonist is its efficacy for the receptor (ability to produce activation). Analogous to effects on affinity, a factor $\beta$ is defined as the ratio of the efficacy of the endogenous agonist in the presence of the modulator to the efficacy in the absence of the modulator. Thus, when $\beta = 10$, the efficacy of the endogenous agonist is increased by a factor of 10 in the presence of the modulator. Finally, the allosteric modulator may itself activate the receptor upon binding; therefore, a term $\tau_B$ is defined as the intrinsic efficacy of the modulator. A total of three parameters, along with the equilibrium dissociation constant of the modulator-receptor complex ($K_B$), define the effects of an allosteric modulator on the receptor. In order to describe guest allostery, it is useful to define allosteric effects in terms of a functional allosteric receptor model. This model combines the allosteric receptor-binding model presented by Stockton et al. [76] and Ehlert [77] with Black and Leff's operational model for agonism. With this model, the response to an agonist A in the presence of an allosteric modulator B (with no direct allosteric agonism by the modulator) is given by [54, 78, 79]

$$\text{Response} = \frac{\tau_A[A]/K_A(1+\alpha\beta[B]/K_B)}{[A]/K_A(1+\alpha[B]/K_B)+\tau_A(1+\alpha\beta[B]/K_B)+[B]/K_B+1} \quad (1)$$

where $\tau_A$ is the endogenous agonist efficacy, $K_A$ and $K_B$ are the respective equilibrium dissociation constants for the agonist- and modulator-receptor complexes, $\alpha$ is the effect of the modulator on agonist affinity, and $\beta$ is the effect of the modulator on agonist A efficacy.

One feature of Eq. (1) is that it allows the nature of the agonist A to interact with the cooperative properties of the allosteric ligand in the terms $\alpha\beta\tau_A[A]/K_A$ $[B]K_B$ and $\alpha[A]/K_A$ $[B]/K_B$; this allows the model to produce probe dependence, i.e., different agonists ($\tau_A$ and $K_A$ values) can produce different effects with the same modulator through varying values of $\alpha$ and $\beta$. This is actually a prominent feature of guest allostery and examples of these effects can be found in chemokine-receptor-mediated responses. For example, the allosteric CXCR4 receptor modular AMD3100 blocks the effects of the natural CXCR4 agonist CXCL12 but does not block the effects of CXCL12 peptide fragments RSVM and ASLW [80]. Similarly, a series of metal-ion chelating compounds have been shown to have differential effects on the binding of two natural agonists for the CCR1 receptor. Specifically, they negatively affect the binding of CCL5 but act as allosteric enhancers of binding of CCL3 illustrating opposite $\alpha$-values [see Eq. (1)] for affinity on CCR5 [81]. Similarly, enhancers of GLP-1 response produce differential sensitization to natural agonists of the GLP-1 receptor. Thus, while the positive allosteric modulator NOVO2 produces virtually no sensitization to the agonist GLP (1-37) ($\alpha\beta = 1.14$), it potentiates the effects of the agonist oxyntomodulin by a factor of 25 ($\alpha\beta = 2.5$) [82].

Another consequence of the functional allosteric model is the potential for modulators to independently affect co-binding ligand affinity and efficacy. Thus, the CCR5 HIV-1 entry inhibitor modulator aplaviroc does not substantially block the binding of the chemokine CCL5 to the receptor but does completely block its functional effect [83]. Similarly, the dual CXCR1 and CXCR2 antagonist reparixin blocks the effects of CXCL8 but not its binding [55, 84]. Similarly, a series of naphthalene-sulfonamide-based antagonists of CCR8 show potent inverse agonist effects (from 1.7 to 23 nM $EC_{50}$) with antagonist values of 100-fold lower potency [85]. A dischotomy in binding and function is also seen with the CCR1 antagonist BX471 which blocks function but cannot be displaced in binding studies with CCL3, CCL5, or CCL7 [86]. Differential effects on CCR5 receptors also have

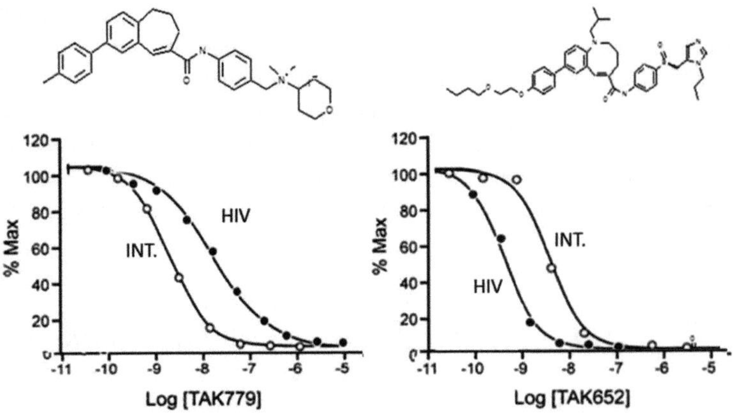


**Fig. 8** Reverse potencies of two allosteric ligands in the blockade of CCL3L1-induced internalization of CCR5 receptors (*curves* labeled INT) and inhibition of HIV-1 entry (labeled HIV). It can be seen that, while TAK779 is more potent as an inhibitor of CCR5 internalization (over HIV), the reverse is true for TAK652. This reflects differing $\alpha$-values for the ligands and guests [see Eq. (1)] and can only be seen with allosteric ligands. Redrawn from Muniz-Medina et al. (2009) [102]

been found with maraviroc and TAK779; specifically while both displace radioactive HIV-1 glycoprotein [$^{35}$S]gp120 with similar potency, maraviroc is 100-fold more potent as an inhibitor of HIV infection [87].

Such probe dependence can have therapeutic ramifications. For instance, HIV-1 infection is mediated by utilization of the chemokine receptor CCR5 by the virus; allosteric HIV-1 entry inhibitors bind to CCR5 to block infection [88–92]. Therefore, allosteric blockade aimed at the CCR5 receptor is an effective method of preventing HIV-1 infection. However, it has also been shown that normal utilization of the CCR5 receptor by chemokines appears to be beneficial in AIDS and delay disease progression [93–97]. In fact, it has been shown that multiple copies of the gene that mediates CCL3L1 production, a natural chemokine agonist for CCR5, confer a higher rate of survival to patients infected with HIV-1 [98] probably through CCL3L1-mediated CCR5 receptor internalization [99–101]. Therefore, a probe-dependent allosteric antagonist of HIV-1 entry that otherwise allows CCL3L1 to function normally would theoretically have an advantage in AIDS therapy [102]. As shown in Fig. 8, there are allosteric antagonists with differential potency against HIV-1 entry and inhibition of CCL3L1-mediated CCR5 internalization; thus while TAK 779 is more potent as an inhibitor of CCL3L1-mediated CCR5 internalization (than HIV-1 entry), the reverse is true for TAK 652. It would be postulated that TAK 652 would have some sparing effect on chemokine function during the process of HIV-1 entry inhibition that would be beneficial [102]; this type of effect would only be possible with allosteric ligands.

Probe-dependent guest allosterism also can have deleterious effects therapeutically when HIV-1 virus mutates to resistant forms. For instance, while the JV1083-PC HIV-1 virus is completely blocked by vicriviroc, maraviroc, and aplaviroc, a laboratory-passaged resistant virus JV1083-VCV$_{res}$ has a reduced susceptibility to maraviroc inhibition, is completely insensitive to aplaviroc, and shows an enhanced infectivity with vicriviroc. These effects are even more pronounced with the virus RU570-PC [103].

### *5.3 Cytosol-Directed Allosterism*

When the allosteric vector is directed toward the cell cytosol (see Fig. 7), then allosteric probe dependence leads to biased agonism. In this regard, the mechanism of biased agonism is identical to guest allostery except the guests in this case will be signaling molecules such as G proteins and β-arrestin. For example, four chemokine agonists for CCR5 have considerably different relative bias factors for CCR5 internalization over IP1 production with CCL3L1 producing 23.7 times more internalization for normalized IP1 production than CCL3 [75]. In terms of probe-dependent allostery, this signifies that CCL3L1 stabilizes a CCR5 receptor conformation that preferentially internalizes to a greater extent than the conformation stabilized by CCL3 [104]. Similarly, the non-peptide allosteric agonist for CXCR3 receptors, VUF10661, has efficacy for [$^{35}$S]GTPγS and cyclic AMP similar to that of the natural agonist CXCL11 but recruits more β-arrestin1 and β-arrestin 2 to the receptor than does the natural chemokine [105]. The agonist AOP-RANTES activates CCR1, CCR3, and CCR5 but has differential effects on receptor recycling; specifically, AOP-RANTES inhibits CCR5 recycling in eosinophils, promotes recycling of CCR1, and has no effect on CCR3 recycling indicating functional selectivity [106]. Similarly the analog 5P14-RANTES produces powerful CCR5 internalization but not activation of G proteins [107].

## 6 Challenges and Opportunities Faced in the Design of Allosteric Ligands for Chemokine Receptors

It is well established that chemokines and their receptors have a key role in the pathogenesis of inflammation, autoimmune diseases, viral infection, cancer, and transplant rejection. The search for small-weight compound with clinical potential is although paved by many challenges and opportunities. Some of them are discussed below.

### *6.1 Orthosteric Versus Allosteric*

The ability to identify subtype selective allosteric modulators suggests that unlike the case of the agonist binding site, where there is significant evolutionary pressure

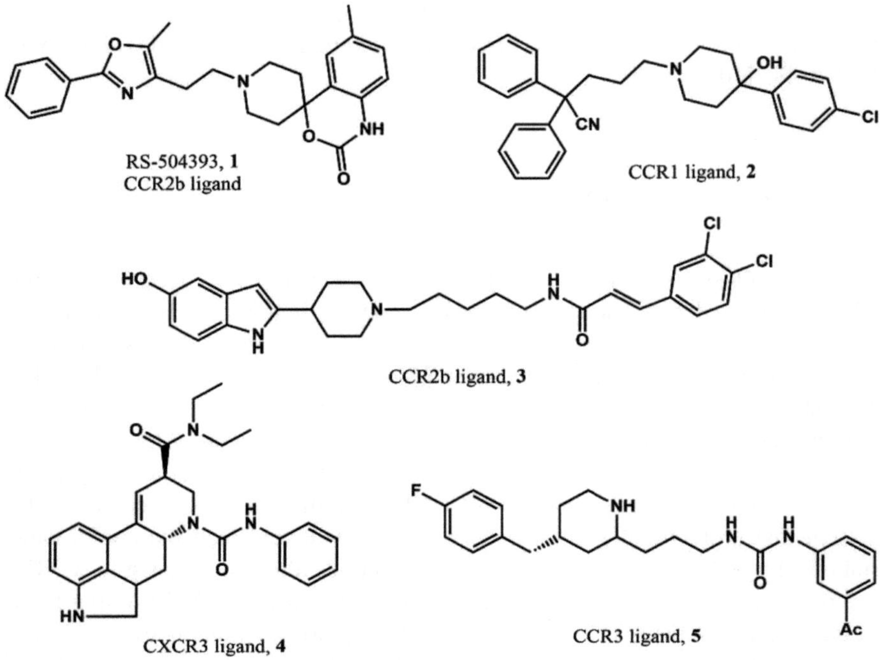


**Fig. 9** Examples of pharmacophores that are shared between allosteric modulators of chemokine receptors, biogenic amine receptors, and neurotransmitter transporters

to conserve residues across families of GPCRs, the allosteric binding sites are under less evolutionary pressure for their conservation [108]. Ligands that bind to allosteric sites and confer activity represent novel chemotypes, structurally unrelated to orthosteric ligands, with high levels of selectivity and improved chemical tractability (vide infra) [109]. Unfortunately an allosteric site in one GPCR might be quite similar to an orthosteric site of another GPCR. Non-peptide chemokine-receptor antagonists are mainly allosteric inhibitors and bind to the biogenic amine-like domains of the receptor [110–115]. Not surprisingly one can find many reports about the chemokine-receptor antagonists, which share the common pharmacophore with the biogenic amine receptors. For CCR2 the spiropiperidine compounds (Fig. 9, **1**) were reported that have affinity for several biogenic amine receptors [111]. The basic nitrogen of spiropiperidine was held responsible for this cross talk. Receptor models indicated that the acidic residue, Glu291$^{7.39}$, from the transmembrane domain 7 of CCR2, is in a position similar to the acidic residue contributed from transmembrane domain 3 of biogenic amine receptors, which may account for the shared affinity of spiropiperidines for these two receptor classes [111]. For CCR1, a potent 4-hydroxypiperidine derivative (Fig. 9, **2**) was reported, which has affinity to the dopamine $D_{2L}$ receptor [116]. The indolopiperidine derivative **3**, a promising CCR2b antagonist, has unwanted affinity for the 5HT and dopamine D2/D3 receptors, with the affinities comparable to that for the

**Fig. 10** Examples of promiscuous inhibitors with a desired cross-reactivity

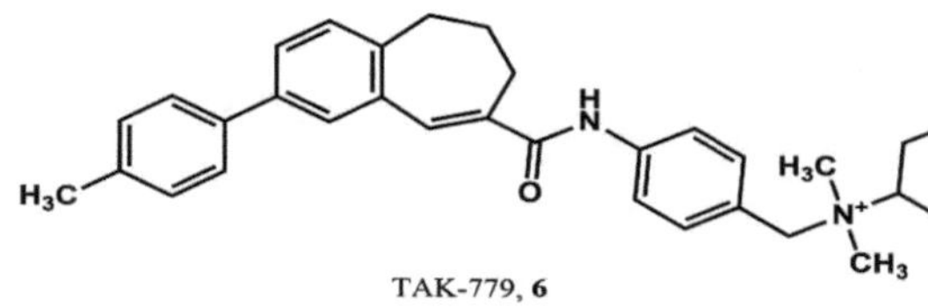

TAK-779, **6**

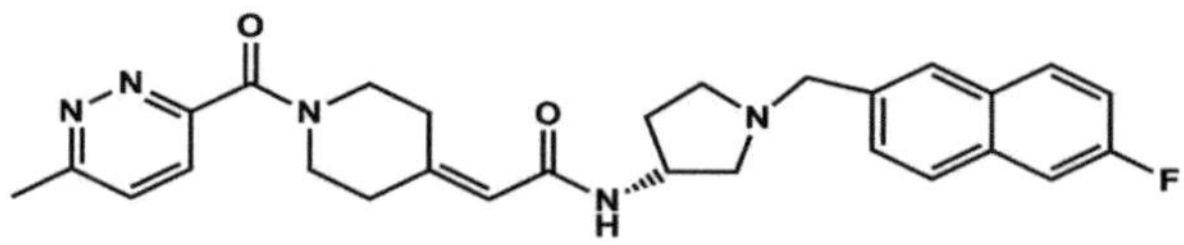

YM-344484, **7**

desired CCR2b receptor [117]. Also, the ergoline-based antagonist of CXCR3 (Fig. 9, **4**) shows submicromolar affinity for $5HT_{2A}$ [118]. A common observation is that a basic nitrogen, which is generally incorporated into constrained aliphatic rings, generates the propensity of a compound to interact with biogenic amine receptors. An additional issue in the development of chemokine allosteric modulators is the possible cross talk of modulators with neurotransmitter transporters. For example, a potent CCR3 antagonist (Fig. 9, **5**) with the substituted piperidine structural motif demonstrated potent inhibitory activity for norepinephrine (NET), dopamine (DAT), and serotonin reuptake transporters (5-HTT) [119].

The biogenic amine-like allosteric binding pocket of chemokine receptors offers, on the other hand, the possibility to design desired dual antagonists that target GPCRs that belong to unrelated families. For example, YM-344484, a dual antagonists of CCR3 and H1R (Fig. 10), was shown to be an attractive approach for the development of novel antiallergic inflammation drug, because eosinophilic CCR3 and H1R are main therapeutic targets in allergic inflammation (e.g., asthma) [120].

The investigation of binding modes of a variety of allosteric modulators of chemokine receptors by homology modeling and docking and site-directed mutagenesis has demonstrated that at least one structurally conserved allosteric site exists between even unrelated chemokine receptors. Amino acids Tyr1.39, Lys2.64, Asn3.35, Tyr6.51, and Glu7.39 seem to be highly conserved in several chemokine receptors [55, 108, 111, 121, 122]. Not surprisingly, negative allosteric modulators that interact with unrelated chemokine receptors were reported. For example, TAK-779 (Fig. 10) was initially described as dual CCR5 and CCR2 antagonist [123]. Later, TAK-779 was shown to efficiently antagonize CXCR3 activation [25, 124]. The benefits of simultaneous antagonism of CCR2, CCR5, and CXCR3 were demonstrated in the murine experimental colitis, where TAK-779 inhibited the recruitment of inflammatory cells like monocytes and macrophages into the mucosa, which indicated that these chemokine receptors may be therapeutic targets for the treatment of inflammatory bowel disease [125].

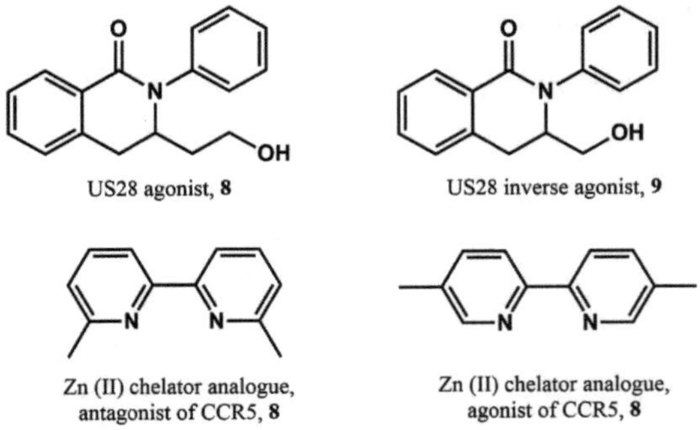


**Fig. 11** Examples of "molecular switches"

## 6.2 *"Molecular Switches"*

Despite increasing numbers of high-resolution structures of chemokine receptors to guide computational efforts in drug design (see below), an additional challenge represents significant hurdle for the rational ligand design. The "molecular switches," which are responsible for the switching between positive and negative efficacy of allosteric modulators, are so far unpredictable as the path from an agonist to an inverse agonist could be as short as a methyl group [56, 126] (Fig. 11).

# 7 Recent Structural Insights into Chemokine Receptors

Within the last decade, there has been an explosion in GPCR structural biology research, resulting in the solution of numerous high-resolution GPCR structures [127]. These breakthroughs have occurred due to a convergence of approaches that facilitated a very high expression of (traditionally low abundance) GPCRs in various cell backgrounds, antibody, and protein engineering techniques to constrain the normally highly flexible nature of GPCRs toward discrete states, new classes of detergents, and improvements in synchrotron microbeam diffraction methods. To date, 3 chemokine receptors have had their structures determined at high resolution: the CXCR4 and CCR5 receptors, using X-ray crystallography, and the CXCR1 receptor using NMR [114, 115, 128]. These structures are beginning to reveal the molecular basis for receptor-ligand interactions and may facilitate novel structure-based drug discovery efforts.

The CXCR1 receptor was solved by NMR in its native state in a lipid bilayer in the absence of ligand [128], whereas the CCR5 and CXCR4 receptors were engineered to improve stability, bound to ligands, and solved by lipidic cubic phase crystallography. Two different sets of structures of the CXCR4 receptor were solved, one bound to the cyclic peptide inhibitor, CVX15, and the other to the small molecule inhibitor, IT1t [114]. CCR5 was solved bound to the small molecule allosteric antagonist, maraviroc [115]. Figure 12 shows a comparison between the three chemokine-receptor structures. Despite sharing the characteristic

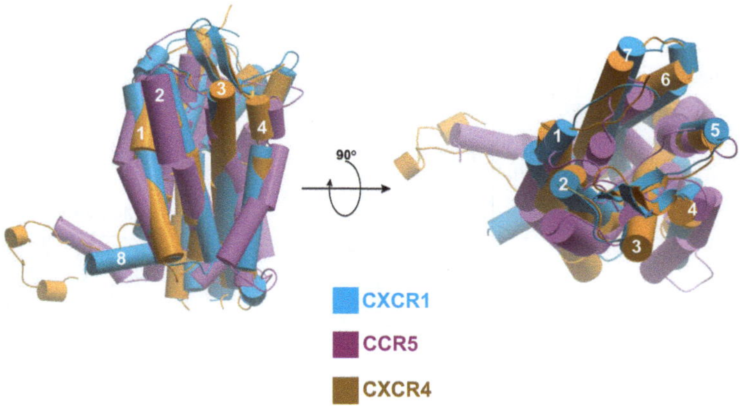


**Fig. 12** Overall structural comparison of three different chemokine-receptor structures: CXCR1 (PDB 2LNL), CCR5 (PDB 4MBS), and CXCR4 (PDB 3ODU)

seven transmembrane-spanning domains typical of GPCRs, some notable differences exist between the structures, likely reflecting sequence variations, the presence or not of bound ligand, and the effect of the conditions used to solve the structures.

One of the most interesting findings in all three chemokine-receptor structures was the identification of a second disulfide bridge between the N-terminal region of the receptor and the top of transmembrane domain (TM) 7; this is in addition to the highly conserved disulfide that links the second extracellular loop to the top of TM3. The presence of the extra disulfide bridge results in the creation of, essentially, a "fourth" extracellular loop that closes the receptor in a ringlike conformation. This loop may play an important and hitherto unappreciated role in the binding of chemokines, for instance, with regard to positioning the appropriate site 1 and site 2 residues of the two-step binding mechanism, as well as in the conformational transition between active states and even in the control of access of small molecule ligands to the helical bundle [129]. Interestingly, the presence of this loop is likely to extend to other chemokine receptors and perhaps other Class A GPCRs as well [129].

In contrast to many other GPCR structures, the ligand-binding regions of the chemokine receptors are larger and more open (Fig. 13); this is particularly so for the CCR5 receptor. This likely reflects the fact that the cognate orthosteric agonists are large peptides and thus require more surface area for establishing appropriate contacts. In the CXCR4 structures, substantial overlap is noted between regions bound by the small molecule, IT1t, and the peptide ligand, CVX15 (Fig. 13); IT1t would thus be classed as an orthosteric antagonist. The peptide, CVX15, forms a disulfide-stabilized $\beta$ hairpin and also makes extensive hydrogen bonds with the second extracellular loop. This ligand fills most of the binding pocket, whereas the IT1t localizes in a region defined largely by TM helices 1, 2, 3, and 7. In the CCR5 structure, however, the binding site for maraviroc is deeper within the

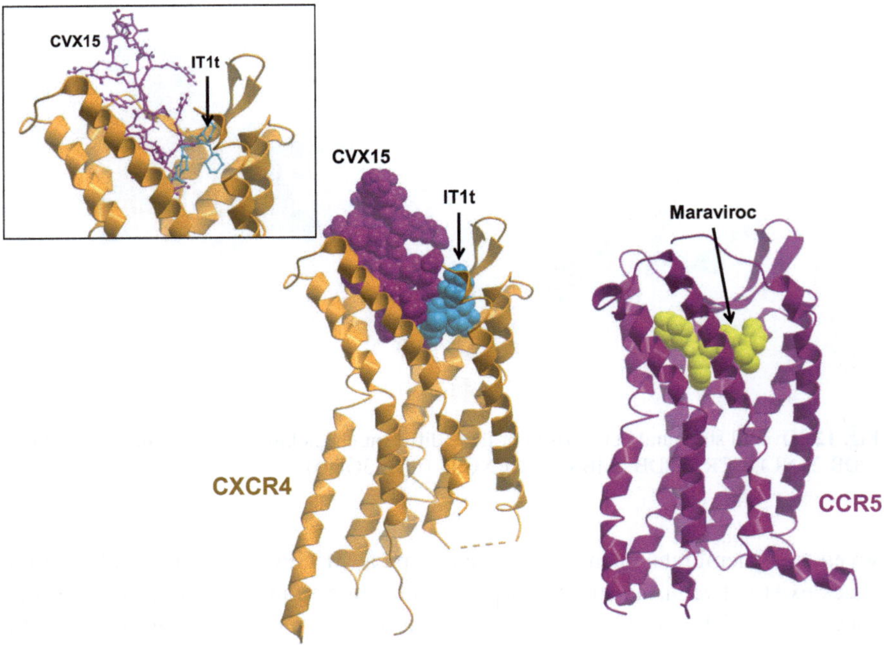


**Fig. 13** Ligand-binding modes in the CXCR4 and CCR5 receptor structures

transmembrane bundle and fills a larger volume than that occupied by IT1t in the CXCR4 structure; this delineates a true allosteric site for maraviroc and related compounds. It also provides some insight into the mechanism of modulation of the receptor by maraviroc as a NAM, namely, that its placement within the allosteric binding pocket most likely leads to disruption of site 2 interactions [115].

On the intracellular surface, the CXCR4 structure lacks the short helix 8, which is a feature common to other GPCR structures. In contrast, both CCR5 and, in particular, CXCR1 receptor structures retain this feature. Perhaps more interestingly, all structures of the CXCR4 receptor revealed similar parallel and symmetric dimers linked via the TM5 and 6 interfaces [114], which may also provide a structural basis for previously reported cooperative binding between orthosteric CXCR4 ligands [130].

## 8 Allosteric Modulators of Chemokine Receptors: From Bench-to-Clinic Challenges

The fundamental role of chemokines and their receptors in inflammation processes provides great opportunity to develop novel therapeutics. The permissive nature, saturability, and probe dependence of allosteric modulation provide unprecedented possibilities to fine-tune therapeutic action of novel drugs and thus largely increase therapeutic and reduce adverse effects. Unfortunately drug discovery efforts in the field of chemokines and their receptors are accompanied by pronounced challenges that influence bench-to-clinic success rate.

Because chemokine systems differ significantly between species, the development of therapeutics targeting this receptor system is hampered by a lack of predictive animal models [1, 4, 5]. The differences in the functions of the immune system between the rodents as first-choice animals for the preclinical drug development and humans are striking. For example, the chemokine CXCL8/IL-8, the ligand of CXCR1 and CXCR2, which is in humans excreted by macrophages, does not exist in mice [131]. The mouse homolog of CXCR2 is activated by two murine chemokines, mCXCL1 and mCXCL2 [132, 133]. The mouse homolog or ortholog of human CXCR1, which shares only 64% identical amino acids with hCXCR1 and is responsive to mCXCL6, was not cloned until 2006 [134, 135]. Also receptor expression levels may differ between species as in, e.g., the expression profile of CX3CR1 [136] and CCR5 [137]. Furthermore, the differences in the selectivity of chemokines from chemokine receptors between species make the transfer even more challenging. For example, mouse CCL5 and mouse CCL7 lack affinity for the mouse CCR1 but are agonists for the human CCR1 [138].

Such cross-species differences thus have important consequences for the validation of chemokines receptors as drug targets and to estimate the efficacy of drug candidates in animal models of various diseases. In, e.g., rheumatoid arthritis, the infiltration of macrophages into the arthritic joint space promotes much of the damage, as these cells produce cytokines that promote inflammation in the diseased joint. Because the infiltration of monocytes and macrophages is driven by the chemokine receptors CCR1 and CCR2, it was suggested that both receptors are equally attractive pharmacological targets. The most commonly used animal models for identification of potential therapeutics for rheumatoid arthritis are adjuvant-induced arthritis (AIA) in rats and collagen-induced arthritis (CIA) in rats and mice [139]. The data from animal studies related to the role of CCR2 in rheumatoid arthritis are inconsistent and even contradictory, by suggesting an anti-inflammatory role of CCR2 in the CIA-model of rheumatoid arthritis [140–142]. Not surprisingly, the CCL2-specific monoclonal antibodies, ABN-912 (Novartis) and MLN1202 (Millennium), and the small molecule antagonist of CCR2, MK0812 (Merck), have not exceeded placebo treatment of rheumatoid arthritis in the Phase II trials. The analysis of synovial fluid from patients with rheumatoid arthritis demonstrated that it contains high levels of alternative CCR1 binding chemokines, CCL15 and CCL23, that are activated by proteolytic processing and exhibit up to 1,000-fold increase in the CCR1-mediated signaling [143]. Consistent with these observations, it was shown that the ability of synovial fluid from rheumatoid arthritis patients to attract monocytes was mediated by CCR1 and not by CCR2 [144]. The new generation of inhibitors targeting CCR1 carries more promises [1, 5, 144]. This example demonstrates the importance of using clinical disease samples because rodent models of arthritis are thought to mimic human rheumatoid arthritis only weakly [5]. The evaluation of potential drug candidates in animal models of inflammatory processes is facing also a species-selectivity problem, which is well illustrated by the next example. The CCR1 antagonist CP-481,715 is highly specific for human CCR1 and thus prevents its evaluation in classical animal models. To circumvent this issue, the humanized CCR1 transgenic mice were generated to evaluate the ability of CP-481,715 to inhibit inflammatory responses in this animals at clinically achievable dose levels [145].

## 9 Conclusions

In general it can be seen that all important unique features of allosteric receptor control (saturation of effect, probe dependence) are embodied in the chemokine-receptor system. While this adds a level of complexity to analyses and approaches to drug discovery, it also introduces a tremendous capacity for pharmacologic control of this physiological system for therapeutic advantage. The bench-to-clinic translation of these opportunities is hampered by the lack of relevant animal models, which continue to make target validation and the drug candidate evaluation challenging.

## References

1. Horuk R (2003) Development and evaluation of pharmacological agents targeting chemokine receptors. Methods 29(4):369–375
2. Kenakin T, Miller LJ (2010) Seven transmembrane receptors as shapeshifting proteins: the impact of allosteric modulation and functional selectivity on new drug discovery. Pharmacol Rev 62(2):265–304
3. Scholten D, Canals M, Maussang D, Roumen L, Smit M, Wijtmans M, de Graaf C, Vischer H, Leurs R (2012) Pharmacological modulation of chemokine receptor function. Br J Pharmacol 165(6):1617–1643
4. Proudfoot AE (2002) Chemokine receptors: multifaceted therapeutic targets. Nat Rev Immunol 2(2):106–115
5. Schall TJ, Proudfoot AE (2011) Overcoming hurdles in developing successful drugs targeting chemokine receptors. Nat Rev Immunol 11(5):355–363
6. Wootten D, Christopoulos A, Sexton PM (2013) Emerging paradigms in GPCR allostery: implications for drug discovery. Nat Rev Drug Discov 12(8):630–644
7. Allen SJ, Crown SE, Handel TM (2007) Chemokine: receptor, structure, interactions, and antagonism. Annu Rev Immunol 25:787–820
8. Jin H, Shen X, Baggett BR, Kong X, LiWang PJ (2007) The human CC chemokine MIP-1β dimer is not competent to bind to the CCR5 receptor. J Biol Chem 282(38):27976–27983
9. Nasser MW, Raghuwanshi SK, Grant DJ, Jala VR, Rajarathnam K, Richardson RM (2009) Differential activation and regulation of CXCR1 and CXCR2 by CXCL8 monomer and dimer. J Immunol 183(5):3425–3432
10. Tan JH, Canals M, Ludeman JP, Wedderburn J, Boston C, Butler SJ, Carrick AM, Parody TR, Taleski D, Christopoulos A (2012) Design and receptor interactions of obligate dimeric mutant of chemokine monocyte chemoattractant protein-1 (MCP-1). J Biol Chem 287 (18):14692–14702

11. Veldkamp CT, Seibert C, Peterson FC, De la Cruz NB, Haugner JC III, Basnet H, Sakmar TP, Volkman BF (2008) Structural basis of CXCR4 sulfotyrosine recognition by the chemokine SDF-1/CXCL12. Sci Signal 1(37):ra4
12. Choe H, Moore MJ, Owens CM, Wright PL, Vasilieva N, Li W, Singh AP, Shakri R, Chitnis CE, Farzan M (2005) Sulphated tyrosines mediate association of chemokines and Plasmodium vivax Duffy binding protein with the Duffy antigen/receptor for chemokines (DARC). Mol Microbiol 55(5):1413–1422
13. Ludeman JP, Stone MJ (2014) The structural role of receptor tyrosine sulfation in chemokine recognition. Br J Pharmacol 171(5):1167–1179
14. Zhu JZ, Millard CJ, Ludeman JP, Simpson LS, Clayton DJ, Payne RJ, Widlanski TS, Stone MJ (2011) Tyrosine sulfation influences the chemokine binding selectivity of peptides derived from chemokine receptor CCR3. Biochemistry 50(9):1524–1534
15. Cowan JE, McCarthy NI, Parnell SM, White AJ, Bacon A, Serge A, Irla M, Lane PJ, Jenkinson EJ, Jenkinson WE (2014) Differential requirement for CCR4 and CCR7 during the development of innate and adaptive αβT cells in the adult thymus. J Immunol 193 (3):1204–1212
16. Sallusto F, Lanzavecchia A, Mackay CR (1998) Chemokines and chemokine receptors in T-cell priming and Th1/Th2-mediated responses. Immunol Today 19(12):568–574
17. Groom J, Luster A (2011) CXCR3 ligands: redundant, collaborative and antagonistic functions. Immunol Cell Biol 89(2):207–215
18. Farber J (1997) Mig and IP-10: CXC chemokines that target lymphocytes. J Leukoc Biol 61 (3):246–257
19. Ohmori Y, Wyner L, Narumi S, Armstrong D, Stoler M, Hamilton T (1993) Tumor necrosis factor-alpha induces cell type and tissue-specific expression of chemoattractant cytokines in vivo. Am J Pathol 142(3):861–870
20. Ohmori Y, Schreiber RD, Hamilton TA (1997) Synergy between interferon-gamma and tumor necrosis factor-alpha in transcriptional activation is mediated by cooperation between signal transducer and activator of transcription 1 and nuclear factor kappa B. J Biol Chem 272 (23):14899–14907
21. Rani MRS, Foster GR, Leung S, Leaman D, Stark GR, Ransohoff RM (1996) Characterization of beta -R1, a gene that is selectively induced by interferon beta (IFN-beta) compared with IFN-alpha. J Biol Chem 271(37):22878–22884
22. Weng Y, Siciliano SJ, Waldburger KE, Sirotina-Meisher A, Staruch MJ, Daugherty BL, Gould SL, Springer MS, DeMartino JA (1998) Binding and functional properties of recombinant and endogenous CXCR3 chemokine receptors. J Biol Chem 273(29):18288–18291
23. Cox MA, Jenh C-H, Gonsiorek W, Fine J, Narula SK, Zavodny PJ, Hipkin RW (2001) Human interferon-inducible 10-kDa protein and human interferon-inducible T cell {alpha} chemoattractant are allotopic ligands for human CXCR3: differential binding to receptor states. Mol Pharmacol 59(4):707–715
24. Heise CE, Pahuja A, Hudson SC, Mistry MS, Putnam AL, Gross MM, Gottlieb PA, Wade WS, Kiankarimi M, Schwarz D, Crowe P, Zlotnik A, Alleva DG (2005) Pharmacological characterization of CXC chemokine receptor 3 ligands and a small molecule antagonist. J Pharmacol Exp Ther 313:1263–1271
25. Verzijl D, Storelli S, Scholten DJ, Bosch L, Reinhart TA, Streblow DN, Tensen CP, Fitzsimons CP, Zaman GJR, Pease JE, de Esch IJP, Smit MJ, Leurs R (2008) Noncompetitive antagonism and inverse agonism as mechanism of action of nonpeptidergic antagonists at primate and rodent CXCR3 chemokine receptors. J Pharmacol Exp Ther 325(2):544–555
26. Xanthou G, Williams T, Pease J (2003) Molecular characterization of the chemokine receptor CXCR3: evidence for the involvement of distinct extracellular domains in a multi-step model of ligand binding and receptor activation. Eur J Immunol 33(10):2927–2936
27. Colvin RA, Campanella GSV, Sun J, Luster AD (2004) Intracellular domains of CXCR3 that mediate CXCL9, CXCL10, and CXCL11 function. J Biol Chem 279(29):30219–30227

28. Colvin RA, Campanella GSV, Manice LA, Luster AD (2006) CXCR3 requires tyrosine sulfation for ligand binding and a second extracellular loop arginine residue for ligand-induced chemotaxis. Mol Cell Biol 26(15):5838–5849
29. Kouroumalis A, Nibbs RJ, Aptel H, Wright KL, Kolios G, Ward SG (2005) The chemokines CXCL9, CXCL10, and CXCL11 differentially stimulate Gαi-independent signaling and actin responses in human intestinal myofibroblasts. J Immunol 175(8):5403–5411
30. Christensen JE, de Lemos C, Moos T, Christensen JP, Thomsen AR (2006) CXCL10 is the key ligand for CXCR3 on CD8+ effector T cells involved in immune surveillance of the lymphocytic choriomeningitis virus-infected central nervous system. J Immunol 176 (7):4235–4243
31. Hensbergen P, Wijnands P, Schreurs M, Scheper R, Willemze R, Tensen C (2005) The CXCR3 targeting chemokine CXCL11 has potent antitumor activity in vivo involving attraction of CD8+ T lymphocytes but not inhibition of angiogenesis. J Immunother 28 (4):343–351
32. Loetscher P, Pellegrino A, Gong J-H, Mattioli I, Loetscher M, Bardi G, Baggiolini M, Clark-Lewis I (2001) The ligands of CXC chemokine receptor 3, I-TAC, Mig, and IP10, are natural antagonists for CCR3. J Biol Chem 276(5):2986–2991
33. Rathanaswami P, Hachicha M, Sadick M, Schall TJ, McColl SR (1993) Expression of the cytokine RANTES in human rheumatoid synovial fibroblasts. Differential regulation of RANTES and interleukin-8 genes by inflammatory cytokines. J Biol Chem 268 (8):5834–5839
34. Berkman N, Robichaud A, Krishnan V, Roesems G, Robbins R, Jose P, Barnes P, Chung K (1996) Expression of RANTES in human airway epithelial cells: effect of corticosteroids and interleukin-4, -10 and -13. Immunology 87(4):599–603
35. Schall T, Jongstra J, Dyer B, Jorgensen J, Clayberger C, Davis M, Krensky A (1988) A human T cell-specific molecule is a member of a new gene family. J Immunol 141 (3):1018–1025
36. Pakianathan D, Kuta E, Artis D, Skelton N, Hebert C (1997) Distinct but overlapping epitopes for the interaction of a CC-chemokine with CCR1, CCR3 and CCR5. Biochemistry 36(32):9642–9648
37. Proudfoot AEI, Buser R, Borlat F, Alouani S, Soler D, Offord RE, Schröder J-M, Power CA, Wells TNC (1999) Amino-terminally modified RANTES analogues demonstrate differential effects on RANTES receptors. J Biol Chem 274(45):32478–32485
38. Weber C, Weber KSC, Klier C, Gu S, Wank R, Horuk R, Nelson PJ (2001) Specialized roles of the chemokine receptors CCR1 and CCR5 in the recruitment of monocytes and TH1-like/ CD45RO+T cells. Blood 97(4):1144–1146
39. Kohout TA, Nicholas SL, Perry SJ, Reinhart G, Junger S, Struthers RS (2004) Differential desensitization, receptor phosphorylation, {beta}-arrestin recruitment, and ERK1/2 activation by the two endogenous ligands for the CC chemokine receptor 7. J Biol Chem 279 (22):23214–23222
40. Blanpain C, Lee B, Vakili J, Doranz BJ, Govaerts C, Migeotte I, Sharron M, Dupriez V, Vassart G, Doms RW (1999) Extracellular cysteines of CCR5 are required for chemokine binding, but dispensable for HIV-1 coreceptor activity. J Biol Chem 274(27):18902–18908
41. Tsamis F, Gavrilov S, Kajumo F, Seibert C, Kuhmann S, Ketas T, Trkola A, Palani A, Clader JW, Tagat JR (2003) Analysis of the mechanism by which the small-molecule CCR5 antagonists SCH-351125 and SCH-350581 inhibit human immunodeficiency virus type 1 entry. J Virol 77(9):5201–5208
42. Samson M, Libert F, Doranz BJ, Rucker J, Liesnard C, Farber C-M, Saragosti S, Lapouméroulie C, Cognaux J, Forceille C (1996) Resistance to HIV-1 infection in Caucasian individuals bearing mutant alleles of the CCR-5 chemokine receptor gene. Nature 382 (6593):722–725

43. de Kruijf P, Lim HD, Roumen L, Renjaän VA, Zhao J, Webb ML, Auld DS, Wijkmans JC, Zaman GJ, Smit MJ (2011) Identification of a novel allosteric binding site in the CXCR2 chemokine receptor. Mol Pharmacol 80(6):1108–1118
44. Andrews G, Jones C, Wreggett KA (2008) An intracellular allosteric site for a specific class of antagonists of the CC chemokine G protein-coupled receptors CCR4 and CCR5. Mol Pharmacol 73(3):855–867
45. Salchow K, Bond M, Evans S, Press N, Charlton S, Hunt P, Bradley M (2010) A common intracellular allosteric binding site for antagonists of the CXCR2 receptor. Br J Pharmacol 159(7):1429–1439
46. Nicholls DJ, Tomkinson NP, Wiley KE, Brammall A, Bowers L, Grahames C, Gaw A, Meghani P, Shelton P, Wright TJ (2008) Identification of a putative intracellular allosteric antagonist binding-site in the CXC chemokine receptors 1 and 2. Mol Pharmacol 74 (5):1193–1202
47. Melchiorre C, Minarini A, Angeli P, Giardina D, Gulini U, Quaglia W (1989) Polymethylene tetraamines as muscarinic receptor probes. Trends Pharmacol Sci 10:55–59
48. Ellis J, Huyler J, Brann MR (1991) Allosteric regulation of cloned m1–m5 muscarinic receptor subtypes. Biochem Pharmacol 42(10):1927–1932
49. Liang J-S, Carsi-Gabrenas J, Krajewski J, McCafferty J, Purkerson S, Santiago M, Strauss W, Valentine H, Potter L (1996) Anti-muscarinic toxins from *Dendroaspis angusticeps*. Toxicon 34(11):1257–1267
50. Gnagey AL, Seidenberg M, Ellis J (1999) Site-directed mutagenesis reveals two epitopes involved in the subtype selectivity of the allosteric interactions of gallamine at muscarinic acetylcholine receptors. Mol Pharmacol 56(6):1245–1253
51. Johnson M, Nisenbaum E, Large T, Emkey R, Baez M, Kingston A (2004) Allosteric modulators of metabotropic glutamate receptors: lessons learnt from mGlu1, mGlu2 and mGlu5 potentiators and antagonists. Biochem Soc Trans 32(5):881–887
52. Chapman RW, Minnicozzi M, Celly CS, Phillips JE, Kung TT, Hipkin RW, Fan X, Rindgen D, Deno G, Bond R (2007) A novel, orally active CXCR1/2 receptor antagonist, Sch527123, inhibits neutrophil recruitment, mucus production, and goblet cell hyperplasia in animal models of pulmonary inflammation. J Pharmacol Exp Ther 322(2):486–493
53. Paton WD, Rang H (1965) The uptake of atropine and related drugs by intestinal smooth muscle of the guinea-pig in relation to acetylcholine receptors. Proc R Soc Lond B Biol Sci 163(990):1–44
54. Kenakin T (2005) New concepts in drug discovery: collateral efficacy and permissive antagonism. Nat Rev Drug Discov 4(11):919–927
55. Bertini R, Allegretti M, Bizzarri C, Moriconi A, Locati M, Zampella G, Cervellera MN, Di Cioccio V, Cesta MC, Galliera E, Martinez FO, Di Bitondo R, Troiani G, Sabbatini V, D'Anniballe G, Anacardio R, Cutrin JC, Cavalieri B, Mainiero F, Strippoli R, Villa P, Di Girolamo M, Martin F, Gentile M, Santoni A, Corda D, Poli G, Mantovani A, Ghezzi P, Colotta F (2004) Noncompetitive allosteric inhibitors of the inflammatory chemokine receptors CXCR1 and CXCR2: prevention of reperfusion injury. Proc Natl Acad Sci U S A 101 (32):11791–11796
56. Thiele S, Malmgaard-Clausen M, Engel-Andreasen J, Steen A, Rummel P, Nielsen M, Gloriam D, Frimurer T, Ulven T, Rosenkilde M (2012) Modulation in selectivity and allosteric properties of small-molecule ligands for CC-chemokine receptors. J Med Chem 55(18):8164–8177
57. Sabroe I, Peck MJ, Van Keulen BJ, Jorritsma A, Simmons G, Clapham PR, Williams TJ, Pease JE (2000) A small molecule antagonist of chemokine receptors CCR1 and CCR3 potent inhibition of eosinophil function and CCR3-mediated HIV-1 entry. J Biol Chem 275 (34):25985–25992
58. Hejnová L, Tuček S, El-Fakahany EE (1995) Positive and negative allosteric interactions on muscarinic receptors. Eur J Pharmacol Mol Pharmacol 291(3):427–430

59. Kalatskaya I, Berchiche YA, Gravel S, Limberg BJ, Rosenbaum JS, Heveker N (2009) AMD3100 is a CXCR7 ligand with allosteric agonist properties. Mol Pharmacol 75 (5):1240–1247
60. Black JW, Leff P (1983) Operational models of pharmacological agonism. Proc R Soc B 220 (1219):141–162
61. Sohy D, Parmentier M, Springael J-Y (2007) Allosteric transinhibition by specific antagonists in CCR2/CXCR4 heterodimers. J Biol Chem 282(41):30062–30069
62. Springael J-Y, Le Minh PN, Urizar E, Costagliola S, Vassart G, Parmentier M (2006) Allosteric modulation of binding properties between units of chemokine receptor homo-and hetero-oligomers. Mol Pharmacol 69(5):1652–1661
63. Springael J-Y, Urizar E, Parmentier M (2005) Dimerization of chemokine receptors and its functional consequences. Cytokine Growth Factor Rev 16(6):611–623
64. Luker KE, Gupta M, Luker GD (2009) Imaging chemokine receptor dimerization with firefly luciferase complementation. FASEB J 23(3):823–834
65. Salanga CL, O'Hayre M, Handel T (2009) Modulation of chemokine receptor activity through dimerization and crosstalk. Cell Mol Life Sci 66(8):1370–1386
66. Wang T, Duan Y (2008) Binding modes of CCR5-targetting HIV entry inhibitors: partial and full antagonists. J Mol Graph Model 26:1287–1295
67. Levoye A, Balabanian K, Baleux F, Bachelerie F, Lagane B (2009) CXCR7 heterodimerizes with CXCR4 and regulates CXCL12-mediated G protein signaling. Blood 113 (24):6085–6093
68. Armour SL, Foord S, Kenakin T, Chen W-J (1999) Pharmacological characterization of receptor-activity-modifying proteins (RAMPs) and the human calcitonin receptor. J Pharmacol Toxicol Methods 42(4):217–224
69. Hay DL, Poyner DR, Sexton PM (2006) GPCR modulation by RAMPs. Pharmacol Ther 109 (1):173–197
70. Mellado M, Rodríguez-Frade JM, Vila-Coro AJ, Fernández S, Martín de Ana A, Jones DR, Torán JL, Martínez-A C (2001) Chemokine receptor homo-or heterodimerization activates distinct signaling pathways. EMBO J 20(10):2497–2507
71. Parenty G, Appelbe S, Milligan G (2008) CXCR2 chemokine receptor antagonism enhances DOP opioid receptor function via allosteric regulation of the CXCR2-DOP receptor heterodimer. Biochem J 412:245–256
72. Dean M, Carrington M, Winkler C, Huttley GA, Smith MW, Allikmets R, Goedert JJ, Buchbinder SP, Vittinghoff E, Gomperts E (1996) Genetic restriction of HIV-1 infection and progression to AIDS by a deletion allele of the CKR5 structural gene. Science 273 (5283):1856–1862
73. Benkirane M, Jin D-Y, Chun RF, Koup RA, Jeang K-T (1997) Mechanism of transdominant inhibition of CCR5-mediated HIV-1 infection by ccr5{Delta}32. J Biol Chem 272 (49):30603–30606
74. Christopoulos A, Kenakin T (2002) G protein-coupled receptor allosterism and complexing. Pharmacol Rev 54(2):323–374
75. Kenakin TP (2012) Biased signalling and allosteric machines: new vistas and challenges for drug discovery. Br J Pharmacol 165(6):1659–1669
76. Stockton J, Birdsall N, Burgen A, Hulme E (1983) Modification of the binding properties of muscarinic receptors by gallamine. Mol Pharmacol 23(3):551–557
77. Ehlert F (1988) Estimation of the affinities of allosteric ligands using radioligand binding and pharmacological null methods. Mol Pharmacol 33(2):187–194
78. Ehlert FJ (2005) Analysis of allosterism in functional assays. J Pharmacol Exp Ther 315 (2):740–754
79. Price MR, Baillie GL, Thomas A, Stevenson LA, Easson M, Goodwin R, McLean A, McIntosh L, Goodwin G, Walker G (2005) Allosteric modulation of the cannabinoid CB1 receptor. Mol Pharmacol 68(5):1484–1495

80. Sachpatzidis A, Benton BK, Manfredi JP, Wang H, Hamilton A, Dohlman HG, Lolis E (2003) Identification of allosteric peptide agonists of CXCR4. J Biol Chem 278(2):896–907
81. Jensen PC, Thiele S, Ulven T, Schwartz TW, Rosenkilde MM (2008) Positive versus negative modulation of different endogenous chemokines for CC-chemokine receptor 1 by small molecule agonists through allosteric versus orthosteric binding. J Biol Chem 283 (34):23121–23128
82. Koole C, Wootten D, Simms J, Valant C, Sridhar R, Woodman OL, Miller LJ, Summers RJ, Christopoulos A, Sexton PM (2010) Allosteric Ligands of the glucagon-like peptide 1 receptor (GLP-1R) differentially modulate endogenous and exogenous peptide responses in a pathway-selective manner: implications for drug screening. Mol Pharmacol 78(3):456–465
83. Watson C, Jenkinson S, Kazmierski W, Kenakin T (2005) The CCR5 receptor-based mechanism of action of 873140, a potent allosteric noncompetitive HIV entry inhibitor. Mol Pharmacol 67(4):1268–1282
84. Casilli F, Bianchini A, Gloaguen I, Biordi L, Alesse E, Festuccia C, Cavalieri B, Strippoli R, Cervellera MN, Bitondo RD (2005) Inhibition of interleukin-8 (CXCL8/IL-8) responses by repertaxin, a new inhibitor of the chemokine receptors CXCR1 and CXCR2. Biochem Pharmacol 69(3):385–394
85. Rummel PC, Arfelt K, Baumann L, Jenkins T, Thiele S, Lüttichau H, Johnsen A, Pease J, Ghosh S, Kolbeck R (2012) Molecular requirements for inhibition of the chemokine receptor CCR8–probe-dependent allosteric interactions. Br J Pharmacol 167(6):1206–1217
86. Vaidehi N, Schlyer S, Trabanino RJ, Floriano WB, Abrol R, Sharma S, Kochanny M, Koovakat S, Dunning L, Liang M (2006) Predictions of CCR1 chemokine receptor structure and BX 471 antagonist binding followed by experimental validation. J Biol Chem 281 (37):27613–27620
87. Garcia-Perez J, Rueda P, Staropoli I, Kellenberger E, Alcami J, Arenzana-Seisdedos F, Lagane B (2011) New insights into the mechanisms whereby low molecular weight CCR5 ligands inhibit HIV-1 infection. J Biol Chem 286(7):4978–4990
88. Atchison RE, Gosling J, Monteclaro FS, Franci C, Digilio L, Charo IF, Goldsmith MA (1996) Multiple extracellular elements of CCR5 and HIV-1 entry: dissociation from response to chemokines. Science 274(5294):1924–1926
89. Doms RW, Peiper SC (1997) Unwelcomed guests with master keys: how HIV uses chemokine receptors for cellular entry. Virology 235(2):179–190
90. Doranz BJ, Lu Z, Rucker J, Zhang T-Y, Sharron M, Cen Y, Wang Z, Guo H, Du J, Accavitti MA (1997) Two distinct CCR5 domains can mediate coreceptor usage by human immunodeficiency virus type 1. J Virol 71(9):6305–6314
91. Picard L, Simmons G, Power CA, Meyer A, Weiss RA, Clapham PR (1997) Multiple extracellular domains of CCR-5 contribute to human immunodeficiency virus type 1 entry and fusion. J Virol 71(7):5003–5011
92. Rucker J, Edinger AL, Sharron M, Samson M, Lee B, Berson JF, Yi Y, Margulies B, Collman RG, Doranz BJ (1997) Utilization of chemokine receptors, orphan receptors, and herpesvirus-encoded receptors by diverse human and simian immunodeficiency viruses. J Virol 71 (12):8999–9007
93. Lori F, Foli A, Matteo PS, Lisziewicz J, Jessen H (1997) Long-term suppression of HIV-1 by hydroxyurea and didanosine. JAMA 277(18):1437–1438
94. Garzino-Demo A, Moss RB, Margolick JB, Cleghorn F, Sill A, Blattner WA, Cocchi F, Carlo DJ, DeVico AL, Gallo RC (1999) Spontaneous and antigen-induced production of HIV-inhibitory β-chemokines are associated with AIDS-free status. Proc Natl Acad Sci U S A 96(21):11986–11991
95. Shieh B, Yan Y-P, Ko N-Y, Liau Y-E, Liu Y-C, Lin H-H, Chen P-P, Li C (2001) Detection of elevated serum {beta}-chemokine levels in seronegative chinese individuals exposed to human immunodeficiency virus type 1. Clin Infect Dis 33(3):273–279

96. Xiang J, George S, Wunschmann S, Chang Q, Klinzman D, Stapleton J (2004) Inhibition of HIV-1 replication by GB virus C infection through increases in RANTES, MIP-1alpha, MIP-1beta, and SDF-1. Lancet 363(9426):2040–2046
97. Ullum H, Lepri AC, Victor J, Aladdin H, Phillips AN, Gerstoft J, Skinhoj P, Pedersen BK (1998) Production of {beta}-chemokines in human immunodeficiency virus (HIV) infection: evidence that high levels of macrophage inflammatory protein-1{beta} are associated with a decreased risk of HIV disease progression. J Infect Dis 177(2):331–336
98. Gonzalez E, Kulkarni H, Bolivar H, Mangano A, Sanchez R, Catano G, Nibbs RJ, Freedman BI, Quinones MP, Bamshad MJ (2005) The influence of CCL3L1 gene-containing segmental duplications on HIV-1/AIDS susceptibility. Science 307(5714):1434–1440
99. Alkhatib G, Locati M, Kennedy PE, Murphy PM, Berger EA (1997) HIV-1 coreceptor activity of CCR5 and its inhibition by chemokines: independence from G protein signaling and importance of coreceptor downmodulation. Virology 234(2):340–348
100. Amara A, Gall SL, Schwartz O, Salamero J, Montes M, Loetscher P, Baggiolini M, Virelizier J-L, Arenzana-Seisdedos F (1997) HIV coreceptor downregulation as antiviral principle: SDF-1{alpha}-dependent internalization of the chemokine receptor CXCR4 contributes to inhibition of HIV replication. J Exp Med 186(1):139–146
101. Mack M, Luckow B, Nelson PJ, Cihak J, Simmons G, Clapham PR, Signoret N, Marsh M, Stangassinger M, Borlat F (1998) Aminooxypentane-RANTES induces CCR5 internalization but inhibits recycling: a novel inhibitory mechanism of HIV infectivity. J Exp Med 187 (8):1215–1224
102. Muniz-Medina VM, Jones S, Maglich JM, Galardi C, Hollingsworth RE, Kazmierski WM, Ferris RG, Edelstein MP, Chiswell KE, Kenakin TP (2009) The relative activity of "function sparing" HIV-1 entry inhibitors on viral entry and CCR5 internalization: is allosteric functional selectivity a valuable therapeutic property? Mol Pharmacol 75(3):490–501
103. Buontempo PJ, Wojcik L, Buontempo CA, Ogert RA, Strizki JM, Howe JA, Ralston R (2009) Quantifying the relationship between HIV-1 susceptibility to CCR5 antagonists and virus affinity for antagonist-occupied co-receptor. Virology 395(2):268–279
104. Kenakin T (1995) Agonist-receptor efficacy II: agonist trafficking of receptor signals. Trends Pharmacol Sci 16(7):232–238
105. Scholten D, Canals M, Wijtmans M, de Munnik S, Nguyen P, Verzijl D, de Esch I, Vischer H, Smit M, Leurs R (2012) Pharmacological characterization of a small-molecule agonist for the chemokine receptor CXCR3. Br J Pharmacol 166(3):898–911
106. Elsner J, Mack M, Brühl H, Dulkys Y, Kimmig D, Simmons G, Clapham PR, Schlöndorff D, Kapp A, Wells TN (2000) Differential activation of CC chemokine receptors by AOP-RANTES. J Biol Chem 275(11):7787–7794
107. Gaertner H, Cerini F, Kuenzi G, Melotti A, Offord R, Rossitto-Borlat I, Nedellec R, Salkowitz J, Gorochov G, Mosier D (2008) Highly potent, fully recombinant anti-HIV chemokines: reengineering a low-cost microbicide. Proc Natl Acad Sci 105 (46):17706–17711
108. Allegretti M, Bertini R, Bizzarri C, Beccari A, Mantovani A, Locati M (2008) Allosteric inhibitors of chemoattractant receptors: opportunities and pitfalls. Trends Pharmacol Sci 29 (6):280–286
109. Lewis JA, Lebois EP, Lindsley CW (2008) Allosteric modulation of kinases and GPCRs: design principles and structural diversity. Curr Opin Chem Biol 12(3):269–280
110. Vaidehi N, Schlyer S, Trabanino RJ, Floriano WB, Abrol R, Sharma S, Kochanny M, Koovakat S, Dunning L, Liang M, Fox JM, de Mendonca FL, Pease JE, Goddard WA III, Horuk R (2006) Predictions of CCR1 chemokine receptor structure and BX 471 antagonist binding followed by experimental validation. J Biol Chem 281(37):27613–27620
111. Mirzadegan T, Diehl F, Ebi B, Bhakta S, Polsky I, McCarley D, Mulkins M, Weatherhead GS, Lapierre J-M, Dankwardt J, Morgans D, Wilhelm R, Jarnagin K (2000) Identification of the binding site for a novel class of CCR2b chemokine receptor antagonists: binding to a

common chemokine receptor motif within the helical bundle. J Biol Chem 275 (33):25562–25571
112. de Mendonça FL, da Fonseca PCA, Phillips RM, Saldanha JW, Williams TJ, Pease JE (2005) Site-directed mutagenesis of CC chemokine receptor 1 reveals the mechanism of action of UCB 35625, a small molecule chemokine receptor antagonist. J Biol Chem 280 (6):4808–4816
113. Gerlach LO, Skerlj RT, Bridger GJ, Schwartz TW (2001) Molecular interactions of cyclam and bicyclam non-peptide antagonists with the CXCR4 chemokine receptor. J Biol Chem 276 (17):14153–14160
114. Wu B, Chien EYT, Mol CD, Fenalti G, Liu W, Katritch V, Abagyan R, Brooun A, Wells P, Bi FC, Hamel DJ, Kuhn P, Handel TM, Cherezov V, Stevens RC (2010) Structures of the CXCR4 chemokine GPCR with small-molecule and cyclic peptide antagonists. Science 330 (6007):1066–1071
115. Tan Q, Zhu Y, Li J, Chen Z, Han GW, Kufareva I, Li T, Ma L, Fenalti G, Li J, Zhang W, Xie X, Yang H, Jiang H, Cherezov V, Liu H, Stevens RC, Zhao Q, Wu B (2013) Structure of the CCR5 chemokine receptor–HIV entry inhibitor maraviroc complex. Science 341 (6152):1387–1390
116. Hesselgesser J, Ng HP, Liang M, Zheng W, May K, Bauman JG, Monahan S, Islam I, Wei GP, Ghannam A, Taub DD, Rosser M, Snider RM, Morrissey MM, Perez HD, Horuk R (1998) Identification and characterization of small molecule functional antagonists of the CCR1 chemokine receptor. J Biol Chem 273(25):15687–15692
117. Witherington J, Bordas V, Cooper D, Forbes I, Gribble A, Ife R, Berkhout T, Gohil J, Groot P (2001) Conformationally restricted indolopiperidine derivatives as potent CCR2B receptor antagonists. Bioorg Med Chem Lett 11(16):2177–2180
118. Thoma G, Baenteli R, Lewis I, Wagner T, Oberer L, Blum W, Glickman F, Streiff MB, Zerwes HG (2009) Special ergolines are highly selective, potent antagonists of the chemokine receptor CXCR3: discovery, characterization and preliminary SAR of a promising lead. Bioorg Med Chem Lett 19(21):6185–6188
119. Watson P, Jiang B, Harrison K, Asakawa N, Welch P, Covington M, Stowell N, Wadman E, Davies P, Solomon K, Newton R, Trainor G, Friedman S, Decicco C, Ko S (2006) 2,4-Disubstituted piperidines as selective CC chemokine receptor 3 (CCR3) antagonists: synthesis and selectivity. Bioorg Med Chem Lett 16(21):5695–5699
120. Suzuki K, Morokata T, Morihira K, Sato I, Takizawa S, Kaneko M, Takahashi K, Shimizu Y (2007) A dual antagonist for chemokine CCR3 receptor and histamine H1 receptor. Eur J Pharmacol 563(1):224–232
121. Hall SE, Mao A, Nicolaidou V, Finelli M, Wise EL, Nedjai B, Kanjanapangka J, Harirchian P, Chen D, Selchau V, Ribeiro S, Schyler S, Pease JE, Horuk R, Vaidehi N (2009) Elucidation of binding sites of dual antagonists in the human chemokine receptors CCR2 and CCR5. Mol Pharmacol 75(6):1325–1336
122. Dragic T, Trkola A, Thompson DAD, Cormier EG, Kajumo FA, Maxwell E, Lin SW, Ying W, Smith SO, Sakmar TP, Moore JP (2000) A binding pocket for a small molecule inhibitor of HIV-1 entry within the transmembrane helices of CCR5. Proc Natl Acad Sci 97 (10):5639–5644
123. Baba M, Nishimura O, Kanzaki N, Okamoto M, Sawada H, Iizawa Y, Shiraishi M, Aramaki Y, Okonogi K, Ogawa Y (1999) A small-molecule, nonpeptide CCR5 antagonist with highly potent and selective anti-HIV-1 activity. Proc Natl Acad Sci 96(10):5698–5703
124. Gao P, Zhou X-Y, Yashiro-Ohtani Y, Yang Y-F, Sugimoto N, Ono S, Nakanishi T, Obika S, Imanishi T, Egawa T, Nagasawa T, Fujiwara H, Hamaoka T (2003) The unique target specificity of a nonpeptide chemokine receptor antagonist: selective blockade of two Th1 chemokine receptors CCR5 and CXCR3. J Leukoc Biol 73(2):273–280
125. Tokuyama H, Ueha S, Kurachi M, Matsushima K, Moriyasu F, Blumberg RS, Kakimi K (2005) The simultaneous blockade of chemokine receptors CCR2, CCR5 and CXCR3 by a

non-peptide chemokine receptor antagonist protects mice from dextran sodium sulfate-mediated colitis. Int Immunol 17(8):1023–1034

126. Kralj A, Wetzel A, Mahmoudian S, Stamminger T, Tschammer N, Heinrich MR (2011) Identification of novel allosteric modulators for the G-protein coupled US28 receptor of human cytomegalovirus. Bioorg Med Chem Lett 21(18):5446–5450
127. Venkatakrishnan A, Deupi X, Lebon G, Tate CG, Schertler GF, Babu MM (2013) Molecular signatures of G-protein-coupled receptors. Nature 494(7436):185–194
128. Park SH, Das BB, Casagrande F, Tian Y, Nothnagel HJ, Chu M, Kiefer H, Maier K, De Angelis AA, Marassi FM (2012) Structure of the chemokine receptor CXCR1 in phospholipid bilayers. Nature 491(7426):779–783
129. Szpakowska M, Bercoff DP, Chevigné A (2014) Closing the ring: a fourth extracellular loop in chemokine receptors. Sci Signal 7(341):pe21–pe21
130. Zhu L, Zhao Q, Wu B (2013) Structure-based studies of chemokine receptors. Curr Opin Struct Biol 23(4):539–546
131. Watanabe K, Iida M, Takaishi K, Suzuki T, Hamada Y, Iizuka Y, Tsurufuji S (1993) Chemoattractants for neutrophils in lipopolysaccharide-induced inflammatory exudate from rats are not interleukin-8 counterparts but gro-gene-product/melanoma-growth-stimulating-activity-related factors. Eur J Biochem 214(1):267–270
132. Harada A, Kuno K, Nomura H, Mukaida N, Murakami S, Matsushima K (1994) Cloning of a cDNA encoding a mouse homolog of the interleukin-8 receptor. Gene 142(2):297–300
133. Bozic CR, Kolakowski L, Gerard NP, Garcia-Rodriguez C, von Uexkull-Guldenband C, Conklyn MJ, Breslow R, Showell HJ, Gerard C (1995) Expression and biologic characterization of the murine chemokine KC. J Immunol 154(11):6048–6057
134. Moepps B, Nuesseler E, Braun M, Gierschik P (2006) A homolog of the human chemokine receptor CXCR1 is expressed in the mouse. Mol Immunol 43(7):897–914
135. Fan X, Patera AC, Pong-Kennedy A, Deno G, Gonsiorek W, Manfra DJ, Vassileva G, Zeng M, Jackson C, Sullivan L, Sharif-Rodriguez W, Opdenakker G, Van Damme J, Hedrick JA, Lundell D, Lira SA, Hipkin RW (2007) Murine CXCR1 is a functional receptor for GCP-2/CXCL6 and interleukin-8/CXCL8. J Biol Chem 282(16):11658–11666
136. Jung S, Aliberti J, Graemmel P, Sunshine MJ, Kreutzberg GW, Sher A, Littman DR (2000) Analysis of fractalkine receptor CX3CR1 function by targeted deletion and green fluorescent protein reporter gene insertion. Mol Cell Biol 20(11):4106–4114
137. Mack M, Cihak J, Simonis C, Luckow B, Proudfoot AE, Jí P, Brühl H, Frink M, Anders H-J, Vielhauer V (2001) Expression and characterization of the chemokine receptors CCR2 and CCR5 in mice. J Immunol 166(7):4697–4704
138. Topham PS, Csizmadia V, Soler D, Hines D, Gerard CJ, Salant DJ, Hancock WW (1999) Lack of chemokine receptor CCR1 enhances Th1 responses and glomerular injury during nephrotoxic nephritis. J Clin Invest 104(11):1549–1557
139. Hegen M, Keith JC, Collins M, Nickerson-Nutter CL (2008) Utility of animal models for identification of potential therapeutics for rheumatoid arthritis. Ann Rheum Dis 67 (11):1505–1515
140. Haringman JJ, Tak PP (2004) Chemokine blockade: a new era in the treatment of rheumatoid arthritis? Arthritis Res Ther 6(3):93–97
141. Quinones MP, Estrada CA, Kalkonde Y, Ahuja SK, Kuziel WA, Mack M, Ahuja SS (2005) The complex role of the chemokine receptor CCR2 in collagen-induced arthritis: implications for therapeutic targeting of CCR2 in rheumatoid arthritis. J Mol Med 83(9):672–681
142. Brühl H, Cihak J, Schneider MA, Plachý J, Rupp T, Wenzel I, Shakarami M, Milz S, Ellwart JW, Stangassinger M, Schlöndorff D, Mack M (2004) Dual role of CCR2 during initiation and progression of collagen-induced arthritis: evidence for regulatory activity of CCR2+ T cells. J Immunol 172(2):890–898
143. Berahovich RD, Miao Z, Wang Y, Premack B, Howard MC, Schall TJ (2005) Proteolytic activation of alternative CCR1 ligands in inflammation. J Immunol 174(11):7341–7351

144. Lebre MC, Vergunst CE, Choi IYK, Aarrass S, Oliveira ASF, Wyant T, Horuk R, Reedquist KA, Tak PP (2011) Why CCR2 and CCR5 blockade failed and why CCR1 blockade might still be effective in the treatment of rheumatoid arthritis. PLoS One 6(7):e21772
145. Gladue RP, Cole SH, Roach ML, Tylaska LA, Nelson RT, Shepard RM, McNeish JD, Ogborne KT, Neote KS (2006) The human specific CCR1 antagonist CP-481,715 inhibits cell infiltration and inflammatory responses in human CCR1 transgenic mice. J Immunol 176 (5):3141–3148

Top Med Chem (2015) 14: 119–186
DOI: 10.1007/7355_2014_75

Published online: 5 October 2014

# Exploring the CXCR3 Chemokine Receptor with Small-Molecule Antagonists and Agonists

**Maikel Wijtmans, Danny Scholten, Wouter Mooij, Martine J. Smit, Iwan J.P. de Esch, Chris de Graaf, and Rob Leurs**

**Abstract** CXCR3 is a CXC chemokine receptor that, together with its three major ligands, CXCL9 (MIG), CXCL10 (IP-10), and CXCL11 (I-TAC), is involved in inflammatory responses, mediated mainly by T cells. In several immune-related diseases, including chronic obstructive pulmonary disease (COPD), inflammatory bowel disease (IBD), rheumatoid arthritis, multiple sclerosis, and atherosclerosis, CXCR3 and/or its ligands are found to be overexpressed, potentially indicating a role for this receptor in these diseases. Animal models have confirmed the therapeutic potential of targeting CXCR3 in the treatment of such diseases. Several peptidic, peptidomimetic, and small non-peptidomimetic CXCR3 ligands have been disclosed in the past 10 years. These ligands have served as chemical tools for the investigation of CXCR3 activation, blocking, and signaling, and some of these ligand series have been developed as potential therapeutic agents against inflammation. Computational modeling studies, facilitated by the recent developments in GPCR structural biology, together with mutagenesis and pharmacological studies, have aided in understanding how these ligands interact with CXCR3.

This chapter will give an overview on how the combination of these chemical, computational, and pharmacological tools and techniques has increased our understanding of the molecular mechanisms by which small-molecule antagonists and agonists bind to CXCR3 compared to the relatively large chemokines. A detailed overview of CXCR3 ligand structure-activity relationships and structure-function relationships will be presented. This comparative analysis reveals that the full spectrum of antagonist and agonist effects on CXCR3 is now within reach by appropriate scaffolds and chemical modifications. Many of these ligands display behavior deviating from simple competition and do not interact with the chemokine

M. Wijtmans (✉), D. Scholten, W. Mooij, M.J. Smit, I.J.P. de Esch, C. de Graaf, and R. Leurs
Division of Medicinal Chemistry, Faculty of Sciences, Amsterdam Institute for Molecules Medicines and Systems, VU University, Amsterdam, De Boelelaan 1083, 1081 HV Amsterdam, The Netherlands
e-mail: m.wijtmans@vu.nl

binding site, providing evidence for an allosteric mode of action. Moreover, the computer-assisted molecular modeling of CXCR3 receptor-ligand interactions is discussed in view of GPCR crystal structures and mutagenesis studies of CXCR3 and other chemokine receptors. Improved insights in the interplay between CXCR3-ligand interactions and CXCR3-mediated signaling pathways potentially open up novel therapeutic opportunities in the area of inflammation.

**Keywords** Allosteric modulation, C–X–C chemokine receptor type 3, G protein-coupled receptor, GPCR crystal structure, Homology modeling, Mutagenesis studies, Receptor–ligand interactions, Small-molecule agonist, Small-molecule antagonist, Structure–activity relationship, Structure–function relationship

**Contents**



# 1 Introduction

## *1.1 CXCR3*

### 1.1.1 History and Pharmacological Aspects

Loetscher and colleagues first cloned the human C-X-C chemokine receptor type 3 (CXCR3) receptor in 1996 from a cDNA library derived from $CD4^+$ T cells. The receptor cDNA sequence encodes for a protein of 368 amino acid in length [1]. CXCR3 is predominantly expressed on activated Th1 cells but also on a proportion of circulating blood T cells and B cells and is expressed on natural killer cells [1–3]. Furthermore, the chemokines CXCL9 (Mig), CXCL10 (IP-10), and CXCL11 (I-TAC) were found to be the endogenous agonists for this receptor, governing the migration of CXCR3-expressing leukocytes to sites of inflammation or infection [1, 4–8]. The expression of CXCR3 chemokines is induced under inflammatory conditions, mainly mediated by interferon $\gamma$ [1, 2]. In addition, CXCL4 and CXCL13 chemokines were also reported to bind and activate CXCR3 at high concentrations [9, 10]. CXCL4 was also shown to bind a splice variant of CXCR3 with an extended N-terminus, called CXCR3B [11].

CXCR3 signaling occurs mainly through activation of the pertussis toxin-sensitive $G\alpha_{i/o}$ class of G proteins, and receptor stimulation is associated with various downstream signaling pathways, including chemotaxis, activation of p44/42 and Akt kinases, mobilization of $Ca^{2+}$ from intracellular stores, and lowering of cyclic AMP levels [1, 12–14]. After receptor activation, the receptor is rapidly desensitized, internalized, and degraded [15–17]. In general, after agonist exposure, a GPCR is phosphorylated at C-terminal serine/threonine residues by G protein-coupled receptor kinases (GRKs), a process also known as desensitization. Subsequently, these phosphorylated residues serve as recognition sites for β-arrestin proteins, generally involved in internalization of GPCRs [18]. Although, it has been shown that β-arrestin is involved in CXCR3 internalization to some extent, it is not the sole determinant in directing CXCR3 receptors away from the cell surface [12, 17]. Also caveolae, another route for receptor internalization, do not seem to be involved in the case of CXCR3. Taken together, the exact mechanism of CXCR3 internalization is still unclear [17]. Nevertheless, CXCR3 is able to efficiently recruit β-arrestin1 and −2 proteins after agonist stimulation [14]. However, its functional implications remain to be explored.

Although GPCRs generally are able to function as monomers, increasing evidence points towards assembly of multiple GPCRs in homo- and/or heteromeric complexes [19]. Chemokine receptors have been found to dimerize or oligomerize with other chemokine receptors [20]. To date, evidence suggests that CXCR3 dimerizes with viral chemokine receptor BILF1 (Epstein-Barr virus) in HEK293T cells [21]. The in vivo consequences are yet unclear, but it might be envisioned that such viruses modulate the host immune defense by altering chemokine receptor signaling, e.g., through dimerization, facilitating its survival [21]. Furthermore, a recent study reported on the formation of heteromeric complexes of CXCR3 and CXCR4 receptors in HEK293T cells [22]. It was found that CXCL10 could be displaced from CXCR3 by CXCR4 chemokine CXCL12 and vice versa. Moreover, CXCR3 agonists were able to increase the dissociation rate of CXCL12 from CXCR4 receptors, both suggesting negative cooperativity through a heteromer interface.

### 1.1.2 Functional Selectivity in CXCR3 Pharmacology

The ligands for CXCR3 have different efficacies and potencies when compared in different in vitro functional assays. CXCL11 seems to be the dominant ligand, as it has the highest affinity, potency, and efficacy in most assays [1, 5–8, 14, 23]. The question why a receptor like CXCR3 would need three different ligands is intriguing, as it suggests redundancy in the chemokine system. However, multiple lines of evidence point at unique nonredundant roles for these ligands in vivo. First of all, CXCL9, CXCL10, and CXCL11 seem to interact with the receptor in a distinct mode, as they need different parts of the receptor for their binding and signaling activities [12, 24]. Next to that, CXCL10 and CXCL11 seem to bind different CXCR3 populations [14, 23], also suggesting distinct functions for these chemokines. This is furthermore supported by recent findings, where functional selectivity or biased signaling for the different CXCR3 chemokines was observed [25]. Whereas CXCL10- or CXCL11-mediated receptor activation both resulted in β-arrestin recruitment to the activated CXCR3 receptors, CXCL9 activation failed to do so, despite effective $G_i$ protein activation [25]. Moreover, and in contrast to CXCL9 and CXCL10, CXCL11 was the only CXCR3 chemokine that provoked elevation of intracellular calcium in myofibroblasts [26]. Altogether, these data are in support of nonredundant functions for CXCR3 chemokines. Second, differential spatiotemporal patterns of expression for the different chemokines and chemokine receptors in our body indicate that these chemokines have distinct roles in vivo [27].

In general, chemokines are thought to activate their receptors according to a two-step model, where the core of the chemokine binds to the N-terminus and extracellular domains of the receptor (step 1) [20]. Subsequently, the N-terminus of CXCL9, CXCL10, or CXCL11 is positioned towards yet to be identified receptor domains, mediating receptor activation (step 2) [20, 24, 28]. This general hypothesis is substantiated by the observation that adding or deleting only a few N-terminal chemokine residues resulted in a change from agonist to antagonist behavior. Indeed, N-terminal truncation of CXCL11 (e.g., CXCL11 4–73) hardly influences binding affinity, yet results in a complete loss of agonism [29]. Similarly, deletion of residues 2–6 from the N-terminus of CXCL10 also resulted in a potent antagonist with high affinity for the receptor [30]. Interestingly, chemokines are also truncated in vivo, where they are processed by proteases to give chemokines with modified affinity and activity. Particularly, CXCL10 and CXCL11 are processed N-terminally by the dipeptidyl peptidase IV CD26, leading to a loss in their chemotactic and calcium signaling activity while retaining their ability to bind the receptor, albeit with reduced affinity [31, 32]. These processed chemokines act as antagonists, as they are able to antagonize activity of full-length CXCL10 and CXCL11 [32].

Altogether, these data indicate that expression and activity of chemokines is tightly regulated in a spatiotemporal manner, giving texture and robustness to the CXCR3 response in vivo, again refuting the notion of redundancy.

Hetero(di)merization might present another way by which selective fine-tuning of receptor signaling can be achieved in the chemokine system [20]. In the case of CCR5, heterodimerization with CXCR4 or CCR2 even led to a shift in coupling from $G_i$- to $G_q$-mediated signaling pathways [33]. Unfortunately, little is known about functional consequences of CXCR3 homo-/heteromerization. However, a recent report from our group has shown that β-arrestin can be specifically recruited to CXCR3 and CXCR4 heteromers, potentially leading to altered receptor desensitization and internalization [22].

### 1.1.3 The CXCR3 Receptor as Potential Drug Target

To date, two small-molecule antagonists targeting chemokine receptors have successfully reached the market. Maraviroc (CCR5, Selzentry®) and AMD-3100 (CXCR4, Mozobil®) are used for treatment of HIV-1 infection and non-Hodgkin lymphoma, respectively. However, they do not target chronic inflammation or autoimmune diseases, conditions that are most often associated with imbalanced expression and signaling within the chemokine system. The druggability of the chemokine system has been debated for quite some time now, mainly due to the high attrition rate of drug candidates in clinical trials [34]. The reason for this remains unclear, but it is often suggested that the complexity ("redundancy") of the system is a key factor. As such, target validation of individual chemokines and/or receptors, linking them to specific diseases, is of vital importance to establish therapeutic potential. Fortunately, data pinpointing specific roles for chemokine receptors in disease models is emerging. In the case of CXCR3, the receptor and one or more of its ligands are found to be highly overexpressed in a variety of inflammatory disorders, including allograft rejection [35–37], atherosclerosis [38], and autoimmune diseases such as rheumatoid arthritis [2, 39], chronic obstructive pulmonary disease [40], multiple sclerosis [41], and systemic lupus erythematosus (SLE) [42]. In addition, the amounts of chemokine mRNA and number of infiltrating CXCR3-expressing leukocytes in tissues from transplant and SLE patients seem to correlate with the severity of disease [43–46]. Moreover, CXCR3 is suggested to play an important role in metastasis of melanoma and colon cancer cells to the lymph nodes and in metastasis of breast cancer cells to the lung [47–50].

Inhibition of CXCR3 by either antibodies or small-molecule antagonists significantly delays disease progression in various mouse models of disease, including atherosclerosis, transplant rejection, and cancer [37, 48, 51–55]. As such, antagonism has been the focus of CXCR3 drug discovery efforts.

Intriguingly, opposing data was reported on the role of CXCR3 in allograft rejection. Where some studies with $CXCR3^{-}/^{-}$ mice reported delayed acute and chronic rejection of cardiac allografts [37] or pancreatic island allografts [56], others revealed that CXCR3 does not play an essential role in cardiac allograft rejection [57, 58]. Therefore, these studies challenge the potential of CXCR3 as a drug target in allograft rejection. Moreover, also contrasting evidence exists for the involvement of CXCR3 in cancer, as multiple lines of evidence point both at a protective and sustaining roles for CXCR3 in cancer. In some reports CXCR3 activation and expression is linked to proliferative signaling and metastasis of tumor cells to tissues with relatively high CXCL9-11 expression [47–49, 59–61], as CXCR3 antagonism is found to decrease tumor growth and metastasis [53, 60, 62–64]. On the other hand, others reported that the presence of CXCR3 or its ligands is associated with slower tumor growth and decreased metastasis [65, 66]. The latter might be explained by the recruitment of antitumor immune cells [67].

An alternative explanation that reconciles these apparent contradictory roles is the presence of different splice variants of CXCR3. Alternative splicing of the CXCR3 gene leads to expression of CXCR3A, CXCR3B (52 aa longer N-terminus compared to CXCR3A), and CXCR3-alt (loss of 3 TM domains as a result of skipping exon 2) [11, 68]. Unfortunately, little is known about the expression patterns and signaling properties of the alternative splice variants CXCR3B and especially CXCR3-alt. Nevertheless, accumulating evidence points at a tumorigenic role for CXCR3A [53, 63] and an angiostatic role for CXCR3B [69]. The angiostasis through CXCR3B might be the result of inhibition of the antiapoptotic protein heme-oxygenase-1 as shown by overexpression of CXCR3B in renal cancer cells [70]. Moreover, other evidence supporting this functional division of the two splice variants is provided by multiple reports that describe a trend for cancers with more invasive phenotypes generally exhibiting decreased CXCR3B mRNA expression compared to CXCR3A, as in the case of skin and prostate cancer [64, 71]. Consequently, the functional outcome would then depend on the relative expression of both splice variants in a given (diseased) tissue. Unfortunately, in the majority of CXCR3 target validation studies, no distinction is made between the different splice variants. In some cases, a distinction is made, yet almost exclusively by measurement of mRNA levels instead of actual protein levels. Altogether, in-depth characterization of the properties of CXCR3A, CXCR3B, and CXCR3-alt is needed, as it aids in the validation of CXCR3 as a therapeutic target.

The use of CXCR3$^{-/-}$ mice has also revealed other interesting effects. For example, CXCR3 plays a role in wound healing of the skin [72, 73]. Activation of CXCR3 on, e.g., fibroblasts seems to contribute to the healing of skin injuries, by recruitment of immune cells to the site of injury, leading to the migration of keratinocytes, and by affecting the reorganization of matrix components including collagen and fibrillin [74]. Absence of CXCR3 or its ligands leads to ineffective and slower healing and hypertrophic scarring [72, 73, 75]. These data suggest that CXCR3 agonism might also be a potential therapeutic avenue in some cases. The discovery and characterization of small-molecule CXCR3 agonists will be discussed in more detail in Sect. 3.1.

## 1.2 Small-Molecule Binding

Given the multifaceted role of CXCR3 in a variety of physiological processes, it is not surprising that many efforts have been devoted to the development of small-molecule CXCR3 modulators. As explained, CXCR3 blockade (rather than activation) has generally been regarded as the therapeutically more relevant approach. Thus, the huge majority of disclosures deal with CXCR3 antagonists from pharmaceutical companies and showcase the challenging balances researchers have to address when developing drug candidates [76–78]. From a molecular point of view, chemokines are relatively large compared to the small ligands (±10–50-fold difference in molecular weight) that generally target this receptor family. Despite the fact that low-molecular weight ligands engage in fewer receptor interactions than the chemokines, many of these small ligands have the ability to disrupt chemokine binding and function with nanomolar potencies [20]. Therefore, it seems likely that these ligands do not act via simple steric competition but rather through an allosteric mechanism (an allosteric site is referred to a binding location for a ligand that is distinct from that of the endogenous ligand). This notion is supported by the increasing number of reports revealing the allosteric nature of many of such compounds binding to chemokine receptors [20, 79]. Indeed, increasing evidence also suggests allosteric binding of small-molecule CXCR3 ligands to the receptor [80]. The first crystal structures of druggable GPCRs have been solved in the past few years [81, 82], including the chemokine CXCR4 [83] and CCR5 [84] chemokine receptors. These structures offer new insights into the molecular details of GPCR-ligand binding and suggest that small molecules can accommodate different (allosteric) binding modes in the relatively large chemokine receptor binding pocket [20]. Chemokine receptor crystal structures, chemokine receptor binding sites, and the elucidation of CXCR3-ligand binding modes by site-directed mutagenesis studies and computational modeling studies are discussed in more detail in Sect. 4.

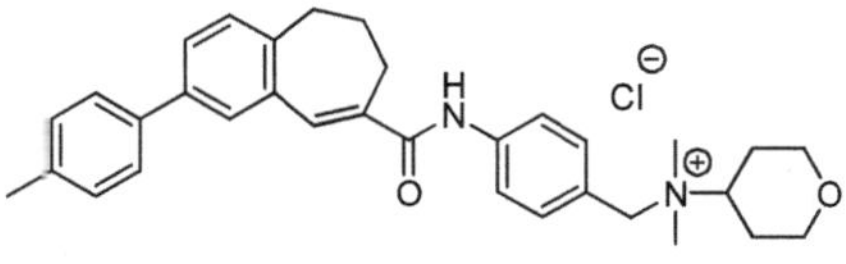

**1**
TAK-779

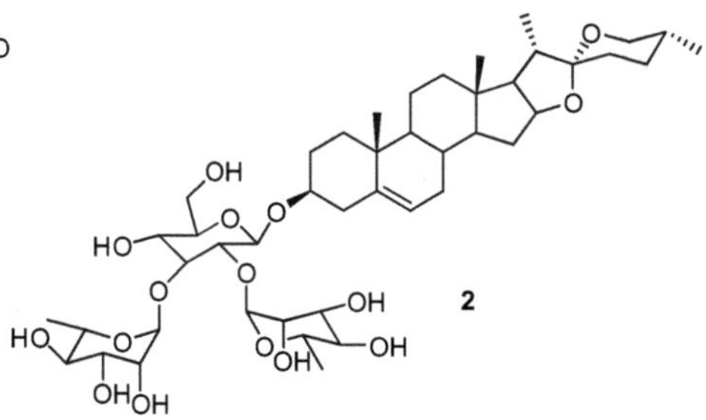

**2**

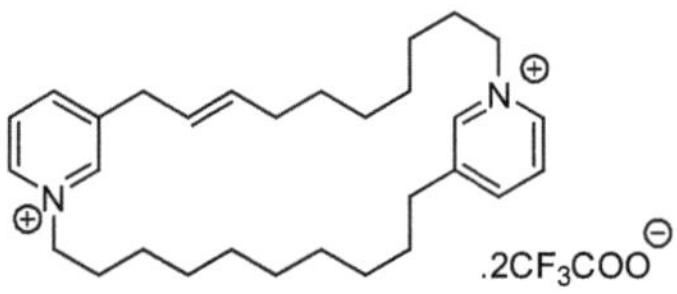

**3**

TAK-779 (**1**, $IC_{50} = 369$ nM, [$^{125}$I]-CXCL10) [51] and a few compounds emerging from a natural product screen, such as sugar-derivatized steroid **2** ($IC_{50} = 0.47$ μM, [$^{125}$I]-CXCL10) and dipyridinium salt **3** ($IC_{50} = 0.69$ μM, [$^{125}$I]-CXCL10) [85], can arguably be considered as the earliest disclosed examples of small-molecule binders of CXCR3. TAK-779 still finds some value as CXCR3 tool compound, but its moderate affinity and low selectivity over other chemokine receptors (notably CCR2 and CCR5) need to be borne in mind. As far as can be deduced from the literature, compounds **2** and **3** seem to have not been followed up upon.

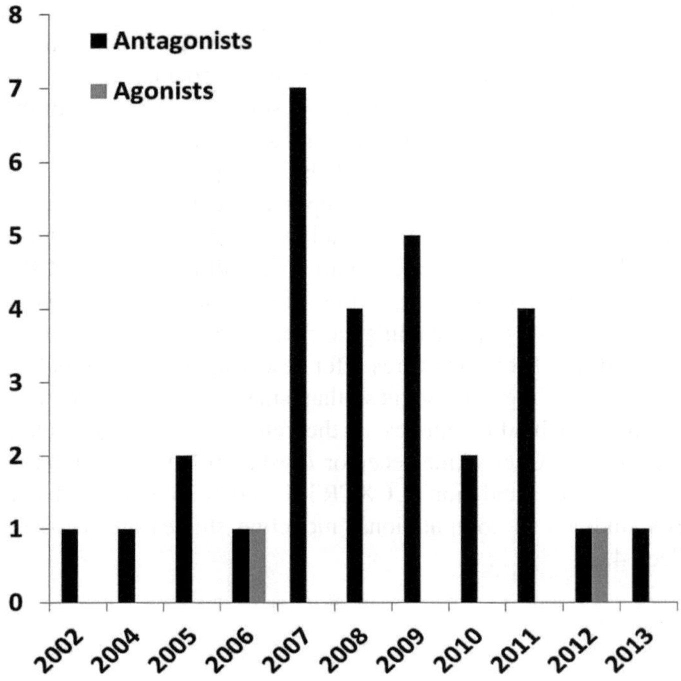


**Fig. 1** Number of medicinal chemistry-oriented publications on small-molecule CXCR3 ligands per year. The date of acceptance is used. A paper was included if it disclosed a new chemotype and/or SAR study on a known chemotype. A distinction is made between antagonists and agonists. All references can be found in the current review

## *1.3 Aim*

The current manuscript aims to review small-molecule modulation of CXCR3 from a molecular point of view. That is, in contrast to our most recent review [78], we pay less attention to proven clinical relevance of published molecules (patents are not included) but instead describe all reported compound classes including ones that would best be classified as "tool compounds". Considerable attention will be paid to articles which appeared after our 2008 review [77], and compact SAR tables are included to further illustrate SAR. In line with our aims, we will also pay special attention to CXCR3 agonists as well as to studies aimed at deciphering CXCR3-ligand binding modes by combining structural models (based on GPCR crystal structures) with ligand SAR and receptor site-directed mutagenesis studies. Collectively, our review aims to illustrate the many venues that have been followed in order to capitalize on small molecules to block or activate CXCR3.

Figure 1, which is an adapted continuation of a graph we published in our 2008 review [77], shows a visual depiction of the progress in CXCR3 ligand research. Two major trends are visible: (1) the publication rate on CXCR3 chemotypes seems to have stalled in recent years, and (2) as expected, much less is published on agonists than on antagonists.

## 2 Antagonists

This paragraph deals with a detailed description of published small-molecule CXCR3 antagonists in order of number of papers on a particular scaffold. Chemotypes with multiple associated publications have been classified in separate subparagraphs. Clarity within a particular table provided, reported affinity/activity values are given in the units used in the corresponding papers.

### 2.1 *(Aza)quinazolinones and Later-Generation Analogues*

#### 2.1.1 (Aza)quinazolinones

The class of (aza)quinazolinones and the resulting offspring of sub-chemotypes is the most widely described collection of CXCR3 antagonists (Amgen) with patents appearing as early as 2001 [86].

The work started with compound **4** as an HTS hit [87]. Several early SAR approaches on hit **4** have been published, with reported affinity values slightly differing likely as a result of the exact assay conditions [87, 88]. Nevertheless, clear trends can be extracted from these reports (Table 1). Replacing the F atom by a cyano group yielded improved ligand VUF5834 (**5**), which could efficiently block CXCR3-mediated calcium release [88]. The decanoyl moiety in **4** could be replaced by a phenylacetyl moiety albeit that substitution with an electron-withdrawing group was required (compare **4** to **6** and **7**). The dimethylamino group could be exchanged for a 3-pyridyl moiety. This is exemplified by compound **8** and **9** (AMG1237845, $IC_{50} = 0.006$ μM), the latter of which also showed good functional activity in a cell migration assay against all three chemokines [35]. Further SAR efforts maintained the OEt present in **9** as $R^3$ for pharmacokinetic reasons while the $CF_3$ was switched for an $OCF_3$ group accompanied by insertion of an additional N-atom in the core bicyclic moiety. This afforded azaquinazolinone **10**, known as AMG487 ($IC_{50} = 0.008$ μM) [87, 90]. The (*R*)-stereomer of AMG487 has the highest affinity [91]. A 4-F,3-$CF_3$ analogue (**11**, NBI-74330) from the same patent[90] was studied by others [89] and found to be more active ($K_i = 1.5$ nM) than AMG487. Our lab also published on the SAR linking **5** to **10** and **11** [92].

A CXCR3 mutagenesis study to detail the binding of **11** has recently been described by our group [80] and highlighted the CXCR3 transmembrane (TM) region as interaction region for the molecule as opposed to N-terminus and extracellular loops which are mainly important for chemokine binding [24]. Particularly, **11** appears to bind mainly in transmembrane site 1 (TMS1) of the TM region, as mutations in this pocket affected affinity of **11** [80].

**Table 1** Optimization of quinazolinone derivatives by several research groups [87–89]

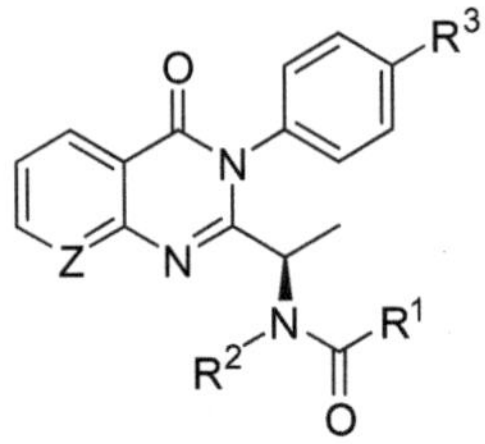


| Compound | | $R^1$ | $R^2$ | $R^3$ | Z | $IC_{50}$[a] |
|---|---|---|---|---|---|---|
| # | In ref. | | | | | |
| **4** | 1 [87], 1c [88] |  | 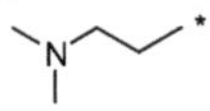  | F | CH | 0.146 μM [87], 3.2 μM [88][b] |
| **5** | 1d [88] |  | 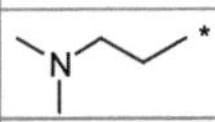  | CN | CH | 0.93 μM[b] |
| **6** | 16 [87] | 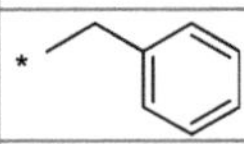 | 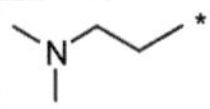  | F | CH | >10 μM |
| **7** | 18 [87] | 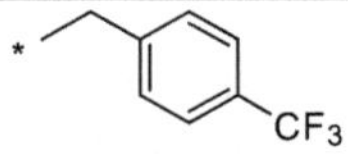  | 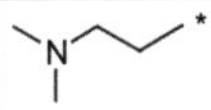  | F | CH | 0.088 μM |
| **8** | 28 [87] | 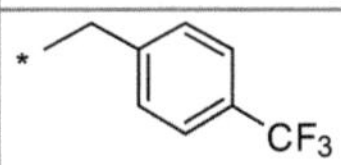  | 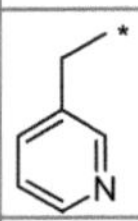  | F | CH | 0.013 μM |
| **9** | 34 [87] | 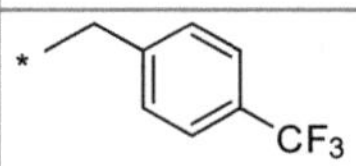  | 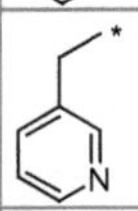  | OEt | CH | 0.006 μM |
| **10** | 47 [87] | 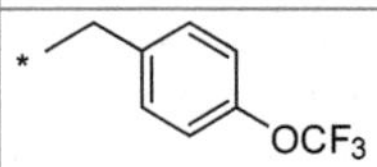  | 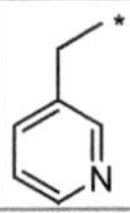  | OEt | N | 0.008 μM |
| **11** | NBI-74330 [89] | 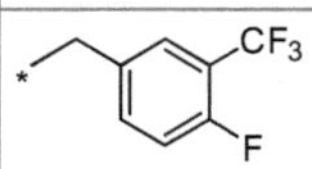  | 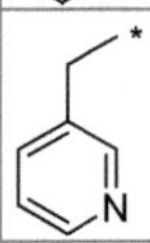  | OEt | N | $K_i = 1.5$ nM |

[a]Assay conditions vary. The reader is referred to the involved references for more details
[b]Racemate

AMG487 is the only ligand, as far as can be deduced from available literature, to have advanced into the clinic. Results of a phase I trial on AMG487 were shared in 2003 [93]. AMG487 was evaluated for safety and pharmacokinetics in 30 healthy males in a randomized, double blind, placebo-controlled dose-escalation study. In general, the compound was well tolerated and adverse events were mild to moderate (25–1,100 mg doses) [93]. In a phase IIa trial, patients suffering from moderate to severe psoriasis received 50 or 200 mg of AMG487 or placebo orally once a day for 28 days. Yet no significant differences in Psoriasis Activity and Severity Index or Physician Global Assessment scores were seen when patient groups were

**Table 2** Optimization of quinazolinone derivatives by Li et al. [97]

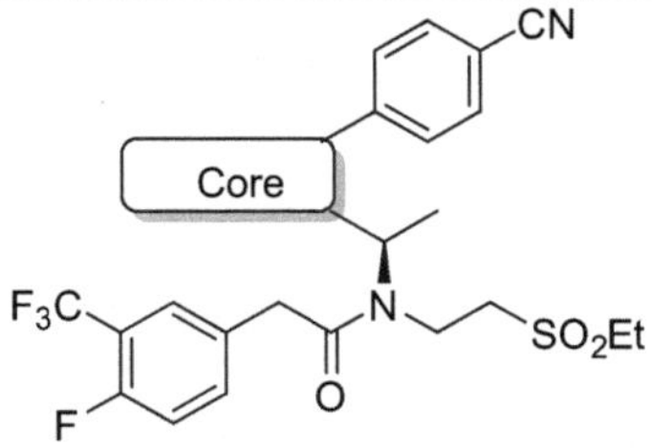


| Compound | | Core | $IC_{50}$ (nM) | |
|---|---|---|---|---|
| # | In ref. | | CXCL10[a] | Migration[b] |
| **12** | 1 | 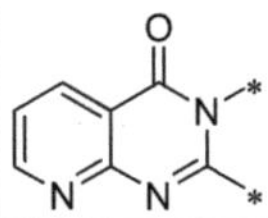 | 11 | 115 |
| **13** | 5 | 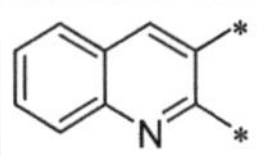 | 0.80 | 72 |
| **14** | 16 | 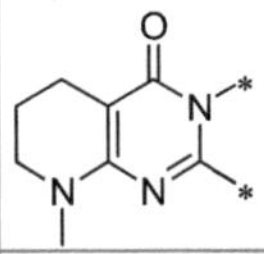 | 4.0 | 88 |
| **15** | 28 | 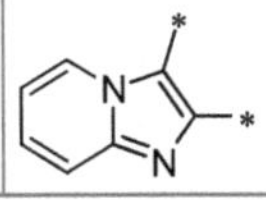 | 3.0 | 72 |

[a][$^{125}$I]-labeled CXCL10 displacement assay to CXCR3 expressed on activated human PBMC in the absence of human plasma
[b]CXCL11-induced cellular migration assay of PBMC

compared [94]. It was hypothesized that high variability in drug exposure may have caused the lack of clinical efficacy [94]. Two subsequent papers addressed the role of several AMG487 metabolites as possible players in the unexpected clinical pharmacokinetic parameters of AMG487 [95, 96]. Not surprisingly, several of the ensuing medicinal chemistry efforts by Amgen seem to have focused on chemically addressing these pharmacokinetic obstacles.

In one such effort, the 8-azaquinazolinone core was inspected for alternative bicyclic systems (Table 2). The 8-azaquinazolinone derivative **12** ($IC_{50} = 11$ nM) exhibited similar affinities and potencies to those of AMG487, but did not induce the time-dependent inhibition of CYP3A4 resulting from *O*-deethylation of AMG487. Therefore, it was decided to explore the SAR of the core with **12** as template [97]. It should be borne in mind, though, that **12** and AMG487 differ in three respects in terms of the periphery of the molecules: −CN instead of −OEt, ethylsulfonyl moiety instead of a pyridine unit, and different substitution pattern of the fluorine-containing aryl moiety. Initially, a series of [6,6]-fused heterocyclic derivatives was synthesized to determine the influence of the carbonyl and the nitrogen atom at the 3-position on the binding of the parental **12** to CXCR3 [97]. It was found that those functional groups are not essential for binding to the receptor. That is, quinoline analogue **13** ($IC_{50} = 0.80$ nM) displayed a more than tenfold increase in affinity for CXCR3 in buffer, although that increase was negligible when the assays were ran in human plasma. A completely aromatic core was not necessary, given that **14** ($IC_{50} = 4.0$ nM) had similar affinity for CXCR3 as **12**. [6,5]-fused heterocycles were also investigated [97]. Most replacements gave compounds with good affinity for the receptor. Notably, the imidazopyridine-derived **15** ($IC_{50} = 3.0$ nM) afforded the highest affinity and showed improved blockade of CXCL11-induced lymphocyte migration compared to **12**. Given the many tolerated core structures, it was concluded by the authors that the main role of the heterocyclic core is to arrange the peripheral substituents in the appropriate orientation [97].

It had been disclosed that the progression of AMG487 in clinical trials proved to be complicated due to the formation of a major pyridine-N-oxide metabolite that was also active on CXCR3 [98]. Arguably as a result of this, the quinazolinone series was further optimized to reduce the risk of formation of a major active metabolite (Table 3) [98]. In these efforts, the quinazolinone core was initially revisited instead of the 8-azaquinazolinone core of AMG487, because quinazolinone derivatives were equally potent and readily available synthetically. First, the aromatic moiety of $R^1$ of **16** ($IC_{50} = 0.006$ μM) was altered with the trifluoromethyl group being conserved during the modifications, since it significantly improved potencies and microsomal stability [98]. Imidazole and pyridine moieties provided similar affinities as the phenyl derivative **16** (not shown). The phenyl unit was maintained, though, and the substituents on this phenyl ring and the linker (*n*) were explored ($R^1$). While a 4-methylsulfonyl group (and less so 4-CN) was not tolerated as a substituent, all other substitution patterns were well tolerated. This may indicate that the electron-withdrawing properties of the trifluoromethyl group cannot fully account for the increased affinity. Elongation of the chain ($n = 2$), however, was not accepted. Next, with the 3-trifluoromethyl-4-fluorophenylacetamide present, the azaquinazolinone core was reexamined [98]. As expected (vide supra), key 8-azaquinazolinone NBI-74330 (**11**) was equally potent to its quinazolinone analogue. However, regioisomeric azaquinazolinones **17** ($IC_{50} = 0.032$ μM) and **18** ($IC_{50} = 0.11$ μM) were less potent than **11**. This somewhat contrasts a previous paper, which indicated that many substitutions within the core ring system are tolerated [97]. The 8-azaquinazolinone core was chosen for further optimization, since it was more polar than the quinazolinone core, and the $R^2$ was studied [98]. As previously reported [87], this area allowed several changes. Alkoxy-ethyl, amino-ethyl, and various heterocyclic moieties led to compounds with good affinity [98]. Notably, compounds such as **19** ($IC_{50} = 0.001$ μM), **20** ($IC_{50} = 0.002$ μM), and **21** ($IC_{50} = 0.001$ μM) stand out. Taken together with previous SAR data (vide supra), it seems that polar groups are preferred in this area but that many types of polar groups are accepted. Compound **21** was taken as a further template, because it was found in in vitro studies that major metabolite formation was avoided for compounds with the ethylsulfonyl moiety [98]. Indeed,

**Table 3** Optimization of quinazolinone derivatives by Liu et al. [98]

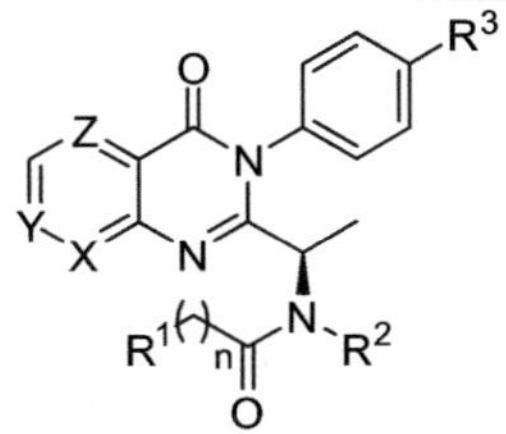


| Compound | | | | | | | | | $IC_{50}$ (μM) | |
|---|---|---|---|---|---|---|---|---|---|---|
| # | In ref. | $R^1$ | $R^2$ | $R^3$ | X | Y | Z | *n* | CXCL10[a] | Migration[b] |
| **11** (NBI-74330) | 21 | 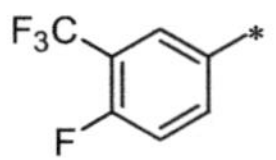 | 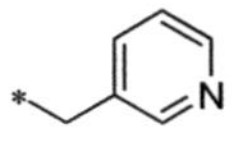 | OEt | N | CH | CH | 1 | 0.005 | 0.023 |
| **16** | 5 | 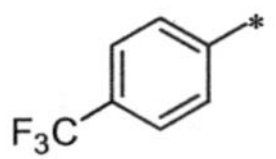 | 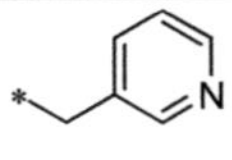 | OEt | CH | CH | CH | 1 | 0.006 | 0.35 |
| **17** | 22 | 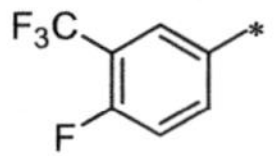 | 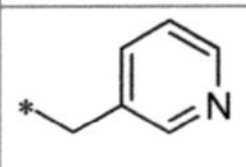 | OEt | CH | N | CH | 1 | 0.032 | – |
| **18** | 23 | 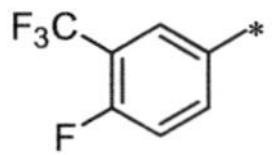 | 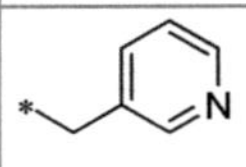 | OEt | CH | CH | N | 1 | 0.11 | – |
| **19** | 30 | 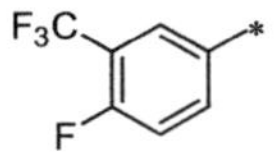 | 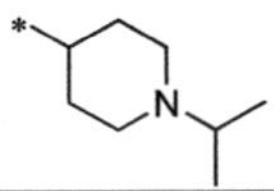 | OEt | N | CH | CH | 1 | 0.001 | 0.005 |

(continued)

**Table 3** (continued)

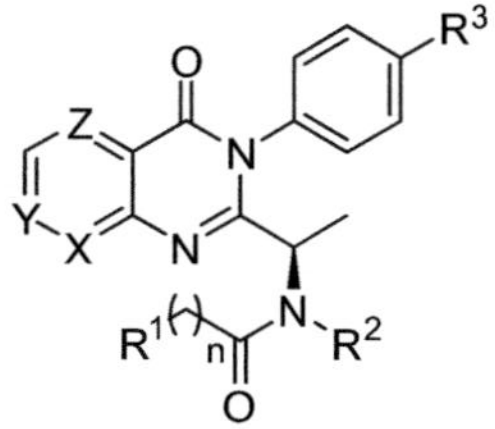


| Compound | | | | | | | | | IC50 (μM) | |
|---|---|---|---|---|---|---|---|---|---|---|
| # | In ref. | $R^1$ | $R^2$ | $R^3$ | X | Y | Z | *n* | CXCL10[a] | Migration[b] |
| **20** | 29 | 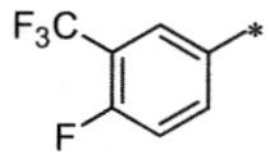 | 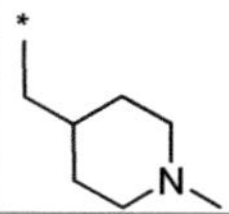 | OEt | N | CH | CH | 1 | 0.002 | 0.020 |
| **21** | 33 | 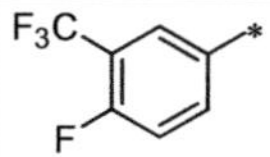 | 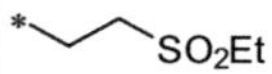 | OEt | N | CH | CH | 1 | 0.001 | 0.022 |
| **22** | 34 | 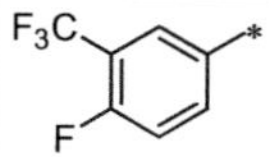 | 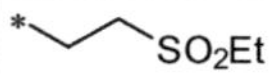 | $OCH_2CF_3$ | N | CH | CH | 1 | 0.001 | 0.015 |

[a]Displacement of [$^{125}$I]-labeled CXCL10 from the CXCR3 receptor expressed on PMBC
[b]CXCL11-induced migration of PBMC in the presence of 100% human plasma

the ethylsulfonyl moiety was already present in previously discussed compounds (vide supra). In an attempt to reduce *O*-dealkylation, a 3,3,3-trifluoroethoxy moiety ($R^3$) was introduced to give **22** ($IC_{50} = 0.001$ μM). The PK profile of **22** was good across several species (including rat, dog, and cynomolgus monkeys), and **22** displayed increased affinity and potency in vitro and in vivo compared to AMG487.

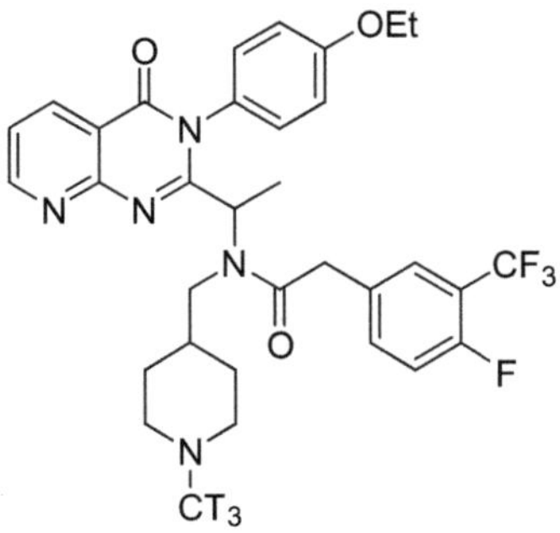


**23** (RAMX3)

A racemic N-$CT_3$ analogue of **20** (i.e., **23** or RAMX3) has been disclosed for use as a CXCR3 radioligand. The paper was accompanied by a brief SAR on additional members of the 8-azaquinazolinone class [99]. This SAR, among others, showed the dramatic loss in potency when the ethoxyphenyl head is removed from AMG487. The radioligand **23** possesses high affinity for CXCR3 ($K_d = 1$ nM). Moreover, and in contrast to CXCL11, unlabelled **23** was not able to completely displace a fluorescently labeled isoform of CXCL11, suggesting an allosteric mechanism of binding for this compound [99].

Evaluation of AMG487 in phase I clinical trials had indicated that the drug accumulated when daily doses above 100 mg were administered [100]. It was hypothesized that a minor de-ethylated metabolite was responsible for the accumulation of the drug, since it was a time-dependent inhibitor of CYP3A4. New compounds were explored to address the potential formation of a similar phenol metabolite [100]. Not surprisingly, for reasons already discussed (N-oxide formation of pyridine), the peripheral ethylsulfonyl group was maintained during these explorations. It was hypothesized that core attachment of the 4-ethoxyphenyl ($R^1$) through a carbon atom would give an oxidatively more stable ligand than attachment through a nitrogen atom. However, selected phenol analogues with a C-linkage were still time-dependent inhibitors of CYP3A4. Thus, it was suggested that the 4-ethoxyphenyl group had to be replaced. To that end, cyano derivatives of **21** were synthesized. Indeed, the general strategy of cyano replacements for the ethoxy group has already been explored previously (Table 2). Cyano-derivative **24** (Table 4) showed a promising potential, because it did not produce any metabolites with CYP3A4 time-dependent inhibitory activity. This compound did test positive in an in vitro chromosomal aberration assay, however. Likewise, most other tested cyano derivatives tested positive in the chromosomal aberration assay, with the

**Table 4** Optimization of quinazolinone derivatives by Chen et al. [100]

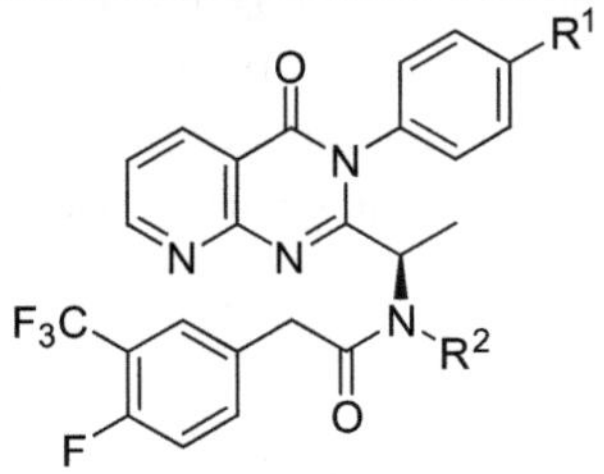


| Compound | | | | $IC_{50}$ (nM) | |
|---|---|---|---|---|---|
| # | In ref. | $R^1$ | $R^2$ | CXCL10[a] | Migration[b] |
| **24** | 15 | CN | 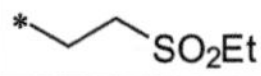 | 11 | n.s. |
| **25** | 19 | CN | 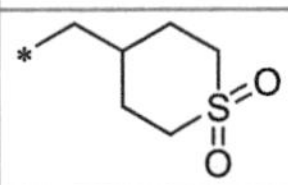 | 12 | 39 |
| **26** | 25 | Cl | 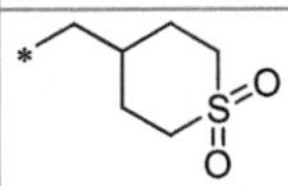 | 6 | 45 |
| **27** | 28 | F | 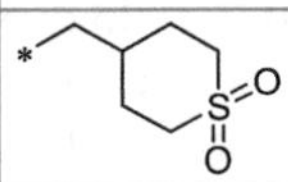 | 7 | 72 |

[a]Displacement of [$^{125}$I]-labeled CXCL10 from the CXCR3 receptor
[b]CXCL11-mediated migration in the presence of 100% human serum
*n.s.* not shown in article

exception of, e.g., **25** ($IC_{50} = 12$ nM). When the cyano moiety as $R^1$ was replaced with a chlorine or a fluorine in certain members, as in **26** ($IC_{50} = 6$ nM), the chromosomal aberration assay activity was abolished. These findings suggest that the cyano group was responsible for the activity in the chromosomal aberration assay. The cyclic sulfone moiety (present in **26** and **27**, $IC_{50} = 7$ nM) generally gave higher potency compared to some ethylsulfone ($R^2$) counterparts. Compounds **26** and **27** were selected for evaluation in multiple species. In general, **26** seemed to have the best PK profile, owing to a lower clearance and longer half-lives across the tested species (with the exception of dogs).

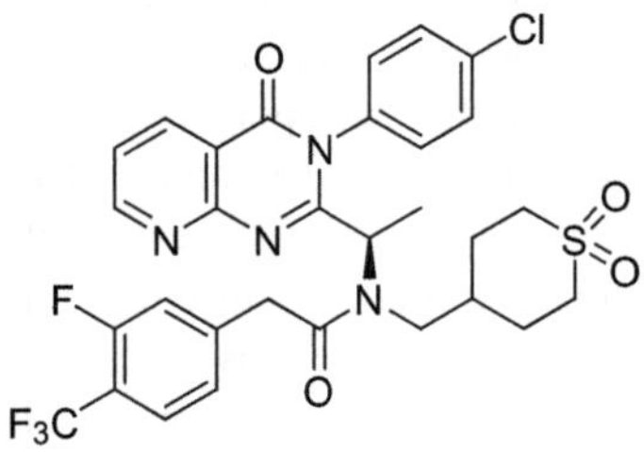


**28**

Interestingly, a paper on process chemistry aspects of the synthesis of **28** (the 4-$CF_3$-3-F regioisomer of **26**) has been published, suggesting that **28** has also been of advanced interest for the Amgen CXCR3 research program. The paper contains valuable details for obtaining members of the (aza)quinazolinone class in enantiopure form [101].

### 2.1.2 Imidazole and Imidazopyrazine Derivatives

Since it was suggested that the core of the quinazolinone series tolerated many changes and served mainly to orient the peripheral groups correctly [97], a series of imidazole derivatives was explored (Table 5) [102]. It was hypothesized that the imidazole group would be a suitable replacement for the quinazolinone moiety, because both groups contain nitrogen atoms in a 1,3 relationship. However, **29** ($IC_{50}$ = 1,800 nM), the parent imidazole derivative, showed poor CXCR3 affinity. It was proposed that a lipophilic moiety mimicking the phenyl part of the quinazolinone core was necessary. Gratifyingly, appending a phenyl group on the 4-position increased the affinity (**30**, $IC_{50}$ = 11 nM). The *N*-alkyl moiety ($R^2$) was then optimized [102]. The SAR for the (aza)quinazolinone series (vide supra) indicated that polar groups are preferred at this site [87, 98]. This trend held for the imidazole series. Similar to the (aza)quinazolinone SAR [98], ethylsulfonyl compounds like **31** ($IC_{50}$ = 0.4 nM) had the best affinity for the CXCR3 receptor. Amines, alkoxy groups, and several pyridyl moieties also displayed good affinity. With the aim to identify a more polar $R^3$-portion, the $R^3$ part was tested for a variety of side chains, such as imidazole-phenyl, triazole-phenyl, and tetrazole-phenyl [102]. All tested side chains had less affinity for the receptor than **31**, although differences were not very dramatic.

As mentioned, metabolic studies of the 4-ethoxyphenyl-substituted analogues demonstrated the formation of phenolic metabolites that are time-dependent inhibitors of CYP3A4, a recurring problem with this class of compounds (vide supra). Thus, replacement of the ethoxy moiety ($R^4$) was sought. The cyano analogue **32** displayed an $IC_{50}$ value of 0.7 nM, and, thus, the cyano was considered a good replacement for the ethoxy moiety [102]. It is noted, however, that a later study by Chen et al. (vide supra) showed that the cyano moiety may cause activity in in vitro chromosomal aberration assays [100]. Exploration of the substitutions at the 4-position of the imidazole ring ($R^1$) showed that replacement of the phenyl moiety by small alkyl groups, like **33** ($IC_{50}$ = 6.6 nM), or by pyridyl resulted in a loss of affinity compared to **32** [102]. However, evaluation of the PK properties indicated that **33** had better solubility and permeability than **32** and was cleared slower than **32**, justifying further inspection of the cyclopropyl class of compounds. Metabolic studies showed that a significant amount of glutathione (GSH) conjugates formed at the imidazole ring of **33** [102]. Consequently, it was tried to circumvent the GSH-conjugate formation by modification of $R^5$. Electron-withdrawing groups difluoromethyl (**34**, $IC_{50}$ = 12 nM) and chlorine (**35**, $IC_{50}$ = 7.8 nM) prevented

**Table 5** Optimization of imidazole derivatives by Du et al. [102]

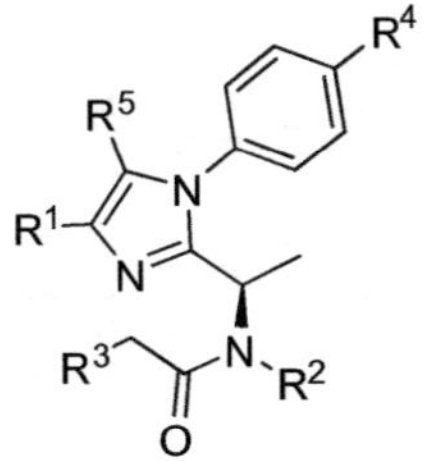


| Compound | | | | | | | $IC_{50}$ (nM) | |
|---|---|---|---|---|---|---|---|---|
| # | In ref. | $R^1$ | $R^2$ | $R^3$ | $R^4$ | $R^5$ | CXCL10[a] | Migration[b] |
| **29** | 2 | H | 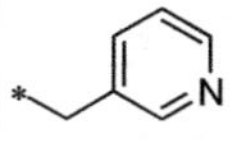 | 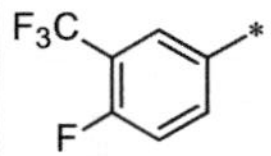 | OEt | H | 1,800 | n.s. |
| **30** | 3 | 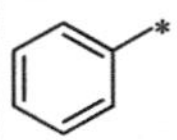 | 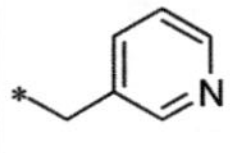 | 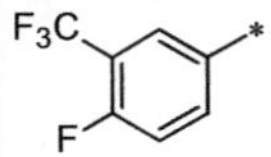 | OEt | H | 11 | n.s. |
| **31** | 16 | 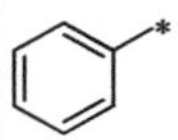 | 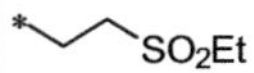 | 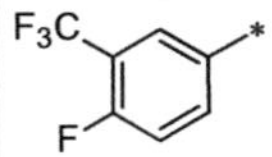 | OEt | H | 0.4 | n.s. |
| **32** | 29 | 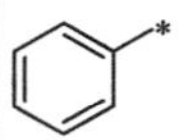 | 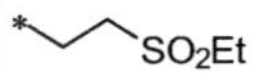 | 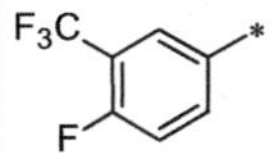 | CN | H | 0.7 | 10 |
| **33** | 32 |  | 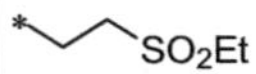 | 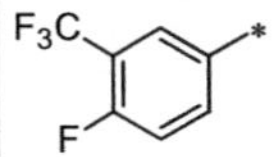 | CN | H | 6.6 | 53 |
| **34** | 40 | 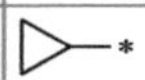 | 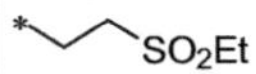 | 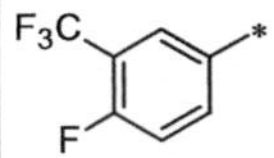 | CN | $CHF_2$ | 12 | 20 |
| **35** | 43 | 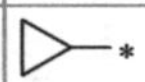 | 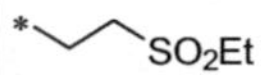 | 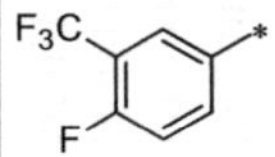 | CN | Cl | 7.8 | 80 |

[a][$^{125}$I]-labeled CXCL10 displacement assay to CXCR3 expressed on IL-2 activated human PBMC in the absence of human plasma
[b]CXCL11-induced cellular migration assay of PBMC

the formation of such adducts while not compromising affinity. Compounds **34** and **35** have oral bioavailability and acceptable PK in rat.

Du et al. continued to improve the potency of the antagonists originating from the (aza)quinazolinone series [103]. Initial efforts on the 8-azaquinazolinone series focused on the C7-position (the CH adjacent to the pyridine unit of the bicyclic system), since a major metabolic pathway was found to be oxidation by aldehyde oxidase at C7. Although in vitro improvement was achieved with small substituents at C7, this did not translate to in vivo profiles. The efforts then turned to a series of imidazopyrimidines (Table 6), exemplified by initial compound **36** ($IC_{50} = 10$ nM). However, neither the affinity of **36** nor the clearance rate of **36** was as good as desired. This prompted the researchers to shift the N by one atom in the core, giving imidazopyrazines. In accord with the affinity increase with small groups at C7 of 8-azaquinazolinone compounds, small groups (e.g., methyl, methoxy, ethyl, and chlorine) at the C8-position ($R_2$) of the imidazopyrazine core significantly increased the affinity compared to $R_2$ being H. Cyclopropyl-derivative **37** had 15-fold increased affinity for the receptor and was the most stable compound in rat microsomes. However, **37** gave significant pregnane X receptor (PXR) activation. PXR activation induces the formation of CYP3A4, which could potentially lead to drug-drug interactions. In order to solve the issue, a more rigid cyclic sulfone moiety was incorporated into the molecule ($R^3$), and the substitution pattern at $R^4$ was reevaluated for PXR activity. This led to compounds **38** and **39** which showed reduced PXR activity. Ligand **38** was more potent in a CXCL11 plasma migration assay and had in vivo efficacy in blocking bleomycin-induced leukocyte migration into the lung.

QSAR studies have been carried out on the early papers from Amgen [87, 102] collectively encompassing several different cores [104]. These studies indicated that highest occupied molecular orbital energy, principal moment of inertia, polar surface area, presence of triple bond, and Kier shape descriptors can be used to predict [$^{125}$I]-CXCL10 $IC_{50}$ inhibition values [104].

## 2.2 *1-Aryl-3-Piperidin-4-yl-Ureas and Later-Generation Analogues*

An HTS campaign (UCB) using a FLIPR-based calcium flux assay delivered cyclooctenyl urea hit **40** ($K_i = 110$ nM) [105]. An early scan of the aromatic moiety revealed fluorinated phenyl groups (e.g., **41**, $K_i = 47$ nM) to be advantageous over the naphthyl group (Table 7). A SAR study at the cycloaliphatic right-hand side ($R^2$), while maintaining the original naphthyl left-hand side, identified the (−) myrtenyl group as a reasonable substitute. With this group at hand, revisiting the aromatic left-hand side yielded a range of affinities with several fluorinated aromatic rings once again standing out. One of the most attractive compounds was **42** ($K_i = 16$ nM) which, compared to hit **40**, had improved affinity, solubility, and

**Table 6** Optimization of imidazopyrazine derivatives by Du et al. [103]

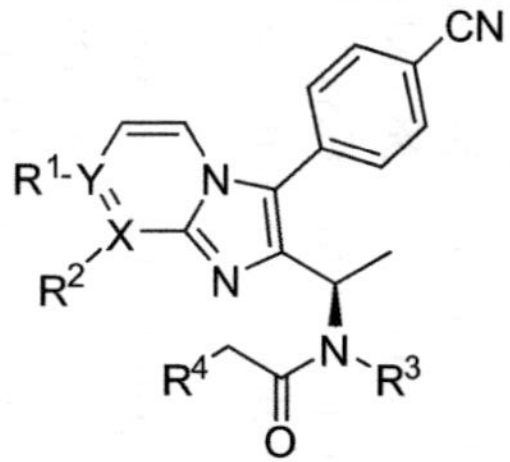


| Compound | | | | | | | | IC50 (nM) | |
|---|---|---|---|---|---|---|---|---|---|
| # | In ref. | $R^1$ | $R^2$ | $R^3$ | $R^4$ | X | Y | CXCL10[a] | Migration[b] |
| **36** | 3 | H | – | 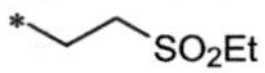 | 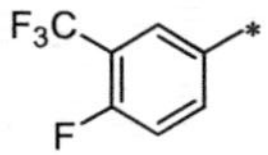 | N | C | 10 | 643 |
| **37** | 18 | – | 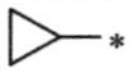 | 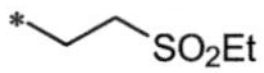 | 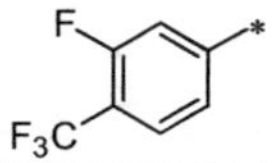 | C | N | 0.9 | 17 |
| **38** | 21 | – |  | 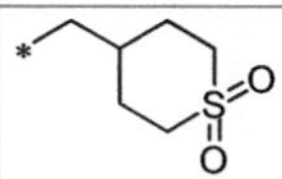 | 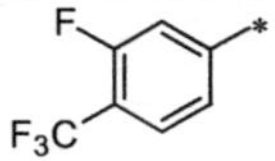 | C | N | 0.9 | 18.9 |
| **39** | 22 | – | 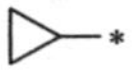 | 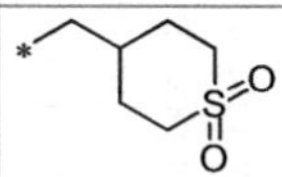 | 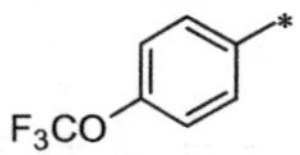 | C | N | 1.1 | 34 |

[a] $[^{125}I]$-labeled CXCL10 displacement assay to CXCR3 expressed on IL-2 activated human PBMC in the absence of human plasma
[b] CXCL11-induced cellular migration assay of PBMC

**Table 7** Optimization of 1-aryl-3-piperidin-4-yl-ureas and analogues by Allen et al. and Watson et al. [105, 106]

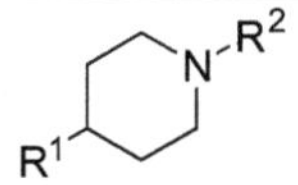

| Compound | | | | $K_i$ (nM) |
|---|---|---|---|---|
| # | In ref. | $R^1$ | $R^2$ | CXCL10/CXCL11[a] |
| **40** | 3 [105] | 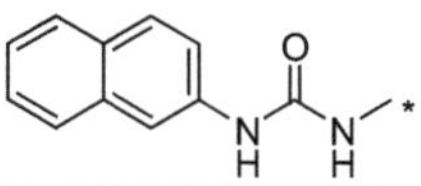 | 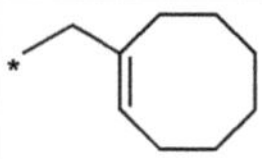 | 110 |
| **41** | 5g [105] | 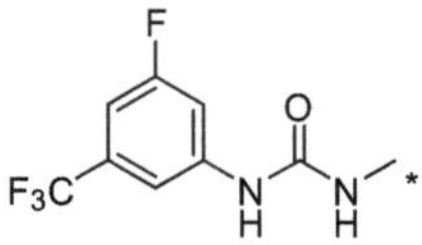 | 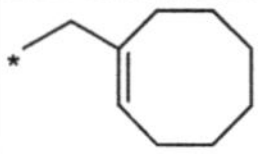 | 47 |
| **42** | 9t [105] | 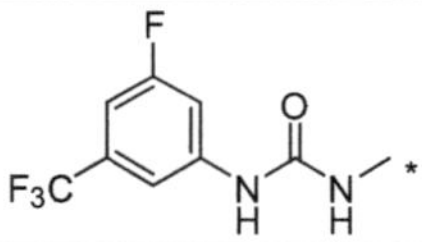 | 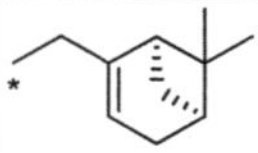 | 16 |
| **43** | 7a [106] | 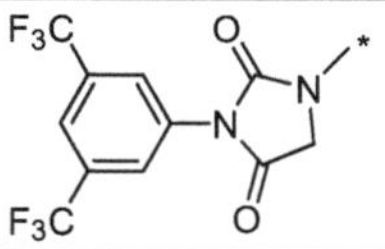 | 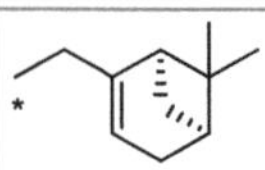 | 26 |
| **44** | 10d [106] | 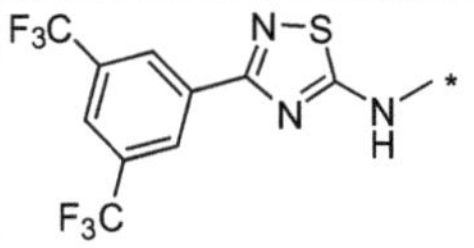 | 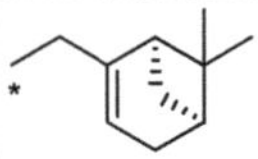 | 270 |

[a]CXCL10/CXCL11 stimulated [$^{35}$S]-GTPγS assay using CXCR3 transfected CHO membranes

Log D. In a follow-up paper [106], the role of the urea portion was investigated further (Table 7). One of the approaches was to use a hydantoin constrainment strategy, which afforded potent compounds (**43**, $K_i = 26$ nM). Since the hydantoin series as a whole suffered from very high microsomal metabolism, replacements were sought. Although these did not match the potency of hydantoin **43**, arylazoles such as **44** ($K_i = 270$ nM) did present advantages in terms of PK properties.

Efforts to further fine-tune the right-hand side of the original series were also initiated (Table 8) [107]. A previously [105] disclosed key myrtenyl analogue **45** ($K_i = 0.026$ μM) was poorly soluble. In contrast, the cyclic sulfone analogue **46** ($K_i = 1.1$ μM) and the piperidine analogue **47** ($K_i = 0.06$ μM) showed good log D and improved solubility, but considerable affinity was lost especially in the case of **46**. It was hypothesized that the lost affinity could be regained by modification of the involved piperidine ring [107]. While methylated piperidine rings did not display an improved affinity, bridged piperidine derivatives like *N*-acetyl homotropene analogue **48** ($IC_{50} = 0.009$ μM) gave good affinity. It was suggested

**Table 8** Optimization of piperidine urea derivatives by Watson et al. [107]

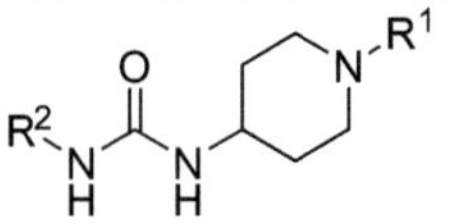

A

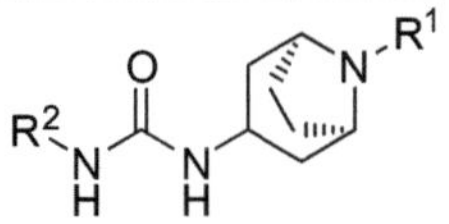

B

| Compound | | | | | $K_i$ (μM) |
|---|---|---|---|---|---|
| # | In ref. | Core | $R^1$ | $R^2$ | CXCL11[a] |
| **45** | 1b | A | 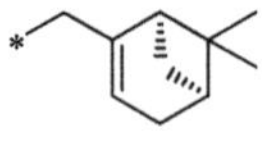 | 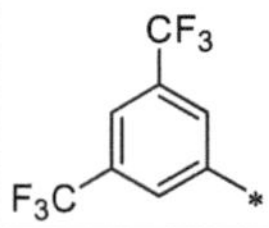 | 0.026 |
| **46** | 3a | A | 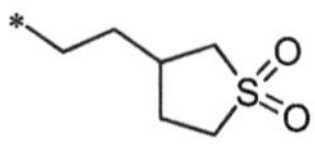 | 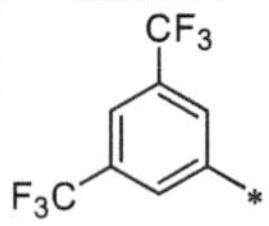 | 1.1 |
| **47** | 3c | A | 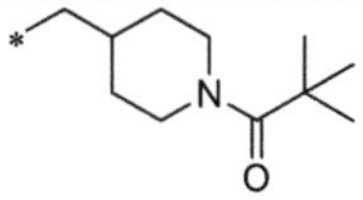 | 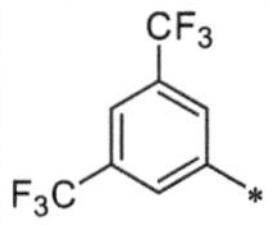 | 0.06 |
| **48** | 9f | A | 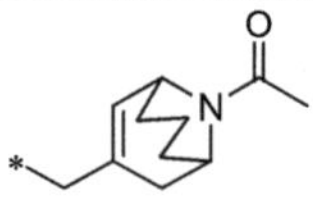 | 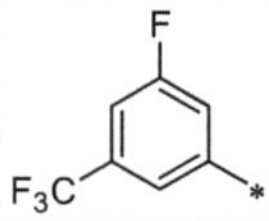 | 0.009 |
| **49** | 9j | A | 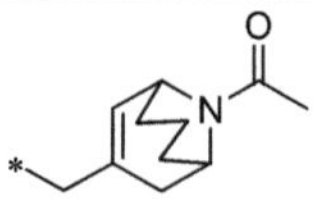 | 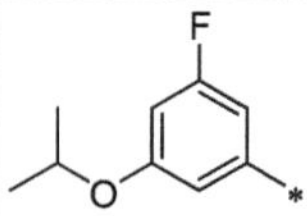 | 0.003 |
| **50** | 15 | B | 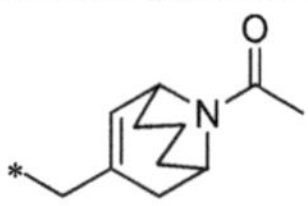 | 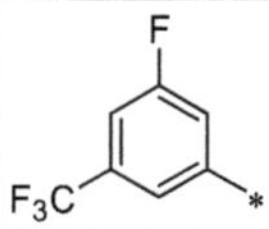 | 0.007 |

[a]CXCL11-activated [$^{35}$S]-GTPγS binding to hCXCR3-transfected CHO cell membranes

that the bridged piperidine fills a similar volume as the myrtenyl group, whereas this is less the case for non-bridged acylpiperidines. Revisiting the left-hand side, it was found that the incorporation of some polar aromatic substituents was tolerated on $R^2$ [107]. The 3-fluoro-5-isopropoxy derivative **49** ($K_i = 0.003$ μM) showed an improved affinity compared to **48**. Several sites of metabolic oxidation were identified around the central piperidine ring of **48** [107]. It was postulated that bridging this ring could lower metabolism and that an *exo*-tropanyl moiety was of interest in this respect. Indeed, **50** ($K_i = 0.007$ μM) showed the desired metabolic stability while maintaining an affinity comparable to **48**.

Knight et al. continued the development of **48** and aimed to create a series of non-urea derivatives with similar affinity and properties [108]. Initial efforts

**Table 9** Optimization of piperidine urea derivatives by Knight et al. [108]

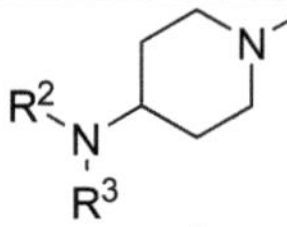


A

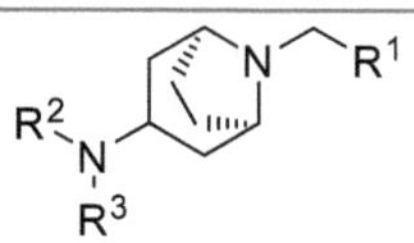


B

| Compound # | Compound In ref. | Core | $R^1$ | $R^2$ | $R^3$ | $K_i$ (nM) CXCL11[a] |
|---|---|---|---|---|---|---|
| **48** | 1b | A | 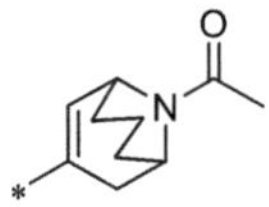 | 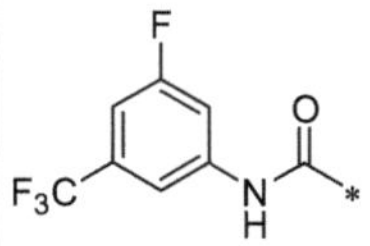 | H | 9 |
| **51** | 6 | A | 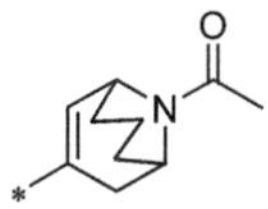 | 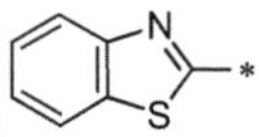 | Me | 95 |
| **52** | 1c | A | 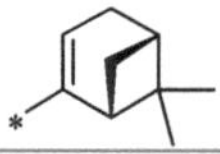 | 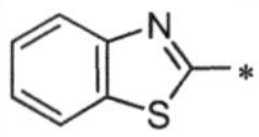 | Me | 126 |
| **53** | 9a | A | 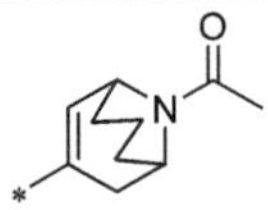 | 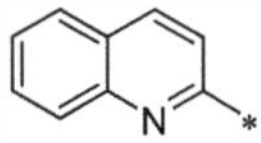 | H | 135 |
| **54** | 14e | A | 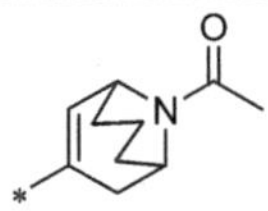 | 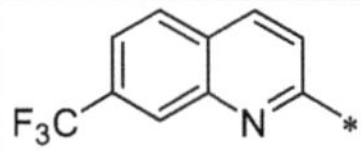 | H | 7 |
| **55** | 24d | B | 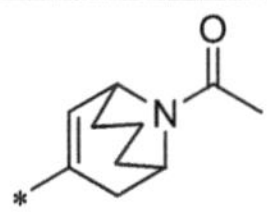 | 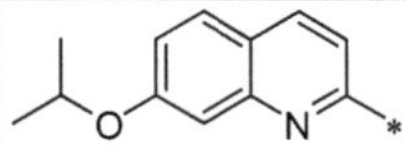 | H | 5 |

[a] CXCL11-activated [$^{35}$S]-GTPγS binding to hCXCR3-transfected CHO cell membranes

delivered azole analogue **51** ($K_i = 95$ nM, Table 9) bearing a homotropene moiety similar to **48** and displaying similar affinity to myrtenyl-parent **52** ($K_i = 126$ nM). However, **51** still had undesirable properties like high plasma protein binding. In order to improve the PK properties, it was decided to differentiate at the $R^2$ moiety [108]. Aminoquinoline analogue **53** ($K_i = 135$ nM) showed good affinity and appropriate substitutions significantly increased the affinity further. For example, 7-trifluoromethyl derivative **54** showed good affinity and had a good clearance of 10 μL/min/mg. As previously discussed (vide supra) [107], replacement of the central piperidine ring for a tropane ring can have benefits in terms of affinities and metabolic stability. When applied here, the resulting compound **55** ($K_i = 5$ nM) had a good affinity, log D of 2.9, and low intrinsic clearance [108].

**Table 10** Optimization of piperazinyl-piperidine derivatives by McGuinness et al. [109]

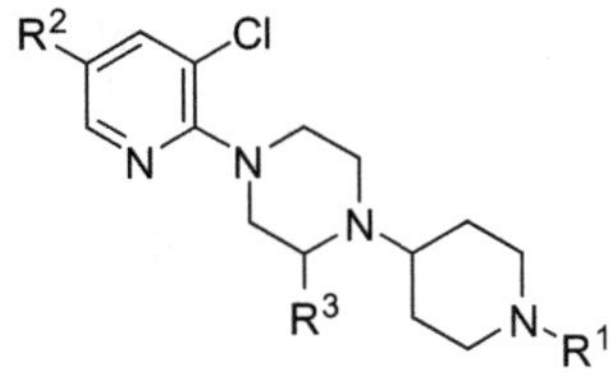

| Compound | | $R^1$ | $R^2$ | $R^3$ | $K_i$ (nM) |
|---|---|---|---|---|---|
| # | In ref. | | | | CXCL10[a] |
| **56** | 1a | 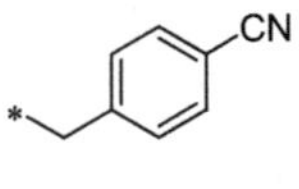 | 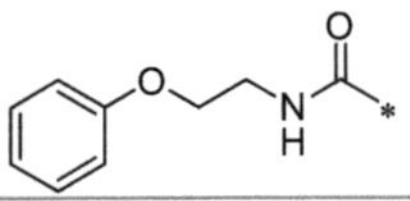 | H | 110 |
| **57** | 1f | 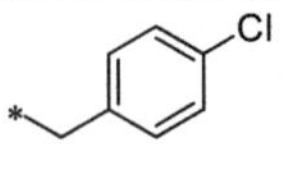 | 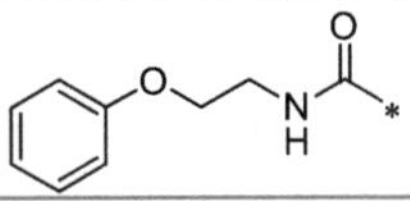 | H | 70 |
| **58** | 1q | 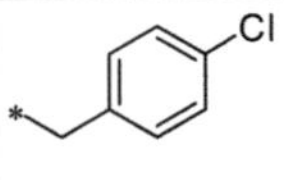 | 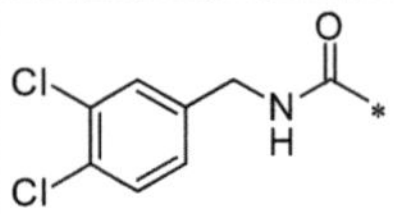 | H | 35 |
| **59** | 12a | 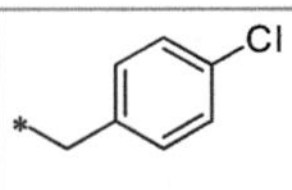 | 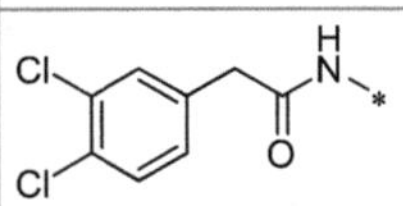 | H | 1,900 |
| **60** | 36 | 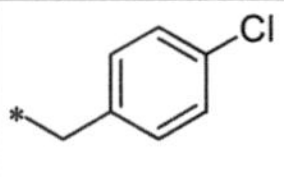 | 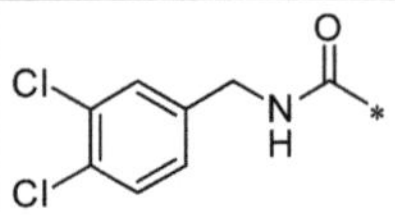 | (*S*)-Me | 16 |
| **61** | 37 | 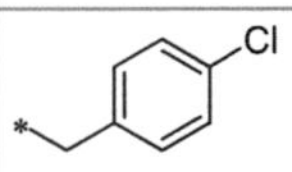 | 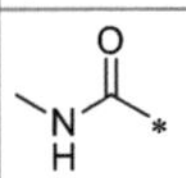 | (*S*)-Me | 32 |

[a]Scintillation proximity binding assay using Ba/F3-CXCR3 membranes and [$^{125}$I]-CXCL10 in buffer

## 2.3 *Piperazinyl-Piperidines*

McGuinness et al. (Merck) identified **56** ($K_i = 110$ nM) as a high-throughput screening hit from an encoded combinatorial library [109]. Further exploration utilized a substantial amount of solid-phase synthesis to rapidly provide a collection of analogues (Table 10). Exploration of the $R^1$-position indicated a preference for a *para*-substituted benzyl moiety given that, e.g., the unsubstituted benzyl analogue, the 3-cyanobenzyl analogue, and the 2-cyanobenzyl analogue all lost affinity. It was found that the 4-chlorobenzyl compound **57** ($K_i = 70$ nM) represented an optimum. This SAR trend was noted to be similar to the benzetimide series described by Bongartz et al. (vide infra) [110]. The authors speculate that both classes may share a similar binding mode with the CXCR3 receptor. With the 4-chlorobenzyl at hand, the left-hand nicotinyl amide was optimized [109]. It was mentioned that complete removal of the amide moiety led to a loss of affinity. Among amide substituents, electron-deficient benzyl rings were preferred (**58**, $K_i = 35$ nM). Replacement of the amide moiety by, for example, a secondary amine, tertiary amine, or urea as well as inversion of the amide (**59**) reduced the affinity for CXCR3. Removal of the 5-chlorine on the pyridine ring or removal of the pyridine nitrogen likewise led to a drop in affinity. SAR was also strict on the piperidinyl-piperazine core [109]. Removal of a nitrogen atom was not tolerated, nor was relocation of nitrogen atoms or ring opening. However, (*S*)-methyl substitution of the piperazine ring increased affinity twofold (**60**, $K_i = 16$ nM). Interestingly, affinity was only slightly reduced when the dichlorobenzyl moiety of $R^2$ of **60** was changed to a methyl group (**61**, $K_i = 32$ nM).

The SAR of the $R^1$ unit was explored in more detail by Shao et al., largely using aromatic and heteroaromatic substitutions (Table 11) [111]. Replacement of the 4-chlorobenzyl moiety of **61** by heteroaromatics or a more polar substituent such as 4-(methanesulfonyl)benzyl resulted in a decrease of affinity, leading the authors to suggest an interaction with a lipophilic region of the receptor binding pocket. However, small lipophilic substituents such as 4-methyl and 4-trifluoromethyl led to a decrease in affinity too. On the other hand, 2,4-dihalo substitution (i.e., 2,4-dichloro analogue **62**, $IC_{50} = 17$ nM) and methyl substitution of the benzylic methylene proved beneficial in increasing the affinity for the receptor (**63**, $IC_{50} = 5$ nM, other diastereomer less active). As McGuinness et al. reported [109], (*S*)-methyl substitution at the 2′-position of the piperazine ring ($R^2$) resulted in an improved affinity (see **60** and **61**). This triggered the examination of methyl substitutions throughout the core [111]. Substitution with (*S*)-methyl on the alternative 5′-position of the piperazine ring was tolerated, while an (R)-methyl resulted in a sixfold loss of affinity. Methylations of the 2″-, 4″-, and 5″-positions of the core piperidine ring were also tolerated, but did not improve the affinity for the receptor significantly. Since many of these methylations will add additional stereochemical complexity, the authors did not pursue these further. Both basic nitrogen atoms (i.e., trialkyl nitrogens) of the piperazinyl-piperidine (X=N, Y=CH) core were essential for CXCR3 affinity [111]. Since inverted piperidinyl-piperazine (X=CH, Y=N) analogues **64** ($IC_{50} = 4{,}500$ nM) and **65** ($IC_{50} = 280$ nM) gave reduced affinity, the hypothesis that both basic nitrogen atoms are crucial in a spatially defined way is further supported. The 5-pyridyl carboxamide ($R^3$) could be reverted, leading to equal or better affinity compared to initial lead compound **61**, but (perhaps for toxicophore reasons) this series was not followed upon [111]. The primary amide **66** ($IC_{50} = 39$ nM) displayed similar affinity as **61**, while an N-ethyl-amide (i.e., **67**, $IC_{50} = 2.3$ nM) gave increased affinity. Having scanned most of the scaffold, the 2′-position substituent ($R^2$) was re-optimized [111]. With the amide $R^3$ being Me, the 2′(*S*)-ethyl analogue **68** ($IC_{50} = 3$ nM) proved to have a tenfold better affinity than **61**. Larger substituents (**69**) or more polar substituents as $R^2$ (**70**) led to a drop in affinity. A tenfold enhancement of affinity was achieved when $R^3$ was substituted with an ethyl moiety (**71**, $IC_{50} = 0.3$ nM; **72**, $IC_{50} = 0.2$ nM) instead of a methyl moiety.

**Table 11** Optimization of piperazinyl-piperidine derivatives by Shao et al. [111]

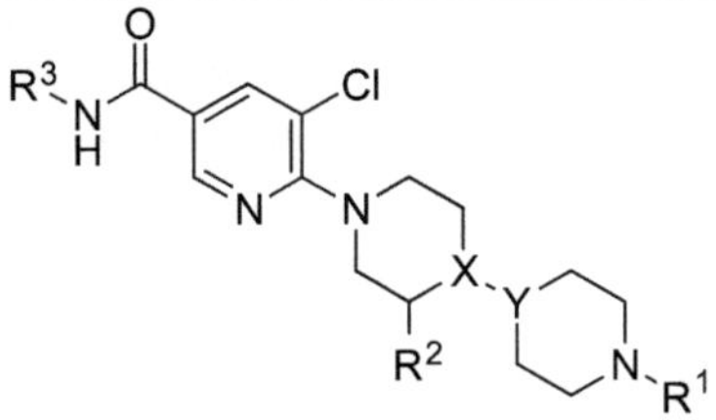


| Compound | | | | | | | |
|---|---|---|---|---|---|---|---|
| # | In ref. | $R^1$ | $R^2$ | $R^3$ | X | Y | $IC_{50}$ (nM)[a] |
| **62** | 6i | 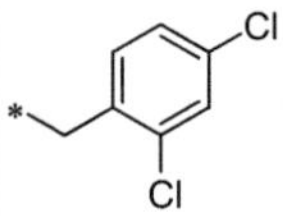 | (*S*)-Me | Me | N | CH | 17 |
| **63** | 6k | 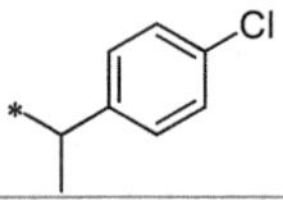 | (*S*)-Me | Me | N | CH | 5 |
| **64** | 16a | 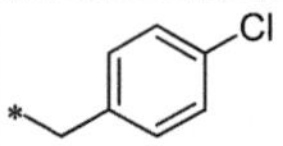 | H | 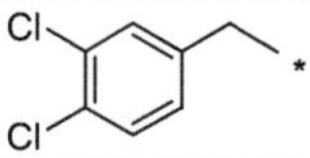 | CH | N | 4,500 |
| **65** | 16d | 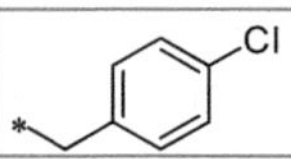 | (*R*,*S*)-Me | Me | CH | N | 280 |
| **66** | 17f | 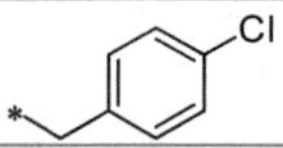 | (*S*)-Me | H | N | CH | 39 |
| **67** | 17g | 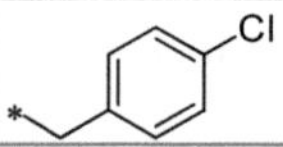 | (*S*)-Me | Et | N | CH | 2.3 |
| **68** | 18a | 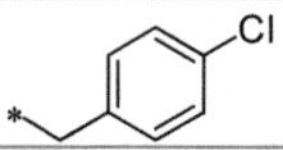 | (*S*)-Et | Me | N | CH | 3 |
| **69** | 18b | 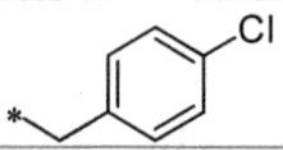 | (*S*)-*i*Bu | Me | N | CH | 260 |
| **70** | 18e | 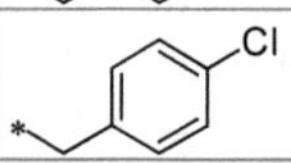 | (*R*,*S*)-$CH_2OH$ | Me | N | CH | 910 |
| **71** | 18i | 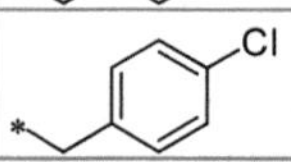 | (*S*)-Et | Et | N | CH | 0.3 |
| **72** | 18j | 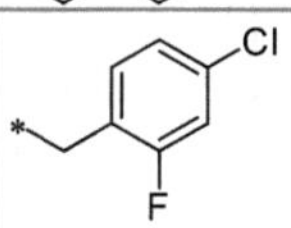 | (*S*)-Et | Et | N | CH | 0.2 |

[a]Details of assay conditions not given in reference

Compounds **71** and **72** represent the first reported sub-nanomolar CXCR3 antagonists. Indeed, at the time, compound **71** had already drawn our attention from the patent literature [112], and we used it in a hybrid-design strategy to probe a potential polycycloaliphatic pocket in CXCR3 (vide infra) [113]. Extensive mutation studies have been done by us on compound **71**, which we called VUF11211 [80]. The binding of **71** to CXCR3 was affected by mutations in the TM region of CXCR3, whereas CXCL11 affinity remained largely unchanged. A binding model was constructed based on homology modeling and the data from the mutagenesis study, revealing a cross-pocket binding mode for **71**, binding to both TMS1 and TMS2, and partially overlapping with the binding mode for **11** [80]. Since the CXCR3 chemokines are binding to the extracellular loops and N-terminus of the receptor, these small molecules probably bind in an allosteric manner (see Sect. 4) [20, 80].

Despite the extraordinary high in vitro affinity of **71**, work continued because **71** exhibited modest PK features in rat and had undesirable hERG affinity [114]. The pyridine ring of **71** was initially replaced by a pyrazine moiety (**73**), and the new core motif was optimized (Table 12), first of all by appending an $NH_2$ group onto the pyrazine ($R^3$). First, the $R^1$-substituent was modified maintaining a primary amide as $R^2$, revealing that many analogues (like **74**) had reasonable affinity for CXCR3 if the $R^1$-substituent was a benzylic group. With the 4-chlorobenzyl fixed, the SAR at the $R^2$-position was probed. Polar substituents such as hydroxyalkyl moieties, sulfonamides, and lactone analogues showed good affinity (i.e., hydroxyalkyl analogue **75**, $IC_{50} = 0.3$ nM). Unfortunately, many of those compounds also displayed affinity for the hERG channel. In an attempt to improve the hERG profile, polar substituents on $R^2$ in combination with polar groups as $R^1$ were investigated. The strategy worked as, e.g., sulfonamide analogues ($R^2$) improved the hERG profile, while affinity was reasonably maintained. Likewise, the hERG affinity decreased for 6-amino-2-chloropyridine-5-carboxamide derivative **76** ($IC_{50} = 3.4$ nM). Moreover, the exposure in rat was improved for **76**. Therefore, the 6-amino-2-chloropyridine-5-carboxamide moiety was fixed as $R^1$, with SAR focusing on other parts of the scaffold. As the previous pyridine series proved good without a 6-amino moiety as $R^3$, this amino group was removed and the $R^2$-position was modified. Primary and secondary amides were both well tolerated, indicating that the 6-amino group was not essential for binding. Subsequently, a more stable replacement was sought for the 3-chloro moiety of the pyrazine, perhaps because it can act as an electrophile. Thus, replacements for $R^4$ were investigated, which showed that methyl and trifluoromethyl analogues were all acceptable. Exemplary cyclopropylamide **77** ($IC_{50} = 1.9$ nM) possesses good affinity, little hERG affinity, and good PK in rat. Additional revisiting of the 4-chlorobenzamide as $R^1$ led to **78** ($IC_{50} = 1.1$ nM), which showed a good combination of CXCR3 affinity and rat PK. The high percentage of hERG inhibition (74% at 10 μM) of **78** could be circumvented by introduction of an (*R*)-methyl moiety at $R^5$ (**79**, $IC_{50} = 1.3$ nM, hERG inhibition of 7% at 10 μM).

Jenh et al. analyzed and reported the properties of **74**, also known as SCH 546738 [54]. The compound exhibited favorable pharmacokinetic properties in rodents and appeared effective as CXCR3 antagonist both in vitro and in vivo. For example, it inhibited the chemotaxis of isolated human T cells towards CXCL9,

**Table 12** Optimization of piperazinyl-piperidine derivatives by Kim et al. [114]

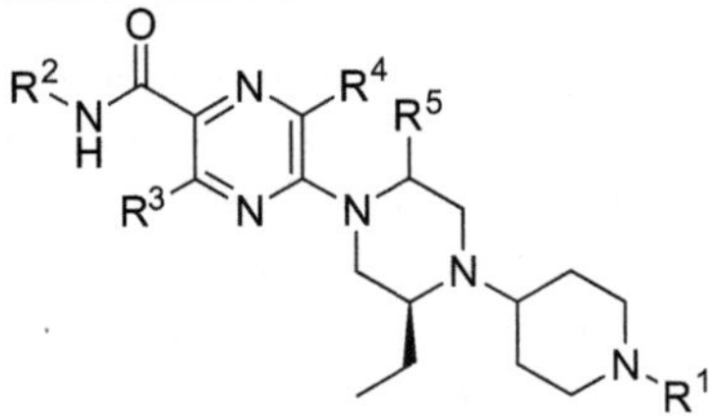


| Compound | | | | | | | $IC_{50}$ (nM) |
|---|---|---|---|---|---|---|---|
| # | In ref. | $R^1$ | $R^2$ | $R^3$ | $R^4$ | $R^5$ | CXCL10[a] |
| **73** | 2 | 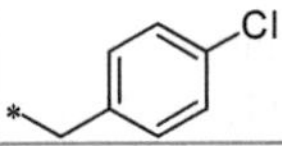 | Et | H | Cl | H | n.s. |
| **74** | 8a | 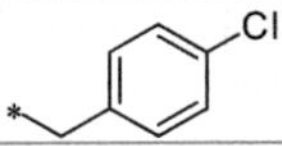 | H | $NH_2$ | Cl | H | 0.8 |
| **75** | 8i | 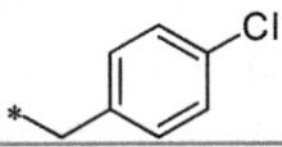 | 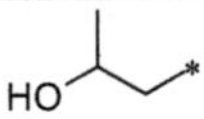 | $NH_2$ | Cl | H | 0.3 |
| **76** | 8r | 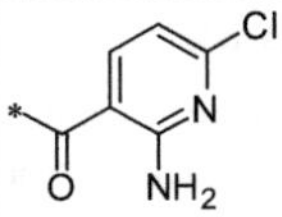 | 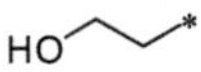 | $NH_2$ | Cl | H | 3.4 |
| **77** | 15f | 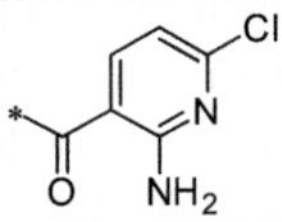 | 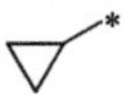 | H | Me | H | 1.9 |
| **78** | 15j | 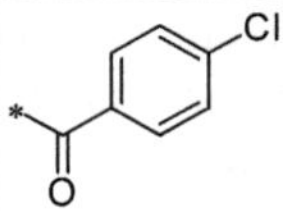 | 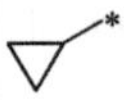 | H | Me | H | 1.1 |
| **79** | 16e | 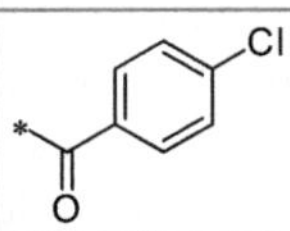 |  | H | Me | (*R*)-Me | 1.3 |

[a][$^{125}$I]-labeled CXCL10 displacement assay performed in Ba/F3 cells expressing human CXCR3
*n.s.* not shown in article

CXCL10, and CXCL11 in a noncompetitive manner. In addition, **74** dose-dependently attenuated collagen-induced arthritis in a mouse model for rheumatoid arthritis. Similarly, it delayed disease onset and attenuated disease severity in murine experimental autoimmune encephalomyelitis, a model for human multiple sclerosis. Moreover, **74** significantly delayed graft rejection in a cardiac allograft model [54].

In the latest disclosed SAR work [115], efforts were directed towards improving PK, hERG, and metabolic parameters. The key approach involved replacing the amide functionality with heterocycles, among which oxadiazoles, imidazoles, and triazoles were probed. Additional balancing involved revisiting other parts of the molecule. Exemplary resulting molecules are **80** ($IC_{50} = 2.8$ nM) and **81** ($IC_{50} = 6.4$ nM). Notably, both had significantly lower hERG affinities (25 and 0% in Rb efflux assay) than several predecessors while displaying good PK properties.

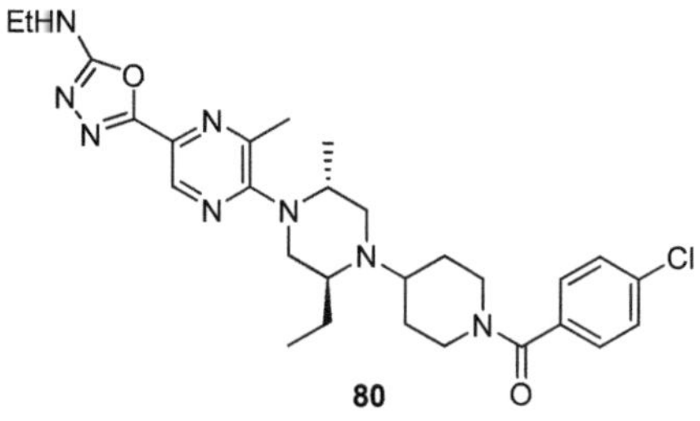


**80**

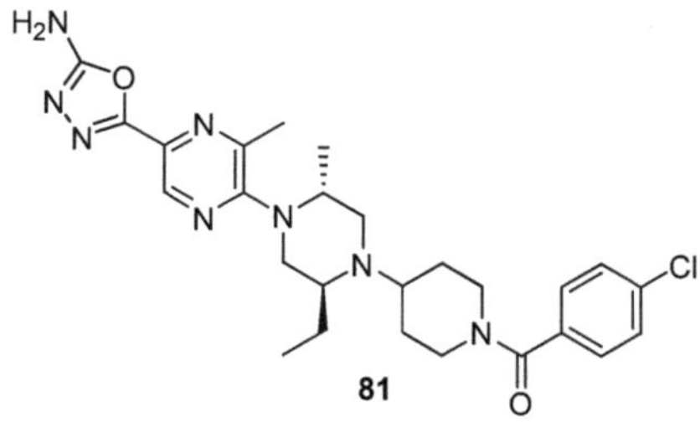


**81**

## 2.4 Ergolines

In 2009, Thoma et al. (Novartis) described the discovery of the lysergic acid-derived inhibitor **82**, a rather unusual type of CXCR3 binder [116]. GPCR selectivity was tested at an early stage, and unlike lysergic acid diethylamide (LSD) and its close derivative **83**, compound **82** did not significantly inhibit serotonin, adrenergic, and dopamine receptors. The selectivity may be explained by different electronic and steric properties of **82** compared to LSD and **83**. For example, **82** (lacking a highly basic N-atom) is neutral under physiological conditions, whereas LSD is protonated. Compound **82** was stable in rat and human microsomes and it had a similar metabolic pattern in all tested species. Preliminary SAR efforts indicated that relatively moderate changes of the core structure strongly affected the affinity [116]. Changes in the amide group ($R^1$), such as removal of the carbonyl moiety, reduced the affinity, as did modification of the urea portion of the molecule ($R^2$). Methylation or benzylation of the indole N ($R^3$) were not of additional value either. In all, a preliminary strict SAR emerged.

In 2011, a more in-depth SAR of the ergoline series was described (also Table 13) [117]. In terms of amide N-substituents ($R^1$), primary and secondary amide analogues lost affinity. Indeed, cyclic tertiary amides, such as **84** ($IC_{50} = 5$ nM), showed the best binding affinities. Various more polar aliphatic heterocycles, such as piperazines and sulfoxide derivatives, had lower affinity for the receptor. With the pyrrolidine amide at $R^1$, variation of $R^2$ was explored [117]. In general, *meta*-substitution of the urea phenyl group was preferred over *ortho*- or *para*-substitution. Aliphatic cyclohexyl urea **85** ($IC_{50} = 24$ nM) and cycloheptyl urea **86** ($IC_{50} = 28$ nM) showed good affinities, while introduction of amides instead of ureas led to a decrease in affinity. In the end, the phenyl urea moiety (as in **84**) remained the best substituent. Unfortunately, **84** only modestly

**Table 13** Optimization of ergolines by Thoma et al. [116, 117]

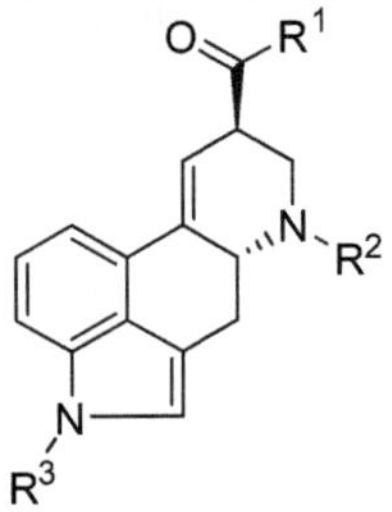

| Compound | | | | | $IC_{50}$ (nM) | |
|---|---|---|---|---|---|---|
| # | In ref. | R[1] | R[2] | R[3] | CXCL11[a] | $Ca^{2+}$ [b] |
| | LSD | 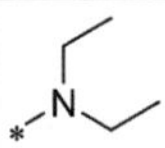 | Me | H | n.s. | n.s. |
| **82** | 1, [116]<br>1a [117] | 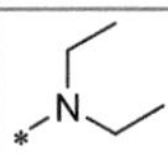 | 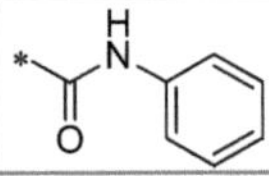 | H | 51 [117] | 18 |
| **83** | 2 [116] | 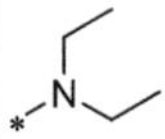 | H | H | >10,000 | n.d. |
| **84** | 1h [117] | 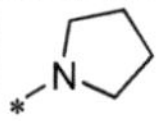 | 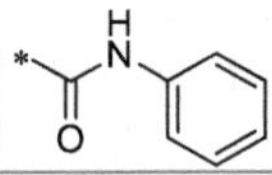 | H | 5 | 4 |
| **85** | 8o [117] | 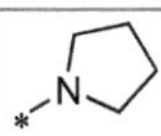 | 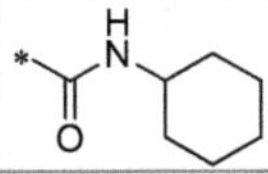 | H | 24 | 11 |
| **86** | 8p[117] | 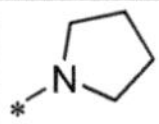 | 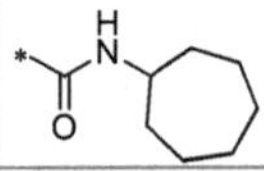 | H | 28 | 14 |
| **87** | 1t [117] | 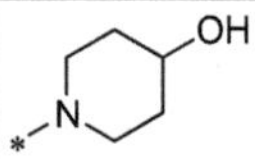 | 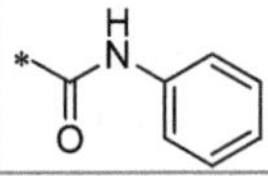 | H | 14 | 5 |
| **88** | 11a [117] | 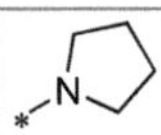 | 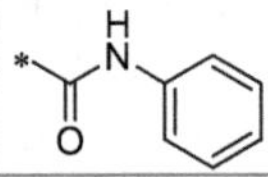 | 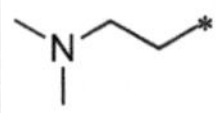 | 2 | 2 |

[a] $[^{125}I]$-labeled CXCL11 displacement assay performed in CHO cells expressing human CXCR3
[b] CXCL11-induced $Ca^{2+}$mobilization assessed in CXCR3-transfected L 1.2 cells
*n.s.* not shown in article, *n.d.* not determined

inhibited CXCL11 binding to CXCR3 in both rat ($IC_{50} = 3{,}300$ nM) and human blood ($IC_{50} = 700$ nM) [117]. Given that the more polar **87** ($IC_{50} = 14$ nM) showed less reduced inhibition in human blood ($IC_{50} = 200$ nM), polarity was postulated as a key factor. More polar groups were sought for at the nitrogen atom of the indole ($R^3$), as polar groups in the amide portion and the urea portion had shown to lead to less potent derivatives. At $R^3$, an ethanol substituent or various basic amino functionalities (i.e., **88**, $IC_{50} = 2$ nM) gave good affinity and potency which was, gratifyingly, largely retained in rat blood ($IC_{50} = 5$ nM). Favorable features of **88** include the bioavailability (97%) and the half-life ($t_{1/2} = 8.9$ h).

A compound named NIBR2130 [118], assumed to be from the ergoline-type class, was recently shown to have only a limited impact on disease outcome in a diabetes type 1 mouse model [118] and on cardiac allograft rejection in mice and rats [57] while having favorable pharmacokinetics and nanomolar affinity for both human and murine CXCR3 (both $IC_{50} = 2.2$ nM, [$^{125}$I]-CXCL11) [57, 118].

## 2.5 *Iminobenzimidazoles*

An HTS of the Abbott corporate compound collection led to the discovery of **89**, a compound with moderate affinity for CXCR3 ($IC_{50} = 3$ μM) [119]. The molecular weight and log P (2.7) of **89** were considered a suitable starting point for hit-to-lead (Table 14). In the acetophenone portion ($R^1$) of the molecule, a cyano group instead of a nitro group was not tolerated, but the bromo and chloro analogues had affinities in the same range as **89**. Removal of the 4-substituent or replacement of the carbonyl moiety of the acetophenone portion with a sulfoxide or alcohol reduced the affinity. Substitution of the benzimidazole core was also explored ($R^2$) [119]. While substitution with a 4-OMe (**90**, $IC_{50} = 3$ μM) showed no change in affinity, substitution at the 5-position and 6-position resulted in compounds with only weak affinity. Because the 2-acyl analogues (collectively referred to as A) like **89** and **90** were poorly soluble in aqueous buffer, the 2-acyl was replaced for an imino group while an additional N-Me was appended as well (referred to as B) [119]. With the parent unsubstituted 2-imino compound **91** ($IC_{50} = 0.8$ μM) at hand, the impact of substitution on the benzimidazole core ($R^2$) and acetophenone moiety ($R^1$) was reevaluated [119]. Notably, substitutions on the 4-position of the benzimidazole improved or maintained affinity compared to **91**, with 4-ethyl analogue **92** ($IC_{50} = 0.03$ μM) displaying the best affinity. Larger substituents and more polar substituents did not perform as well, leading the authors to suggest that the binding subpocket for C4-groups is small and lipophilic. As in previous evaluations (vide supra), substitution on the 5- and 6-position did not improve the affinity.

The focus of Hayes et al. also turned to replacing the N-Me of the iminobenzimidazole core ($R^1$ in Table 15) [120]. The beneficial ethyl moiety on the 4-position ($R^2$) was removed in order to better examine the impact of N-modification. Thus, the N3-position of **91** was substituted with increasingly larger groups, which was generally tolerated. A notable improvement in functional antagonism (FLIPR assay) was found for amide analogue **93** ($IC_{50} = 0.4$ μM). The reverse amide analogue **94** ($IC_{50} = 0.4$ μM) was less potent in a functional assay, but was nevertheless used for further optimization. Introduction of a chlorine atom as $R^2$ increased affinity, and with this substituent the optimal linker length between the iminobenzimidazole nitrogen atom and the amide nitrogen atom was found to be propylene. Several aryl, heteroaryl, and aliphatic amides were probed to identify a more optimal amide substituent. Amide analogue **95** ($IC_{50} = 0.02$ μM) emerged

**Table 14** Optimization of benzimidazole derivatives by Hayes et al. [119]

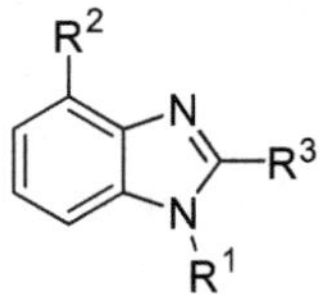


A

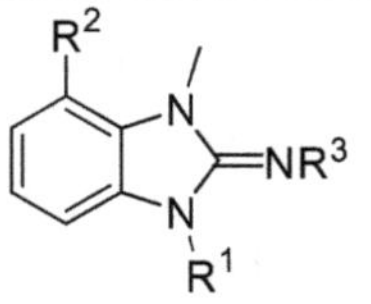


B

| Compound | | | | | | $IC_{50}$ (μM) |
|---|---|---|---|---|---|---|
| # | In ref. | Core | $R^1$ | $R^2$ | $R^3$ | CXCL10[a] |
| **89** | 1 | A | 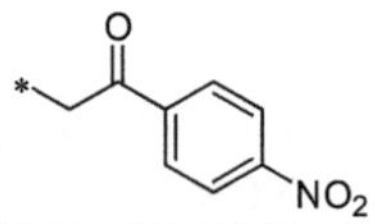 | H | COMe | 3 |
| **90** | 4e | A | 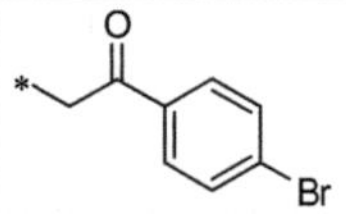 | OMe | COMe | 3 |
| **91** | 12d | B | 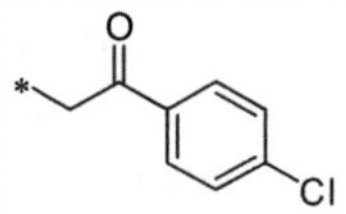 | H | H | 0.8 |
| **92** | 12p | B | 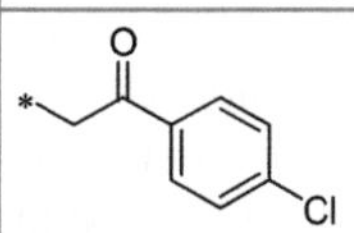 | Et | H | 0.03 |

[a] [$^{125}$I]-labeled CXCL10 displacement assay performed in Chinese hamster ovary (CHO) cells expressing human CXCR3

from this, as it had good affinity for both human and murine CXCR3. However, the half-life of **95** in mouse liver microsomes was short due to *N*-demethylation as the main metabolic event in vitro. To prevent such *N*-demethylation, the *n*-propyl linker was constrained into a ring (**96**, $IC_{50} = 0.015$ μM). Another compound with a ring-containing linker (2-pyrrolidine compound **97**, $IC_{50} = 0.008$ μM) showed good potency across species, but was nonetheless rapidly metabolized (apparently not to the demethylated analogue) in mouse liver microsomes.

## 2.6 *VUA Compounds: Targeting a Hypothesized Polycycloaliphatic Pocket*

Our group has published multiple articles on tool compounds that make use of the recurring [105, 121] polycycloaliphatic motif in CXCR3 ligands. Several venues were followed.

**Table 15** Optimization of benzimidazole derivatives by Hayes et al. [120]

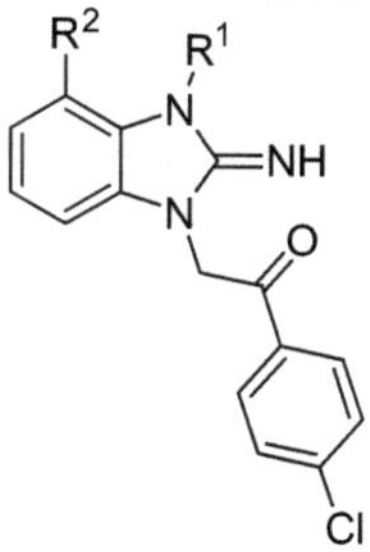


| Compound # | Compound In ref. | $R^1$ | $R^2$ | $IC_{50}$ (μM) CXCL10[a] |
|---|---|---|---|---|
| **93** | 11b | 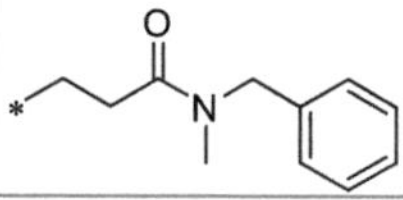 | H | 0.4 |
| **94** | 14d | 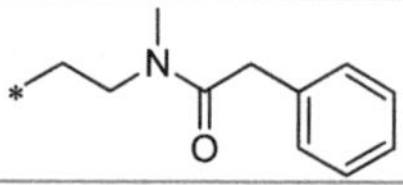 | H | 0.4 |
| **95** | 14n | 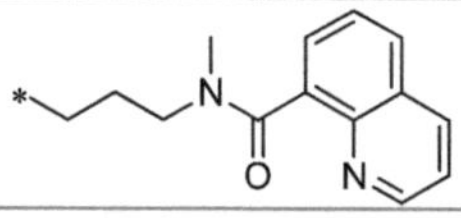 | Cl | 0.02 |
| **96** | 14o | 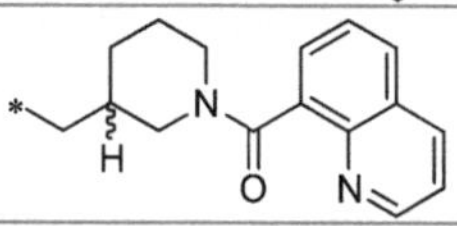 | Cl | 0.015 |
| **97** | 14r | 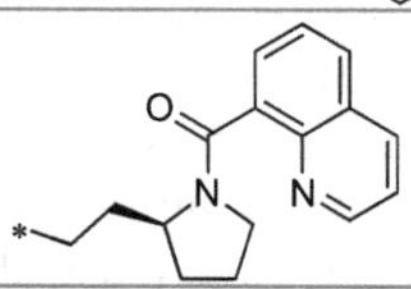 | Cl | 0.008 |

[a] [$^{125}$I]-labeled CXCL10 displacement assay performed in Chinese hamster ovary (CHO) cells expressing human CXCR3

In our earliest efforts, we decided to use the piperazinyl-piperidine series disclosed by Merck to explore the CXCR3 binding site and the receptor's apparent preference for polycycloaliphatic groups [113]. More specifically, **71** and **72** were selected as a starting point from the patent literature [112], because their picomolar affinities may allow the removal of a substantial portion of the molecule followed by the appending of a polycycloaliphatic group. In a first scan, the benzylaminopiperidine part of **71** was equipped with various (poly)cycloaliphatic groups ($R^1$, Table 16) [113]. Neither introduction of monocyclic aliphatic rings, such as pyrrolidine and N-cyclohexyl rings, nor the introduction of bicyclic aliphatic rings, such as (−)-myrtenyl and tropine groups, resulted in an acceptable affinity. One exception is the 2-adamantane analogue **98** ($pK_i = 6.8$). SAR around this 2-adamantane unit was very strict, with all further manipulations (shifting to the

**Table 16** Optimization of polycycloaliphatic aminopiperidines by Wijtmans et al. [113]

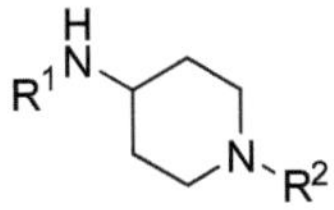


| Compound | | | | $pK_i$ |
|---|---|---|---|---|
| # | In ref. | $R^1$ | $R^2$ | CXCL10[a] |
| **98** | 19 | 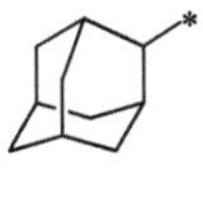 | 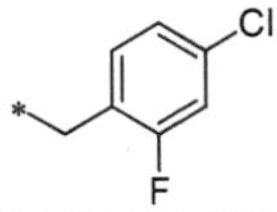 | 6.8 |
| **99** | 28 | 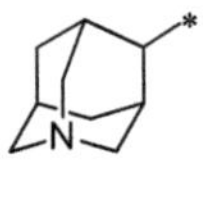 | 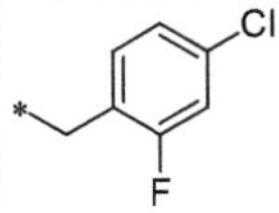 | 5.1 |
| **100** | 37 | 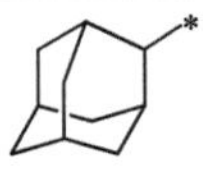 | 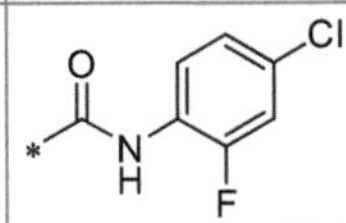 | <5 |
| **101** | 38 |  | 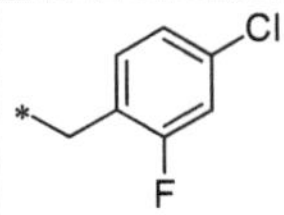 | 6.4 |

[a][$^{125}$I]-labeled CXCL10 displacement assay performed in HEK293 cells expressing human CXCR3

1-position, methylation, linker elongation, incorporation of an N-atom to give **99** ($pK_i = 5.1$)) leading to reduced affinity. SAR was also strict around the 4-aminopiperidine core [113]. Changes such as substitution on the benzylic methylene, methylene insertion, introduction of a (un)saturated ester, and incorporation of a urea group to give **100** ($pK_i < 5$) all gave reduced affinity. For further efforts, another polycycloaliphatic group was sought to overcome the high crystallinity and poor solubility of many of the tested adamantane compounds. The (*R*)-isobornyl analogue **101** ($pK_i = 6.4$) displayed a somewhat reduced affinity compared to **98**, but did have the desired reduced crystallinity and was therefore selected for in-depth SAR on the benzyl portion ($R^2$). In general, the (*R*)-isobornyl series exhibited similar SAR trends as the adamantane series, but none of the tried substitutions matched the affinity of parent **101**. In contrast to the achiral 2-adamantane series, compounds with an isobornyl group as $R^1$ could be used to probe the "polycycloaliphatic pocket" with stereochemical subtleties of the bornyl group. However, little effect of stereochemistry was observed.

In second approach targeted at using the "polycycloaliphatic pocket", we built on a medium-throughput screen of 3,360 pharmacologically active compounds performed in our labs [122]. The screen resulted in 90 hits that displace [$^{125}$I]-CXCL10 for more than 50% at a concentration of 10 μM. One of those hits was IPAG (**102**), a sigma-receptor ligand. Although it had a moderate $K_i$ of 4,000 nM

**Table 17** Optimization of polycycloaliphatic ammonium salts by Wijtmans et al. [122]

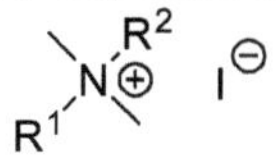

| Compound | | | | p$K_i$ |
|---|---|---|---|---|
| # | In ref. | R[1] | R[2] | CXCL10[a] |
| **103** | 8a | 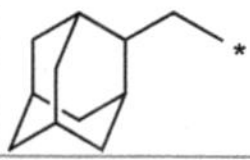 | 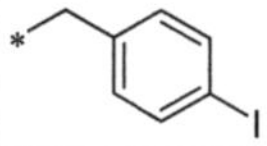 | 6.1 |
| **104** | 8f | 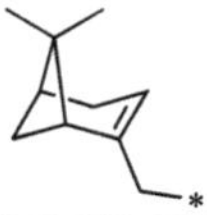 | 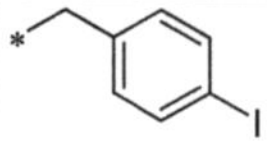 | 6.6 |
| **105** | 9f | 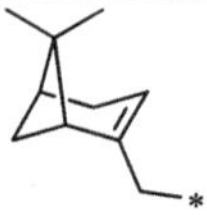 | 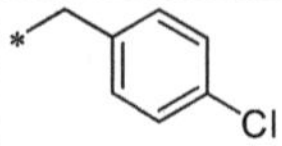 | 6.7 |
| **106** | 9j | 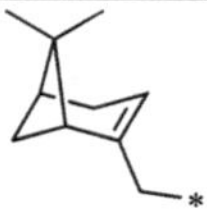 | 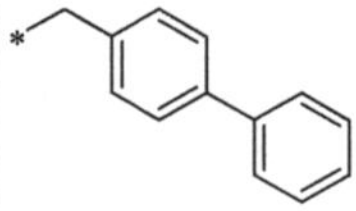 | 6.5 |
| **107** | 10c | 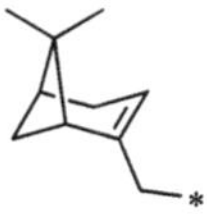 | 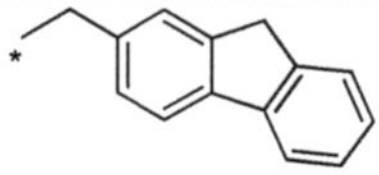 | 6.3 |
| **108** | 10q | 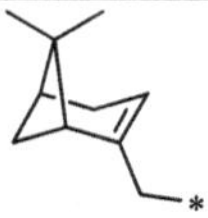 | 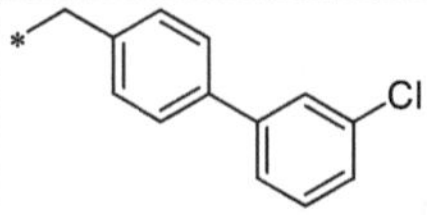 | 6.9 |

[a] [$^{125}$I]-labeled CXCL10 displacement assay performed in HEK293 cells expressing human CXCR3

for CXCR3, its adamantane substructure drew attention to the hypothesized "polycycloaliphatic pocket." In order to simplify the generation of chemical diversity (Table 17), the guanidine unit was replaced with a tertiary amine or a quaternary ammonium cation. The distance between the aryl and adamantanyl units was kept comparable to IPAG by the insertion of two methylene spacers.

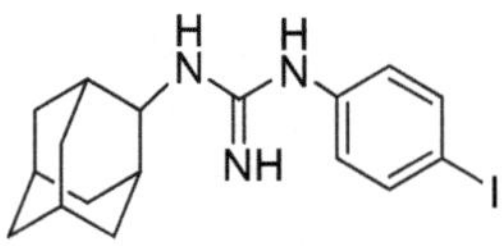

**102** (IPAG)

The tertiary amine analogue of IPAG led to a reduced affinity, but insertion of a permanent cation by methylation of the nitrogen (i.e., **103**, p$K_i = 6.1$) substantially increased the affinity. It was hypothesized that the ability of the core to engage in electrostatic interactions plays a dominant role in this, so the permanent cation was maintained in subsequent SAR. During probing of the left-hand side portion ($R^1$), the most notable observation was that introduction of a myrtenyl group resulted in an enhanced affinity (**104**, p$K_i$ = 6.6) compared to adamantane analogue **103**, which is opposite to what we discovered for the 4-aminopiperidine series (vide supra) [113]. This suggests that the IPAG derivatives bind in a different manner than the 4-aminopiperidine series. The aromatic right-hand side ($R^2$) tolerated *para*-substitution with chlorine (**105**, p$K_i$ = 6.7), bromine, iodine but less so with fluorine. As a whole, the $R^2$ SAR suggested that room for growth was available at the *para*-substitution of the benzyl substituent. Indeed, a biphenyl compound was found to have a good affinity for the receptor (**106**, p$K_i$ = 6.5) [122]. SAR at the biphenyl core itself was pretty strict, with, e.g., shifting of the phenyl ring, insertion of an oxygen atom, or constrainment to give **107** (p$K_i$ = 6.3) not being of surplus value. The biphenyl moiety lends itself well to peripheral SAR, though. *Meta*-substitution was preferred, but protic polar *meta*-substitutions, such as hydroxyl and amino groups, reduced affinities. Indeed, substitution with the more lipophilic chlorine on the *meta*-position restored the affinity. In fact, *meta*-chloro analogue **108** was the compound with the best affinity of this series (p$K_i$ = 6.9).

This series served as a stepping stone for the discovery of a novel class of CXCR3 agonists (Sect. 3.2).

## 2.7 *Miscellaneous*

Several chemotypes have been the subject of one medicinal chemistry paper only. These have been collected in the current paragraph.

### 2.7.1 Benzetimide Derivatives

Bongartz et al. (Johnson & Johnson) screened a database of compounds for their inhibitory activity on cAMP in CXCL11-stimulated hCXCR3-transfected CHO cells [110]. Hit compound **109** inhibited [$^{35}$S]-GTPγS binding with an $IC_{50}$ value of 0.78 μM. Separation of the enantiomers showed that the (−) stereoisomer and the (+) stereoisomer had comparable CXCR3 antagonistic effects. Compound **109** resembles the muscarinic receptor antagonist benzetimide (where the Br is an H atom). For that reason, the anticholinergic activity of **109** was assayed as it may carry the risk of off-target effects. The (+) stereoisomer of **109** showed nanomolar affinity for the muscarinic receptors M1, M2, and M3, whereas the (−) stereoisomer of **109** showed only marginal affinity to M1. This is in accord with the notion that only the (+) stereoisomer of benzetimide (dexetimide) shows anticholinergic activity [110]. Acknowledging **109** as a good starting point for CXCR3 ligands, the N-substitution of the piperidine ring ($R^1$) was explored (Table 18). Most substitution patterns for the benzyl ring were not of surplus value. Only a 3-fluoro-4-chloro

**Table 18** Optimization of benzetimide derivatives by Bongartz et al. [110]

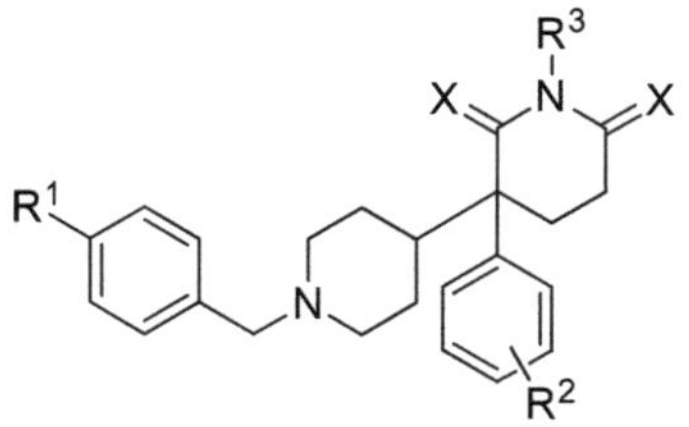

| Compound | | | | | | | $IC_{50}$ (μM) |
|---|---|---|---|---|---|---|---|
| # | In ref. | Stereochemistry | $R^1$ | $R^2$ | $R^3$ | X | [$^{35}$S]-GTPγS[a] |
| **109** | 1 | ± | 4-Br | H | H | O | 0.78 |
| **110** | 12 | – | 3-F,4-Cl | H | H | O | 0.34 |
| **111** | 18 | ± | 4-Br | 2,4-F | H | O | 0.12 |
| **112** | 21 | ± | 4-Br | 2-OMe-5-$SO_3H$ | H | O | 0.17 |
| **113** | 41a | – | 4-Br | H | Acetyl | H, H | 0.11 |
| **114** | 47b | + | 4-Br | H | α-Acetamide | H, H | 0.06 |
| **115** | 48a | – | 4-Br | H | Phenylurea | H, H | 0.03 |

[a][$^{35}$S]-GTPγS binding assay

substitution pattern, as in **110** ($IC_{50} = 0.34$ μM), exhibited better affinity for the receptor than **109**. However, it was decided to continue with the original bromo-substituted moiety and the $R^2$-substituted phenyl group was modified. Small *para*-substituents like fluorine where tolerated, as were various substituents on the *meta*-position, such as amine and carboxylic acid groups. The *ortho*-position was found to be a good additional anchor point, as suggested by the relatively high affinities of **111** ($IC_{50} = 0.12$ μM) and **112** ($IC_{50} = 0.17$ μM). When the glutarimide group was explored, retaining a carbonyl group (X=O) proved essential for affinity. The imide hydrogen ($R^3$) was replaced by alkyl groups without improvement in affinity. Interestingly, testing of enantiopure acetyl (**113**, $IC_{50} = 0.11$ μM), α-acetamide (**114**, $IC_{50} = 0.06$ μM), or phenylurea derivatives (**115**, $IC_{50} = 0.03$ μM) led to significantly enhanced activity even with the two glutarimide carbonyls (X) removed. In contrast to **113** and **115**, for **114** the (+) enantiomer proved most active. Compounds **113**, **114**, and **115** showed no antimuscarinic activity, and it was mentioned that those compounds had comparable affinity for mouse CXCR3.

### 2.7.2 N-Benzyl Benzenesulfonamides

Crosignani et al. (Merck Serono) reported on the screening of 90,000 compounds employing a high-throughput screening method [123]. As a result, compound **116**

**Table 19** Optimization of *N*-benzyl benzenesulfonamides by Crosignani et al. [123]

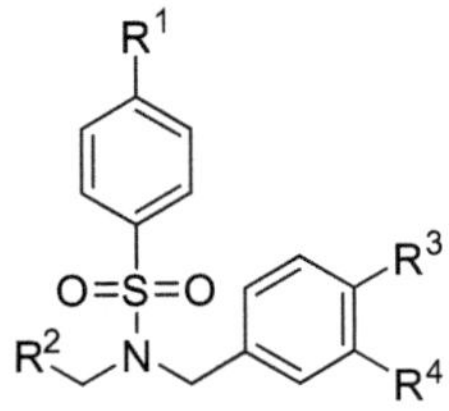


| Compound | | | | | | $IC_{50}$ (nM) |
|---|---|---|---|---|---|---|
| # | In ref. | $R^1$ | $R^2$ | $R^3$ | $R^4$ | Chemotaxis[a] |
| **116** | 1 | Cl | 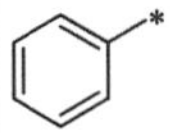 | 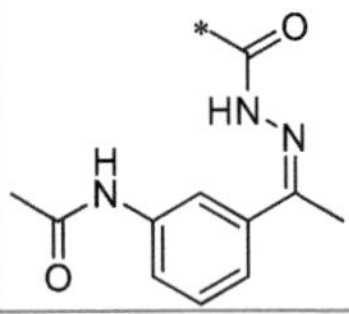 | H | 538 |
| **117** | 2 | Cl | 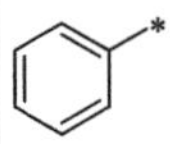 | 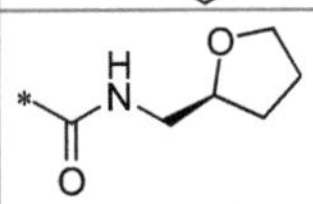 | H | 2,275 |
| **118** | 27 | CN | 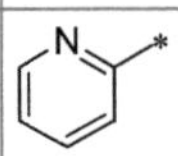 | 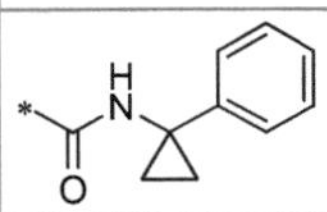 | H | 13 |
| **119** | 13 | Cl | 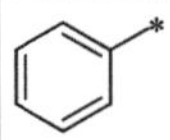 | 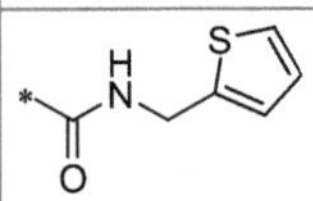 | H | 192 |
| **120** | 14 | Cl | 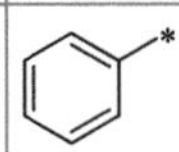 | 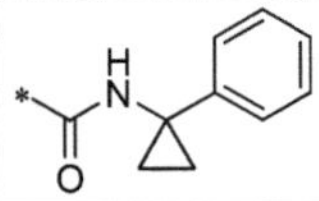 | H | 238 |
| **121** | 28 | Cl | 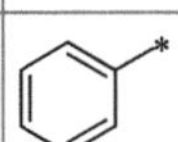 | 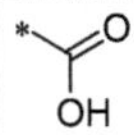 | H | 2,940 |
| **122** | 47 | CN | 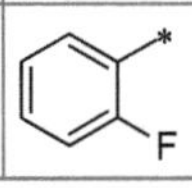 | 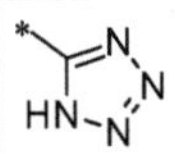 | F | 192 |

[a]Chemotaxis assay with CXCR3-overexpressing L1.2 cells and CXCL10 as chemoattractant

was discovered (Table 19). While the compound was selective against other tested GPCRs, it had low solubility, low permeability in the Caco-2 assay, high clearance in human and rat liver microsomes, and was unstable in acidic media. Some of those properties may be contributed to the acylhydrazone moiety present in **116**. Encouragingly, additional efforts delivered structure **117** indicating that the acylhydrazone moiety could be replaced. Assay values reported in this paper are mostly those of chemotaxis assays, strictly taken not binding values, but the authors show that the correlation with binding data is good. A first optimization was performed for $R^1$, $R^2$, and $R^3$. SAR on $R^1$ was tight and it was mentioned that *ortho*- and *meta*-substitution were not tolerated. A *para*-chlorine functioned well, but methoxy and cyano moieties were allowed too. The $R^2$-position tolerated many (hetero)aromatic groups. Especially compounds containing a 2-pyridyl ring had good potencies (i.e., **118**, $IC_{50} = 13$ nM). The authors state that an unsubstituted methylene linker should be present between the (hetero)aromatic ring and the core structure and that aliphatic or cycloaliphatic groups were not tolerated as $R^2$. For the $R^3$ moiety, lipophilic groups could be used as exemplified by **118** and **119** ($IC_{50} = 192$ nM). The thiophene analogue **119** had slightly better affinity than a 1-phenylcycloprop-1-yl analogue (**120**, $IC_{50} = 238$ nM). The instability in liver microsomes of compounds in the series had to be addressed. Carboxylic acid analogue **121** ($IC_{50} = 2{,}940$ nM) was found to be stable in human and rat liver microsomes, although the potency was reduced. This led the authors to believe log D was possibly involved in the metabolic stability. In order to boost potency, several carboxylic acid bioisosteres, such as tetrazole and oxadiazolone, were tested. Exploration of the SAR around these structures followed similar trends compared to the amide subseries. Yet the tetrazoles generally had higher microsomal stability than the amides. The compound with the best PK profile was **122** (chemotaxis $IC_{50} = 192$ nM).

### 2.7.3 Camphor Sulfonamide Derivatives

A high-throughput screening described by Wang et al. (GlaxoSmithKline) led to the identification of **123** ($pIC_{50} = 6.6$) [121]. The ligand bears a characteristic polycycloaliphatic moiety (camphor), a recurring motif in CXCR3 compounds [105, 113]. The left-hand aryl moiety ($R^1$) was first explored (Table 20) [121]. Moving the trifluoromethyl substituent on the pyridine ring to other positions was not allowed. Replacement of the trifluoromethyl substituent by, for example, bromo, methyl, or cyano also led to a loss of affinity. Adding a second substituent could be beneficial, as evidenced by **124** ($pIC_{50} = 6.8$). Several compounds with aromatic (phenyl) or heteroaromatic (i.e., pyridazinyl) rings as $R^1$ showed significantly reduced affinity, except for installing a pyrimidine, which gave only slightly reduced affinity (**125**, $pIC_{50} = 6.5$). The piperazine core of the scaffold was subjected to an extensive SAR. Most manipulations (substituted piperazines, homopiperazine, other (a)cyclic diamines) gave reduced affinity. Noteworthy, though, (*S*)-methylation on the 3-position of the piperazine core ($R^3$) resulted in a compound with a better affinity. Finally, the camphor portion of the molecule was probed ($R^2$). The (*S*)-isomer **123** exhibited a better affinity than its (*R*)-isomer counterpart. Modification of the ketone of **123** was reasonably (but not fully) allowed. For example, while the alcohol **126** ($pIC_{50} = 6.8$) showed better affinity than its ketone counterpart **123**, the amine **127** ($pIC_{50} = 6.1$) displayed decreased affinity. Some of the best affinity-improving moieties in each region were combined to give, for example, **128** ($pIC_{50} = 7.1$) and **129** ($pIC_{50} = 7.5$).

**Table 20** Optimization of camphor sulfonamide derivatives by Wang et al. [121]

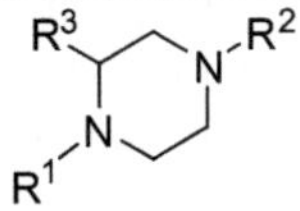

| Compound | | | | | $pIC_{50}$ |
|---|---|---|---|---|---|
| # | In ref. | R[1] | R[2] | R[3] | CXCL10[a] |
| **123** | 1a | 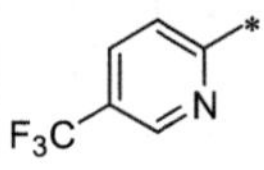 | 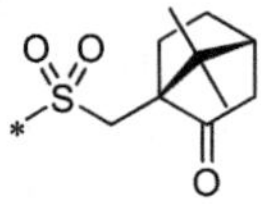 | H | 6.6 |
| **124** | 5k | 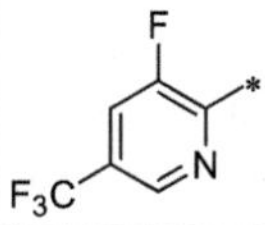 | 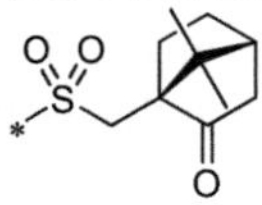 | H | 6.8 |
| **125** | 5q | 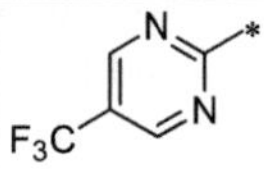 | 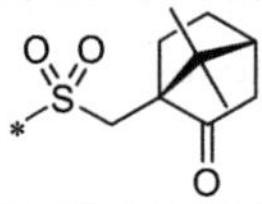 | H | 6.5 |
| **126** | 8a | 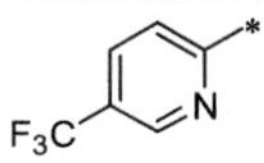 | 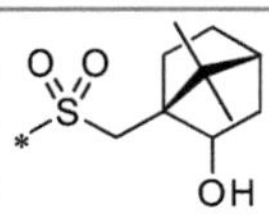 | H | 6.8 |
| **127** | 13a | 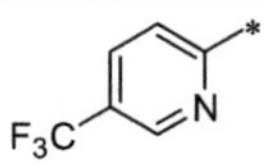 | 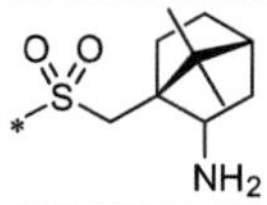 | H | 6.1 |
| **128** | 18a | 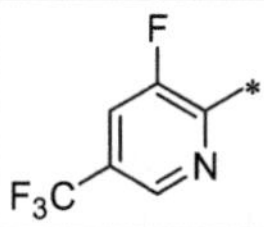 | 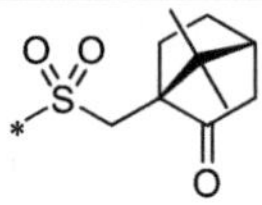 | (*S*)-Me | 7.1 |
| **129** | 18h | 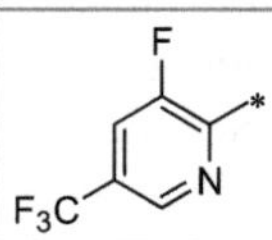 | 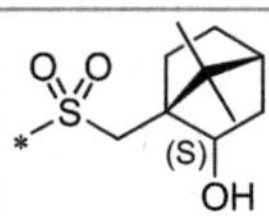 | (*S*)-Me | 7.5 |

[a]CXCL10 induced calcium mobilization assay performed on a CHO-K1 cell line expressing CXCR3 and Gα16

### 2.7.4 4-*N*-Aryl-[1,4]diazepane Ureas

Pharmacopeia researchers screened 90 libraries containing over 4 million compounds using Encoded Combinatorial Libraries on Polymeric Support (ECLiPS™) [124]. From this emerged the 4-*N*-aryl-[1,4]diazepane-urea chemotype on which SAR was carried out (Table 21). The R[1] group benefited from a 2,4-dichlorophenethyl moiety, as is evident from comparing, e.g., **130** ($IC_{50} = 0.51$ μM) and **131** ($IC_{50} = 0.06$ μM). On the $R^2$ side, several substituted phenyl groups as well as a few heteroaromatics were tolerated, with 3-Cl-Ph (as in

**Table 21** Optimization of 4-*N*-aryl-[1,4]diazepane-urea derivatives by Cole et al. [124]

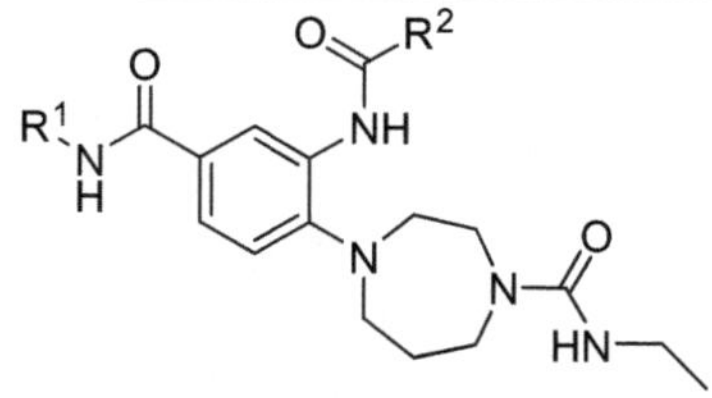


| Compound | | | | $IC_{50}$ (μM) |
|---|---|---|---|---|
| # | In ref. | $R^1$ | $R^2$ | CXCL11[a] |
| **130** | 6v | 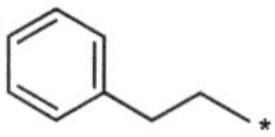 | 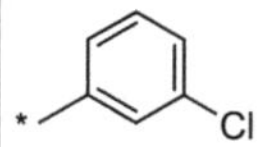 | 0.51 |
| **131** | 6c | 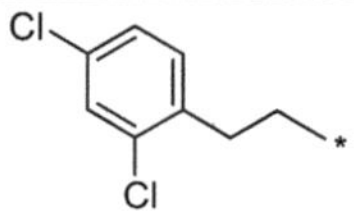 | 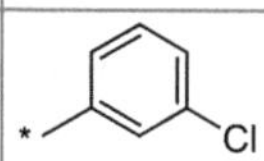 | 0.06 |
| **132** | 6b | 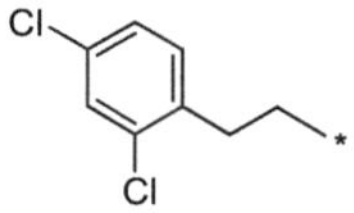 | 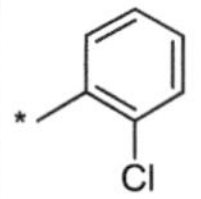 | 0.70 |

[a]Reduction in CXCL11-stimulated calcium release for a cell line (HEK293) overexpressing recombinant human CXCR3 and chimeric G protein Gqi5

**131**) and 3-F-Ph being optimal. Shifting the Cl atom by one position led to a drop in affinity (**132**, $IC_{50} = 0.70$ μM). The ethylurea unit was probed with a selected set of groups, but this did not lead to significantly improved affinities. A decrease in diazepane ring size or its ring opening was not allowed. Compound **131** was used by the same research group in their investigation into small-molecule CXCR3 agonists [125]. A QSAR model has been constructed on this scaffold that consists of molecular descriptors that encode information about the structure, branching, electronic effects, chains, and rings and account for cooperative effects between functional groups [126].

### 2.7.5 Tetrakis-(Diisopropyl-Guanidino) Zinc Phthalocyanine

A metal complex (Zn-DIGP, **133**) has been claimed as a CXCR3 binder, albeit with low affinity ($K_i = 29$ μM, [$^{125}$I]-CXCL10) [127]. The $IC_{50}$ of **133** for inhibition of CXCL10-CXCR3 activation amounted to 3.8 μM. The authors invoke this ability to interfere with CXCL10-CXCR3 signaling as a potential explanation for the antimetastatic activity of **133**.

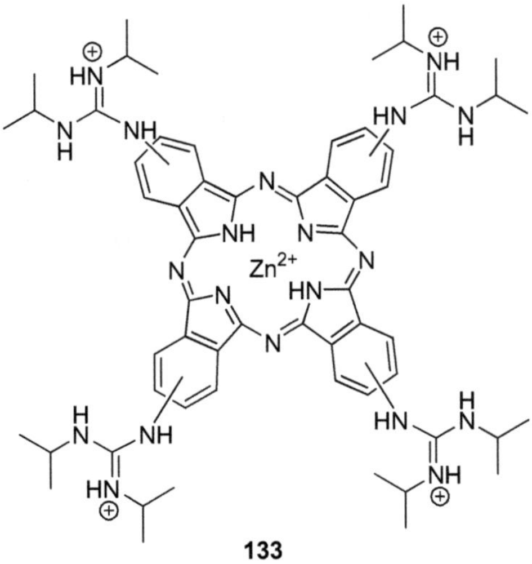


## 3 Agonists

As mentioned in the introduction of this chapter, the field of small-molecule CXCR3 agonists has been much less explored. Arguably, this is because relatively few therapeutic indications have been disclosed for CXCR3 agonists. Moreover, general trends seem to suggest that it is intrinsically complex to find small-molecule agonists for the chemokine receptor family [128]. In total, three distinct agonist small-molecule chemotypes for wild-type CXCR3 have been published (two of which in the same paper in 2006, the other in 2012), and these represent important tool compounds to interrogate the signaling events by CXCR3. Indeed, it will be shown how these agonists have attracted interest from researchers to facilitate fundamental CXCR3 research.

### *3.1 Agonists Emerging from the Pharmacopeia Screen*

During what appears to be the same screen of 90 libraries [124] used for identification of the 4-*N*-aryl-[1,4]diazepane ureas, such as **131** (vide supra), Pharmacopeia researchers also picked up a few agonist chemotypes [125]. The paper only provides qualitative SAR and SFR statements, which will be recapped here. In general, it was noted that irrespective of the scaffold, an agonist chemotype preferably has a single basic amino acid (thus bearing some peptidomimetic character) and a hydrophobic peripheral group [125]. While three agonists are described in more detail in the paper, they are best classified into two chemotypes: a fused piperidinyl diazepanone and two tetrahydroisoquinolines.

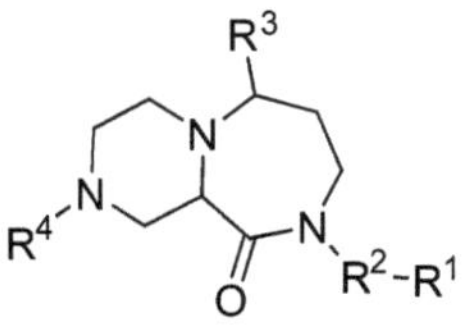


**134**

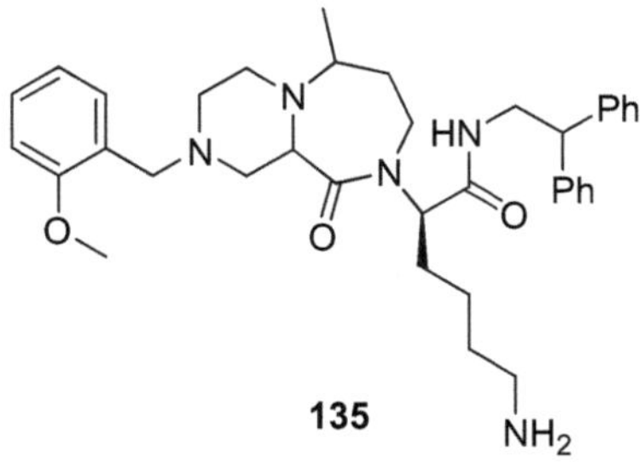


**135**

### 3.1.1 Fused Piperidinyl Diazepanone

One agonist chemotype is that of a fused piperidinyl diazepanone (general structure **134**). While this probed library contained ca. 80,000 members, CXCR3 agonists emerged only from a sublibrary containing an *ortho*-methoxybenzyl moiety as $R^4$. Within this sublibrary of 1,575 compounds (which contained lysines as well as acidic and uncharged amino acids, presumably all on $R^2$), only two compounds were identified as agonists, both bearing a (D)-Lys substituent. Thus, the structure-function relationship (SFR) around this fused piperidinyl diazepanone scaffold, as far as can be deduced, seems strict. The structure of one of the identified agonists was given: **135** ($IC_{50} = 65$ nM, [$^{125}$I]-CXCL10). In a calcium flux assay, **135** presented itself as an agonist with similar efficacy as CXCL11, albeit with lower potency ($EC_{50} = 800$ nM, $\alpha = 1$). Similarly, **135** provoked chemotaxis of human T cells with a maximum response around 1000 nM. Interestingly, the maximum effect in chemotaxis induced by **135** is ±50% of that elicited by CXCL11 ($\alpha = 0.5$). Pease et al. confirmed the agonistic properties of **135** in a murine L1.2 pre-B cell migration assay. Again, **135** showed lower efficacy and potency compared to CXCL11 [129]. In addition, *mutational* studies combined with ab initio modeling on **135** binding to CXCR3 suggested that residue D112$^{2.63}$ acts as the counterion for the positive charge of the arginine moiety in **135** [129]. This suggests that **135** binds to TMS1 of the TM region of CXCR3.

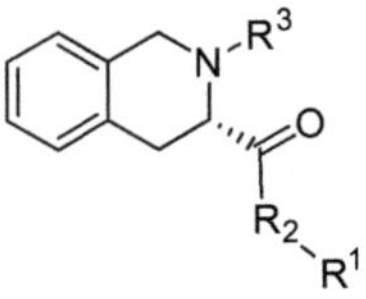

**136**

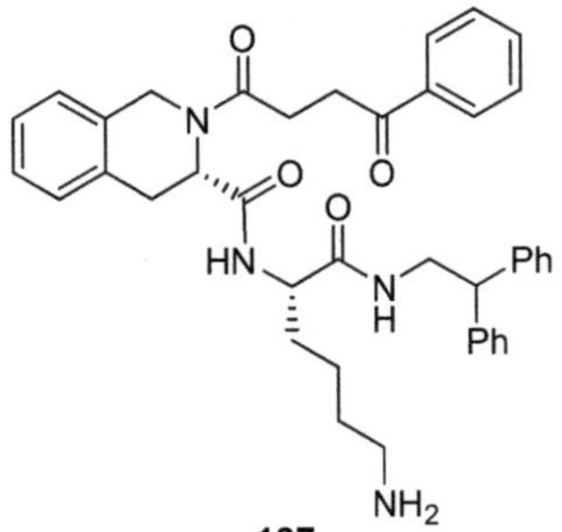

**137**

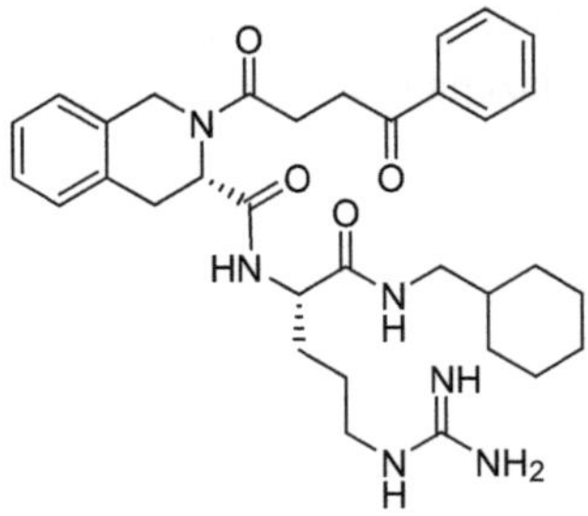

**138**

### 3.1.2 Tetrahydroisoquinolines

The second agonist chemotype emerging from the Pharmacopeia screen [125] is that of a tetrahydroisoquinoline (general structure **136**). This library, with ca. 30,000 members, only provided agonists when $R^3$ was a 3-benzoyl-propionyl side chain. Even so, with that particular $R^3$, only 7 out of the 2520 embedded compounds gave CXCR3 agonism. All of these 7 compounds possessed a basic amino acid as $R^2$ (arginine in 6 out of 7 cases, lysine 1 out of 7). Interestingly, both (L) and (D) amino acids were present in the original library, but the 7 hits all contained the (L) stereomers of the basic amino acid. In terms of $R^1$, it was qualitatively described how a SAR was visible and suggested a preference for a hydrophobic moiety as $R^1$. In all, also for this chemotype, the SFR seems strict. Of the found 7 agonists, the structures of two were disclosed: **137** ($EC_{50} = 3.3$ μM, calcium flux assay) and **138** ($IC_{50} = 42$ nM, [$^{125}$I]-CXCL10; $EC_{50} = 1.1$ μM, calcium flux assay). They are very much alike but differ from each other in the exact nature of the hydrophobic and basic groups.

No synthesis for **137** and **138** was described in the paper. Therefore, our group developed a synthesis strategy for **137** [14]. During these efforts, we also made a handful of derivatives of **137** (Wijtmans et al., unpublished data) which are in line with the qualitative statements from the Pharmacopeia paper [125]. That is, when the $R^3$ group of **137** was changed for an acetyl moiety and/or the $R^2$ for a glycine, the affinity dropped by at least by 1.5 log unit (activity not tested).

The synthesis of **137** enabled detailed pharmacological and mutation studies [14]. Stroke and colleagues showed that both **137** and **138** acted as agonists in a calcium flux and chemotaxis assay. In later publications from our group and Pease et al., these molecules were described in more detail [14, 25, 129].

Our group reported a detailed pharmacological characterization of **137** (called VUF10661), which acts as agonist in different assays, including [$^{35}$S]-GTPγS ($EC_{50} = 0.6$ μM), cAMP ($EC_{50} = 0.5$ μM), β-arrestin recruitment ($EC_{50} = 1.0$ μM), and receptor internalization ($EC_{50} \sim 3$ μM) [14]. **137** also produced PTX-sensitive impedance responses ($EC_{50} = 0.8$ μM) in a label-free impedance assay, comparable to CXCR3 chemokines [25]. Interestingly, **137** showed differential behavior in some assays compared to CXCL11, suggesting functional selectivity. For example, CXCR3 stimulation with **137** resulted in a maximum migration of L1.2 pre-B cells that was about 50% lower than when CXCL11 was used, while in a β-arrestin recruitment assay the efficacy of **137** was 167% compared to CXCL11 [14]. In G protein-dependent assays like [$^{35}$S]-GTPγS and cAMP the efficacy of CXCL11 and **137** were identical. Moreover, **137** likely binds to specific subset of CXCR3 conformations (with distinct functional properties), as it was unable to completely displace [$^{125}$I]-CXCL11, yet completely displaced [$^{125}$I]-CXCL10 radioligand in a whole-cell binding assay. Moreover, in saturation binding experiments, **137** was able to decrease the $B_{max}$ but left the affinity of both chemokine radioligands unchanged [14]. Preliminary mutagenesis studies highlight a binding site for **137** in the TM region of CXCR3 (Scholten et al., unpublished data). Altogether these data indicate that **137** operates in a noncompetitive, potential allosteric way at CXCR3.

In the L1.2 cell migration assay compound **138** also exhibited potency and efficacy comparable to **135** [129]. However, **138** was not able to fully displace [$^{125}$I]-CXCL10 from the receptor, suggesting differential binding to CXCR3 (e.g., different receptor populations). Computational modeling revealed the possibility that **138** mimics residues 35–39 of the 30s loop of CXCL10 and might suggest a similar CXCR3 activation mechanism for this small-molecule agonist. In keeping with this, Pease et al. showed that CXCR3 activation by either CXCL10 or **138** was affected by the same mutations, which did not affect CXCL11 action [129]. Similar to **135**, D112$^{2.63}$ from TMS1 is suggested to act as the counterion for the positive charge in **138**.

O'Boyle has shown how these tool compounds can shed more light on the physiological role of CXCR3 [130]. Based on the molecular mass given [130], they used **138**, which they refer to as PS372424. They expanded on the signaling repertoire of this class of agonists, as by showing ERK phosphorylation induced by **138** comparable to CXCL11. Furthermore, the compound caused significant and sustained internalization of CXCR3 receptors from the cell surface. Again, the compound was able to direct cell migration of T cells over a bare filter (comparable to previous studies [14, 129]). Interestingly, and in contrast to CXCL11, **138** was not able to induce transendothelial migration, but instead antagonized the migration of T cells towards CXCL11 but also CXCL12 and CCL5. The latter two are ligands for CXCR4 and CCR5 receptors, respectively, also expressed on activated T cells [130]. The authors suggest that the inability of **138** to induce transendothelial migration is due to the lack of glycosaminoglycan (GAG) binding, which is probably needed for a ligand concentration gradient serving as a vectorial cue for the immune cells, also shown with a study on a non-GAG-binding mutant of CXCL12 [131]. However, it cannot be ruled out that these ligands are functionally selective agonists that are less efficacious in activating cell migration compared to chemokines. As **138** does not appear to bind to the murine CXCR3 receptor, human T cells were introduced in NOD.Cg-*Prkdc*$^{scid}$ Il2rgtm1$^{Wjl/SzJ}$ mice, as a model to mimic human arthritic inflammation [130]. In this model, migration of T cells was observed towards air pouches injected with solutions containing CXCL11 alone or synovial fluid of rheumatoid arthritis patients (RASF), containing a broad spectrum of chemokines. Interestingly, **138** significantly antagonized migration towards both solutions, whereas selective CXCR3 antagonism by **11** (NBI-74330) or a CXCR3-blocking antibody was not able to block migration towards RASF. In the article, O'Boyle and colleagues show that next to CXCR3 desensitization, cross-desensitization of other chemokine receptors on the human T cells, including CCR5, is likely the mechanism of action for this small-molecule CXCR3 agonist that produces a functional antagonistic response in vivo [130]. The work by O'Boyle presents an interesting new avenue for the treatment of CXCR3-linked disease and potentially for immune diseases in general. Instead of pursuing selective chemokine receptor antagonists, selective small-molecule agonists might be developed that functionally antagonize the chemokine-induced responses by receptor (cross-)desensitization and internalization, without producing a migratory response themselves.

## 3.2 *Biaryl Ammonium Salt Agonists*

Our CXCR3 biaryl ammonium antagonist class [122] (vide supra) harbored a very subtle agonism trigger [132]. This represented a clear departure from the pioneer agonists **135**, **137** and **138** which were all peptidomimetic in nature and needed, e.g., the basic amino acid substructure. Our ammonium chemotype, in contrast, does not possess any peptidomimetic character. The trigger revealed itself when the *ortho*-position of the "outer" aryl ring was probed with a Cl-substituent (Table 22) [132]. The resulting compound **139** proved to be a partial agonist ($pEC_{50} = 5.8$, $\alpha = 0.73$, [$^{35}$S]-GTP$\gamma$S assay). Since the analogous *meta*-compound **108** did not show any agonism, we investigated the SFR of the biaryl substructure using 26 compounds. The focus was on the *ortho*-position of the "outer" ring but also that of the "inner" ring was probed. Interestingly, though, regioisomeric compound **140** did not show agonism. Collectively, variation of $R^1$ led to the whole spectrum of efficacies, but only a large halogen atom on the *ortho*-position of the "outer" ring provided *full agonists* of CXCR3, as illustrated by **141** (VUF11222, $pEC_{50} = 6.1$, $\alpha = 0.95$, [$^{35}$S]-GTP$\gamma$S assay) and **142** (VUF11418, $pEC_{50} = 6.0$, $\alpha = 0.99$, [$^{35}$S]-GTP$\gamma$S assay). Once again, shifting this large halogen atom to the *meta*-position abolished all agonism while reasonably maintaining affinity (**143**). Stereochemical and regiochemical exploration of the myrtenyl moiety did not qualitatively change these results. All this clearly underscores the *ortho*-position of the "outer" ring as the activity switch. We have since confirmed this switch with a thiophene as "outer" ring as well (Wijtmans et al., unpublished results). Some efforts were directed towards elucidating the switch using a combination of QSAR, QM, and NOESY NMR techniques [132]. This analysis suggests key roles for a dihedral angle within the biaryl system of ca. 60° and for appropriate electrostatic potential of the biaryl rings.

Compounds **141** and **142** were investigated in more detail, mostly with the *meta*-chloro antagonist (**108**) as in-class reference. Both **141** and **142** showed agonist

**Table 22** Exploration of polycycloaliphatic ammonium salt agonists by Wijtmans et al. [132]

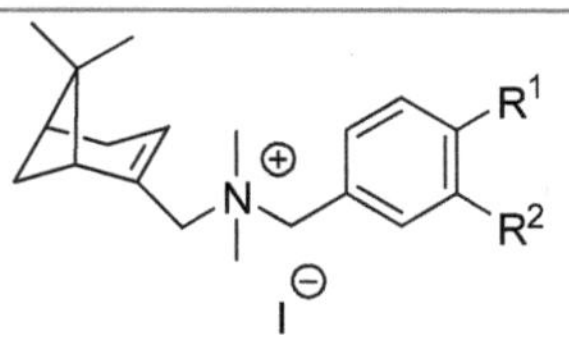


| Compound | | | | p$K_i$ | | |
|---|---|---|---|---|---|---|
| # | In ref. | $R^1$ | $R^2$ | CXCL10[a] | $pEC_{50}$[b] | α[c] |
| **139** | 6 | 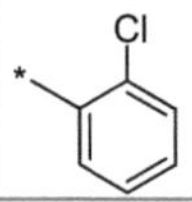 | H | 7.0 | 5.8 | 0.73 |
| **108** | 27 | 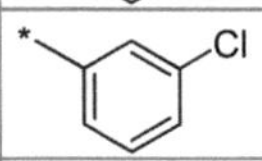 | H | 6.6 | –[d] | 0.05 |
| **140** | 31 | 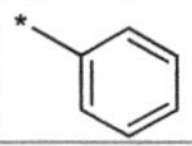 | Cl | 6.2 | –[d] | 0.06 |
| **141** | 38 | 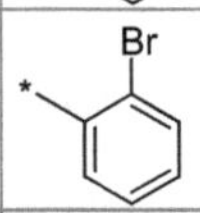 | H | 7.2 | 6.1 | 0.95 |
| **142** | 39 | 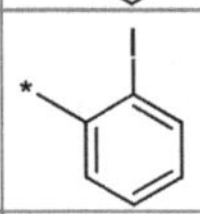 | H | 7.2 | 6.0 | 0.99 |
| **143** | 40 | 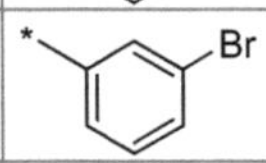 | H | 6.7 | –[d] | 0.08 |

[a] [$^{125}$I]-labeled CXCL10 displacement assay performed in HEK293 cells expressing human CXCR3

[b] [$^{35}$S]-GTPγS functional assay with membranes prepared from HEK293 cells stably expressing the CXCR3 receptor

[c] α represents the relative efficacy of a ligand compared to the endogenous agonist CXCL11 (which is set at α = 1.0)

[d] Could not be determined due to the too low functional assay window

responses in a second functional assay: a cAMP-dependent CRE-luciferase reporter gene assay, whereas **108** did not [132]. Selective CXCR3 antagonist **11** (NBI-74330) inhibited these agonist responses. Moreover, no effects were observed in a [$^{35}$S]-GTPγS assay for these compounds on cells lacking CXCR3 expression. Altogether these data indicate that the induced responses are specifically mediated by the activation of CXCR3 receptors present on these cells. Preliminary mutagenesis studies indicate a potential binding mode for these compounds within the TM region of CXCR3 (Scholten et al., unpublished data).

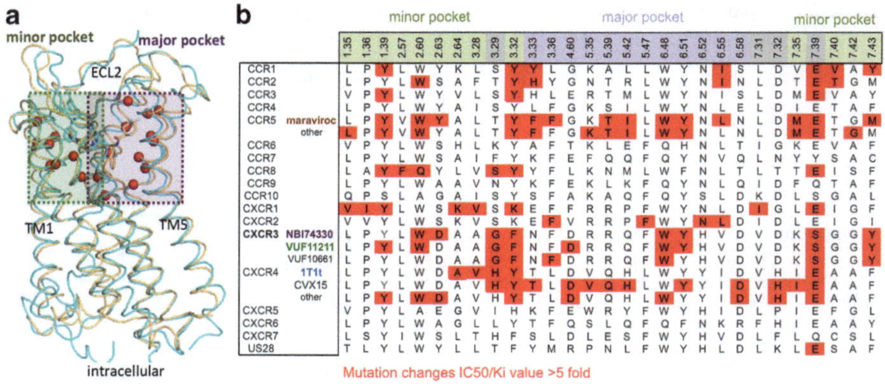


**Fig. 2** (**a**) Alignment of GPCR crystal structures, including structures of the CXCR4 (*cyan*) and CCR5 (*orange*) chemokine receptors [83, 84], highlighting the positions of amino acid residues (Cα atoms depicted by *red spheres*) that play a role in ligand binding of chemokine receptors based on mutation studies [20, 80], as presented in more detail in panel. (**b**) Alignment of residues in the transmembrane binding pockets of chemokine receptors (enumerated according to the Ballesteros-Weinstein residue numbering scheme [133]). Residues lining the minor pocket (TMS1), major pocket (TMS2), and interface are marked *green*, *purple*, and *gray*, respectively. Residues are marked red per receptor when mutation of that particular residue is reported to affect affinity or antagonism of any ligand. CCR5 residues that interact with maraviroc and CXCR4 residues that interact with 1T1t and CVX15 in crystal structures are marked red on additional lines, while CXCR3 mutation effects are presented for **11** (NBI74330), **71** (VUF11211), and **137** (VUF10661) individually

# 4 CXCR3-Ligand Binding: From GPCR X-Rays to Presumed CXCR3 Binding Modes

## *4.1 GPCR Crystal Structures and Chemokine Receptor-Ligand Interaction Modeling*

After the first GPCR crystal structure of bovine rhodopsin in 2000, the first crystal structures of druggable GPCRs have been solved only in the past 7 years [81, 82]. The three-dimensional structures of 25 different GPCRs have been determined, including the CXCR4 and CCR5 chemokine receptors [83, 84] (Figs. 2a and 3a, b), members of other class A GPCR subfamilies (opioid, aminergic, peptide, adenosine, and lipid receptors) [81, 82], and recently also the first crystal structures of class B [134], C [135], and F [136] GPCRs. These GPCR crystal structures give new opportunities to push the limits of structure-based rational ligand discovery and design and offer higher resolution templates for modeling the structures of GPCRs for which crystal structures have not yet been solved [137]. It should however be noted that modeling of GPCRs with low homology to the currently available GPCR crystal structures still remains a difficult task in which experimental data are of utmost importance to restrict the number of possible models. Several of the challenges of GPCR structural modeling have been demonstrated in recent community-wide competitions to predict GPCR crystal structures (GPCR DOCK [138, 139], including modeling challenge to predict ligand-bound crystal structures of the CXCR4 chemokine receptor [139]), and GPCR structure modeling methods and applications have been described in recent reviews [79, 137].

CXCR4 crystal structures have been elucidated with a large cyclic peptide CVX15 and with the small-molecule antagonist IT1t [83] (Fig. 3a), while recently a crystal structure complex of CCR5 and the small-molecule HIV entry inhibitor maraviroc was solved (Fig. 3b) [84]. The CXCR4 crystal structures have been solved as parallel dimers that interact at the extracellular side of helices V and VI [83], in a similar way as observed in μ-opioid receptor crystallized dimers [140]. Comparison of other dimer/tetramer GPCR crystal structures however suggests the existence of different dimer interfaces for different GPCR homodimers, and complementary biochemical and biophysical studies indicate that GPCR oligomerization interfaces may depend on receptor conformations that can be stabilized by specific ligands [19]. While the overall seven transmembrane helical fold is conserved between GPCRs, the CCR5 and CXCR4 chemokine receptor crystal structures show differences compared to other GPCRs, including a more outward orientated second extracellular loop (ECL2) and a different conformation of the top of transmembrane (TM) helix 2. Chemokine receptors (as well as opioid receptors) contain a $S/T^{2.56}XP^{2.58}$ sequence motif that stabilizes a different helical kink in TM2 compared to other GPCR crystal structures and orients residues 2.60 and 2.63 ($W94^{2.60}$ and $D97^{2.63}$ in CXCR4 and $W86^{2.60}$ and $Y89^{2.63}$ in CCR5) towards the minor ligand binding site TMS1 [20]. These structural differences create a wider, more open ligand binding site between TM1, 2, 3, and 7 (TMS1 or "minor pocket") [20] in chemokine receptors compared to most other class A GPCRs. While the ligands in most other class A GPCR co-crystal structures primarily occupy a "major pocket" surrounded by TM3, 4, 5, 6, and 7 (TMS2) [20], the CXCR4 and CCR5 crystal structures show that small chemokine receptor modulators can target TMS1 exclusively (CXCR4-IT1t, Fig. 3a) [83] or TMS1 and TMS2 simultaneously (CCR5-maraviroc, Fig. 3b) [84].

The GPCR DOCK 2010 competition demonstrated that the computational prediction of chemokine receptor-ligand interactions [139] is particularly challenging because of the existence of multiple potential binding sites and ligand binding modes in chemokine receptors [79, 141]. Furthermore the symmetry in both chemokine ligands and chemokine receptor binding sites [20, 79] makes it difficult to prioritize plausible ligand binding mode hypotheses, even when experimental ligand structure-activity relationship (SAR) and receptor mutagenesis data are available. The CXCR4 and CCR5 crystal structures (Fig. 3a, b) and site-directed mutagenesis studies (Fig. 2b) indicate that acidic residues (e.g., $D/E^{2.63}$, $D/E^{4.60}$, $D/E^{6.58}$, $E^{7.39}$) and aromatic residues (e.g., $Y^{1.39}$, $W^{2.60}$, $Y/F^{3.32}$, $W^{6.48}$, $Y/F^{6.51}$, $Y^{7.43}$) present in the TMS1 and/or TMS2 of many chemokine receptors play important role in binding the basic and aromatic/hydrophobic moieties of small-molecule ligands [20, 79]. Mutation of these acidic and/or aromatic residues has an effect on chemokine binding to and/or potency for some but not all receptors. This suggests that small ligands and chemokines bind overlapping yet differential

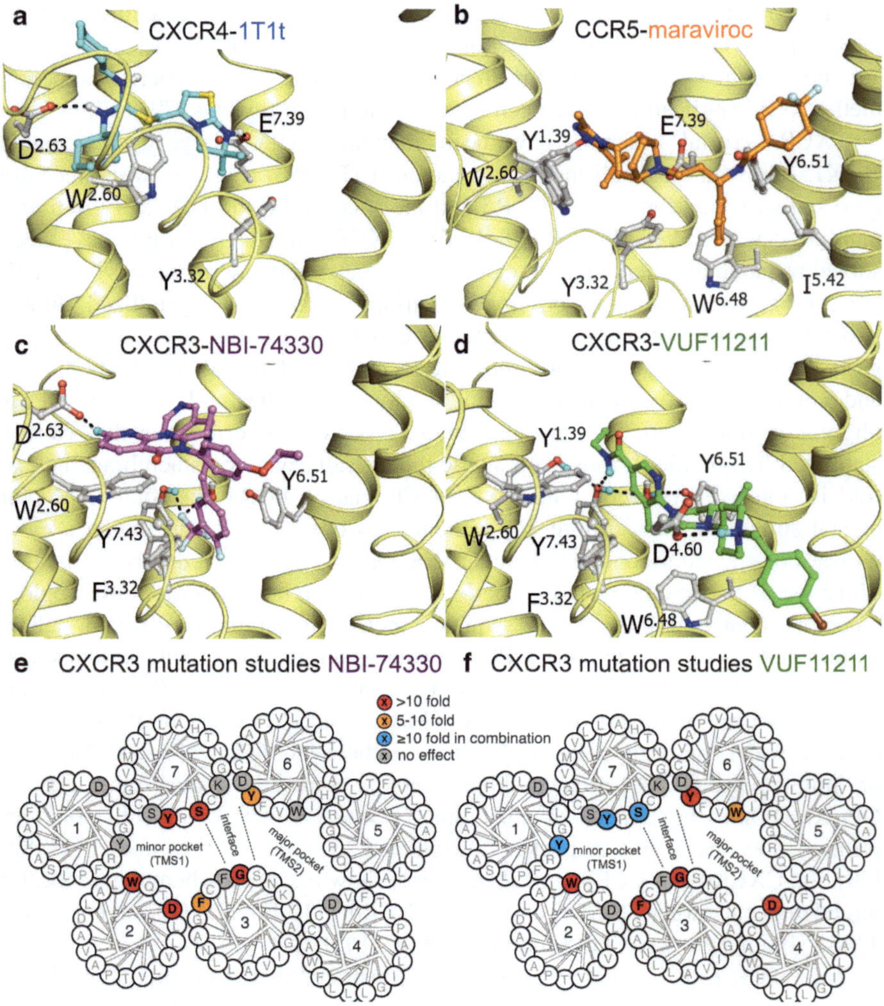


**Fig. 3** Comparison of binding modes of **145** (1 T1, *cyan*, **a**) in the CXCR4 crystal structure, maraviroc **144** (*orange*, **b**) in the CCR5 crystal structure, and **11** (NBI-74330, *magenta*, **c**) and **137** (VUF11211, *green*, **d**) bound CXCR3 homology models. TM helices around the TMS1 and TMS2 binding sites (see Fig. 2) are shown in *yellow*. Side chains of proposed interacting residues are shown in *gray*. Hydrogen bonds/polar interactions are shown as *dashed blue lines*. Helical wheel diagrams are shown for a top view of the TM domains of CXCR3 with effects of mutations highlighted for (**e**) **11** (NBI-74330) and (**f**) **137** (VUF11211). Residues that show a 10-fold or more decrease in affinity upon mutation are indicated in *red*, and mutations that result in a decrease in affinity between 5- and 10-fold are indicated in *orange*. Residues that give a significant decrease (10-fold or more) in affinity when mutated together are shown in *blue*. Other residues that were mutated but did not give a significant change in affinity are colored *gray*

binding sites [20]. For example, while mutation of D97$^{2.63}$ and E288$^{7.39}$ affects CXCL12 binding to CXCR4 [142, 143], mutation of Y60$^{1.39}$, W109$^{2.60}$, D112$^{2.63}$, F131$^{3.32}$, D186$^{4.60}$, W268$^{6.48}$, Y271$^{6.51}$, D278$^{6.58}$, or Y308$^{7.43}$does *not* affect CXCL11 binding to CXCR3 [80].

The following paragraphs show how the combination of CXCR3 mutagenesis studies, CXCR3 ligand SAR, and computational modeling studies can be used to map CXCR3-ligand binding sites and predict the three-dimensional structure of CXCR3-ligand complexes (Sect. 4.2), and give a perspective on the use of such structural models for CXCR3 (structure-based) virtual screening (Sect. 4.3).

## 4.2 In Silico-Guided CXCR3 Mutation Studies to Elucidate CXCR3-Ligand Binding Modes

### 4.2.1 Chemokine and Small Ligand Binding Regions in CXCR3

Chimera and point mutation studies have shown that multiple extracellular domains of CXCR3 are required for chemokine binding and/or receptor activation, including the N-terminus, and regions in the second and three extracellular loops (ECL1-3) [24, 144]. M1-V16 and sulfated Y27 and Y29 in the N-terminus and D282 in ECL3 are required for both CXCL10 and CXCL11 binding, while charged residues in the top of TM2 (D112$^{2.63}$), ECL2 (D195, E196), the top of TM6 (D278$^{6.58}$), ECL3 (E293), and the top of TM7 (D297$^{7.32}$) are required only for CXCL10 binding, but not CXCL11 binding [24, 129, 144]. Recent mutagenesis studies guided by CXCR4 crystal structure-based CXCR3 homology models have shown that residues that play a role in binding of small ligands **11** (NBI-74330), **71** (VUF11211), and **137** (VUF10661) are primarily located in the TM binding site (Fig. 3c, d). Almost all mutations in TMS1 and TMS2 that affect binding of small ligands do not significantly affect CXCL11 affinity [80], suggesting that the TM domains do not play an important role in CXCL11 binding. It should be noted however that chemokines are considerably larger and that most of their binding affinity is determined by interactions with the N-terminus and ECLs of the receptor [20]. The N-terminus of the chemokine is thought to interact with the receptor TM bundle for receptor activation [20], and its binding site may overlap with the binding pockets of small molecules in TMS1 and TMS2. Moreover, alignment of the peptidomimetic CXCR3 agonists **135** and **138** with the CXCL10 chemokine and mutation studies suggest that these small molecules mimic the 30s loop of CXCL10 and target both ECL2 and TMS1 [129, 145]. In general, radiolabeled chemokines are used to investigate the effect of mutations on allosteric ligand affinity. However, the effect on allosteric ligand binding is most likely a combination of both ligand affinity and allosteric cooperativity towards the chemokine radioligand. This cooperativity might also change depending on the specific mutation, potentially resulting in under- or overestimated influences of the mutation on the binding affinity of small-molecule ligands.

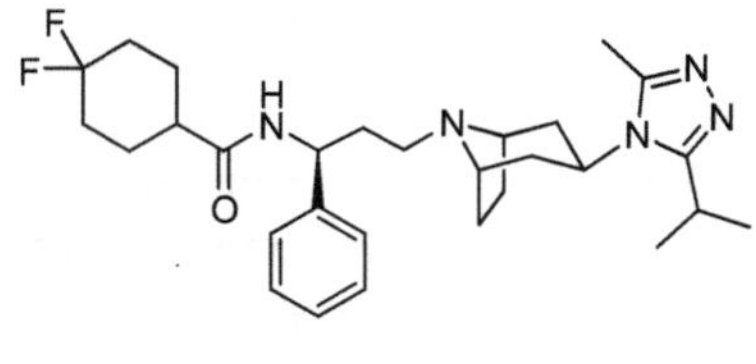

**144**
Maraviroc

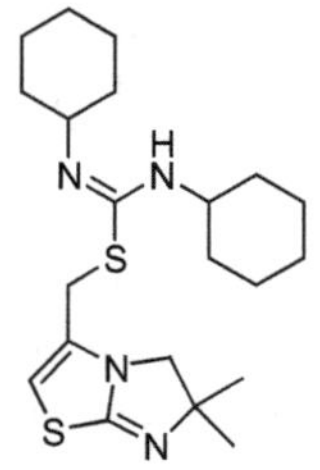

**145**
IT1t

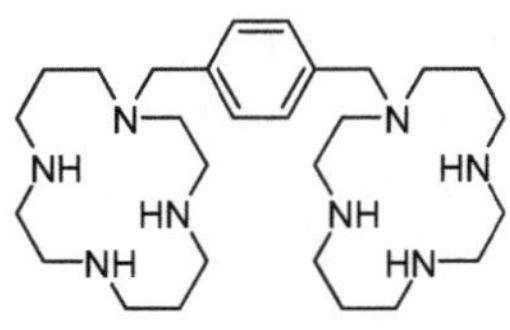

**146**
AMD3100

### 4.2.2 CXCR3 Ligand-Specific Anionic Interaction Sites

Maraviroc (**144**)-bound CCR5 and 1T1t (**145**)-bound CXCR4 crystal structures [83, 84] as well as CCR1, CCR2, CCR3, CCR5, CCR8, CXCR1, CXCR4, and US28 mutation studies [20] (Fig. 2b) show that $E^{7.39}$ acts as an important acidic ionic anchor for the basic moieties of small ligands in most chemokine receptors. Unlike other chemokine receptors, CXCR3 does not contain an acidic (glutamate) residue at position 7.39 in TM7, but a small polar $S304^{7.39}$ serine residue (Fig. 2b). Mutation of $S304^{7.39}$ into a glutamate residue (in combination with the $K300^{7.35}A$ mutation) results in a significant increase in affinity of the CXCR4 ligand AMD-3100 (**146**) for CXCR3, demonstrating its role in CXCR3/CXCR4 ligand selectivity [146]. In addition to several acidic residues in N-terminal region and extracellular loops ECL1, ECL2, and ECL3, there are four acidic residues in the TM helical binding site of CXCR3: $D112^{2.63}$ in TM2, $D186^{4.60}$ in TM4, $D278^{6.58}$ in TM6, and $D297^{7.32}$ in TM7. $D112^{2.63}$ is specific for CXCR3, CXCR4 ($D97^{2.63}$), and CXCR5 ($E308^{2.63}$) and forms an ionic interaction with 1T1t in one of the CXCR4 crystal structures (Fig. 3a). $D186^{4.60}$ and $D278^{6.58}$ are present in several chemokine receptors and form ionic interactions with basic moieties of the CVX15 peptide ligand in the other CXCR4 crystal structure [83].

The important role of basic nitrogen atoms in piperazinyl-piperidines (Sect. 2.3) in CXCR3 binding has been demonstrated in SAR studies [111] (e.g., **61** vs. **65**, Tables 10 and 11), and recent CXCR3 mutation studies guided by a CXCR4 crystal structure-based CXCR3 homology model [80] have identified $D186^{4.60}$ as an ionic interaction site of the piperidine moiety of **71** (VUF11211) (Fig. 3d, f). Piperidinyl diazepanone (Sect. 3.1.1) and tetrahydroisoquinoline (Sect. 3.1.2) peptidomimetic agonist ligands contain a basic amino acid that is required for CXCR3 binding [125], and mutation studies have indicated that E196 in ECL2 plays a role in CXCR3 binding of **135** [129] and **137** (VUF10661) [147]. Azaquinazolinone **11** (NBI-74330) does not possess a highly basic moiety, but SAR studies show (Sect. 2.1) that the 8-azaquinazolinone nitrogen atoms and associated positive partial charge on the 7-position are important for CXCR3 binding affinity (e.g., see **11** vs. **17** and **18**, Table 3) [87, 98]. CXC3 homology model-guided mutation studies indicate that the negatively charged carboxylate group of

D112$^{2.63}$ plays an important role in **11** (NBI-74330) and suggest that an electropositive aromatic –CH group of the ligand forms a weak hydrogen bond to this residue (Fig. 3c, e) [80]. Similar H-bonds between N-heteroaromatic –CH groups and oxygen atoms play, for example, an important role in intermolecular bonding of N-heteroaromatic ring systems [148] and kinase-ligand interactions (i.e., CH–O hydrogen bonds with hinge backbone carbonyl oxygen atoms) [149, 150]. D46N, D52$^{1.31}$ N, D195N, D278$^{6.58}$ N, E293$^{7.28}$Q, and D297$^{7.32}$ N mutations do *not* significantly affect binding of any of the abovementioned small ligands [80]. The ligand-specific roles of different acidic residues in CXCR3 demonstrate the ligand binding mode diversity in chemokine receptors and make it challenging to predict the (main) anionic interaction sites of the basic moieties of other CXCR3 ligands, e.g., aryl-3-piperidin-4-yl-ureas (Sect. 2.2), ergolines (Sect. 2.4), iminobenzimidazoles (Sect. 2.5), polycycloaliphatic aminopiperidines (Sect. 2.6), benzetimides (Sect. 2.7.1), and polycycloaliphatic ammonium salts (Sects. 2.6 and 3.2).

### 4.2.3 Overlapping, but Differential (Aromatic) Binding Pockets for Different CXCR3 Ligand Chemotypes

Several conserved aromatic residues line the minor (Y$^{1.39}$, W$^{2.60}$, Y/F$^{3.32}$, Y$^{7.43}$) and major (W$^{6.48}$, Y/F$^{6.51}$) subpockets of CXCR3 and other chemokine receptors (Figs. 2 and 3). Mutation studies indicate that the roles of these residues in CXCR3 binding are ligand dependent. W109$^{2.60}$, F131$^{3.32}$, and Y308$^{7.43}$ are important residues for binding all three CXCR3 ligands **11** (NBI-74330), **71** (VUF11211), and **137** (VUF10661). Mutation of Y271$^{6.51}$ on the other hand only affects CXCR3 binding affinity for **11** and **71** (but does not affect **137** affinity), while mutation of W268$^{6.48}$ only affects binding affinity for **71** (but does not affect **11** or **137** affinity), and point mutation of Y60$^{1.39}$ and F135$^{3.36}$ only affect binding affinity for **137** (but do not affect **11** or **71** affinity) [80, 147]. These and other (see Sect. 4.2.1) ligand-dependent mutational effects (Fig. 3e, f) and ligand SAR data (vide infra) suggest that these small CXCR3 ligands have overlapping but differential binding modes (Fig. 3c, d). While **11** (NBI-74330) primarily occupies the minor binding pocket and minor and major pocket interface between TM2, TM3, TM6, and TM7 (Fig. 3c, e), combined mutation and modeling studies suggest that **71** (VUF11211) occupies both minor and major pockets between TM1-7 (Fig. 3d, f). For both ligands, the G128$^{3.29}$ mutant diminishes CXCR3 binding affinity, suggesting that the binding site volume at the interface between major and minor pockets is restricted. This binding mode is in line with SAR studies that show that rigidification of the benzyl moiety of **71** either by ring closure or intramolecular hydrogen bonding could maintain ligand affinity, indicating the importance of directionality for the chlorobenzyl moiety [111, 114]. The tight fit of the rigid **71** ligand in the CXCR3 pocket (Fig. 3d) furthermore explains the steep SAR and preference for small apolar substituents over larger or polar substituents of the piperazine ring (e.g., **68** vs. **69–70**, Table 11) [111, 112] that are proposed to bind in a small subpocket between TM5 and TM6. SAR studies have also identified hydrophobic groups in other ligand chemotypes that are important for CXCR3 binding, including the benzene substituent of imidazole **30** (vs. **29**, Table 5), the adamantyl group of polycyclic aliphatic aminopiperidine **98** (vs. **99**, Table 16), the cyclopropyl substituent of imidazopyrazine **37** (vs. **36**, Table 6), and the ethyl substituent of benzimidazole **92** (vs. **91**, Table 14), that may target similar hydrophobic subpockets in CXCR3 as **11** and/or **71**.

The recent CCR5 crystal structure shows that the hydroxyl groups of conserved tyrosine residues $Y^{1.39}$ and $Y^{6.51}$ can also form H-bond interactions with polar functional groups in the ligand (Fig. 3b). CXCR3 modeling studies in combination with CXCR3 mutation and ligand SAR studies suggest that these tyrosine aromatic residues may form a H-bond network with $S304^{7.39}$ and $Y308^{7.43}$ and can act as (alternative) H-bond interaction partners of polar functional groups in CXCR3 ligands **11** and **71** (Fig. 3c–f). SAR studies indicate that the (geometry of the) amide moiety of **71** (e.g., **58** vs. **59**, Table 10) and electron-withdrawing character of the trifluoromethyl group of **11** (e.g., **6** vs. **7**, Table 1) are important determinants for CXCR3 binding, and the CXCR3 models suggest that these functional groups may interact with this H-bond network in the CXCR3 binding site. Electron-withdrawing groups are also important determinants of CXCR3 binding by piperidine ureas (e.g., **54** vs. **53**, Table 9).

## *4.3 Perspectives for (Structure-Based) Virtual Screening for CXCR3 Ligands*

As described in Sects. 4.1 and 4.2, refined chemokine receptor models have successfully been used to guide site-directed mutagenesis studies and design new compounds. Despite the challenges in chemokine receptor-ligand modeling [79, 139], customized chemokine homology models based on bRho and ADRB2 crystal structures as well as de novo receptor models have already been successfully used to identify new ligands for several chemokine receptors [79], including CCR3 [151], CCR4 [152], CCR5 [153], and CXCR4 [154]. *Retrospective* virtual screening experiments have been used to validate CXCR3 homology models and virtual screening methods to discriminate known receptor ligands from decoy molecules with similar physicochemical properties [155]; *prospective* virtual screening studies to discover new CXCR3 ligands have *not* yet been reported. The recent crystal structures of CXCR4 [83, 156] and CCR5 [84] have opened up new opportunities for structure-based discovery and design novel chemokine receptor ligands, as exemplified by successful structure-based virtual screening studies against the CXCR4 crystal structure [156, 157] and CXCR4 crystal structure-based homology models of CXCR7 [158]. It should be noted that hit rates (i.e., the percentage of experimentally confirmed ligands among all tested in silico hits) and binding affinity and/or potency of hits identified in structure-based virtual studies for chemokine receptors are somewhat lower than the hit rates reported for other GPCRs, like aminergic receptors

ADA1A [159], ADRB2 [160], DRD3 [161, 162], and $H_1R$ [163]. This can on one hand be explained by the challenges in computer-aided prediction of chemokine receptor-ligand interactions compared to aminergic GPCRs, for which more crystal structure templates are available and the protein-ligand interaction binding mode is generally more well defined (including a key ionic interaction with the conserved $D^{3.32}$ residue [164]). On the other hand the binding sites of aminergic GPCRs are considered more druggable than the binding sites of chemokine receptors. The TM binding pockets of aminergic receptors contain a combination of hydrophilic and hydrophobic regions that are compatible with the features of drug-like small molecule and favor water displacement upon ligand binding, whereas the open binding region of chemokine receptors (Fig. 3), with its limited number of energetically unfavorable (unhappy) water molecules, is more challenging from a drug design perspective [165]. In addition to the minor and major TM binding sites (Fig. 2), alternative binding sites in chemokine receptors may be targeted by small molecules, including dimer interfaces and intracellular G protein binding site region (as, e.g., proposed for CXCR2 [166–168]). The identification of (chemokine) receptor selective regions in such alternative, relatively shallow, binding pockets (compared to the more occluded TM binding sites) for efficient ligand design is also expected to be difficult. Despite these challenges, the availability of more homologous structural templates, successful virtual screening campaigns for several chemokine receptors, and increased understanding of CXCR3-ligand binding (Sect. 4.2), has made structure-based virtual discovery and design of small CXCR3 modulators more and more feasible.

## 5 Conclusion

This review has addressed, from a molecular point of view, all the progress made in discovery and development of small-molecule CXCR3 ligands. The present efforts have made the full spectrum of efficacies (from antagonists to full agonists) within reach for detailed biological probing of the role of CXCR3 and its ligands. Antagonist papers amount to 28 total and describe chemotypes that vary widely. Most of these papers have been published by the pharmaceutical industry aiming to utilize CXCR3 to address therapeutic needs. While no clinical successes can be reported here yet, the studies represent a valuable arsenal of tool compounds that can be used to study the CXCR3 receptor. Small-molecule CXCR3 agonists have been much less described (only three chemotypes, two of which peptidomimetic), while they too are useful tools especially for emerging concepts such as biased agonism. A unified pharmacophore model for CXCR3 ligands seems challenging to construct as recent combined mutagenesis and CXCR3-ligand modeling studies indicate that the binding modes of different ligand chemotypes are different and only partially overlapping. Gratifyingly, the progress in GPCR crystal structural biology (including CCR5 and CXCR4 chemokine receptor crystal structures) and emerging 3D CXCR3-ligand interaction models have improved our understanding of ligand-dependent molecular determinants of CXCR3 binding. These new structural insights into CXCR3-ligand binding mode (diversity), in combination with (experimentally supported) virtual screening methods, can be used to guide future CXCR3 ligand discovery and design.

## References


1. Loetscher M, Gerber B, Loetscher P, Jones SA, Piali L, Clark-Lewis I, Baggiolini M, Moser B (1996) Chemokine receptor specific for IP10 and mig: structure, function, and expression in activated T-lymphocytes. J Exp Med 184(3):963–969. doi:10.1084/jem.184.3.963
2. Qin S, Rottman JB, Myers P, Kassam N, Weinblatt M, Loetscher M, Koch AE, Moser B, Mackay CR (1998) The chemokine receptors CXCR3 and CCR5 mark subsets of T cells associated with certain inflammatory reactions. J Clin Invest 101(4):746–754. doi:10.1172/JCI1422
3. Bonecchi R, Bianchi G, Bordignon PP, D'Ambrosio D, Lang R, Borsatti A, Sozzani S, Allavena P, Gray PA, Mantovani A, Sinigaglia F (1998) Differential expression of chemokine receptors and chemotactic responsiveness of type 1 T helper cells (Th1s) and Th2s. J Exp Med 187(1):129–134. doi:10.1084/jem.187.1.129
4. Liu L, Callahan MK, Huang D, Ransohoff RM (2005) Chemokine receptor CXCR3: an unexpected enigma. Curr Top Dev Biol 68:149–181. doi:10.1016/S0070-2153(05)68006-4
5. Cole KE, Strick CA, Paradis TJ, Ogborne KT, Loetscher M, Gladue RP, Lin W, Boyd JG, Moser B, Wood DE, Sahagan BG, Neote K (1998) Interferon-inducible T cell alpha chemoattractant (I-TAC): a novel non-ELR CXC chemokine with potent activity on activated T cells through selective high affinity binding to CXCR3. J Exp Med 187(12):2009–2021. doi:10.1084/jem.187.12.2009
6. Weng Y, Siciliano SJ, Waldburger KE, Sirotina-Meisher A, Staruch MJ, Daugherty BL, Gould SL, Springer MS, DeMartino JA (1998) Binding and functional properties of recombinant and endogenous CXCR3 chemokine receptors. J Biol Chem 273(29):18288–18291. doi:10.1074/jbc.273.29.18288
7. Tensen CP, Flier J, Van Der Raaij-Helmer EM, Sampat-Sardjoepersad S, Van Der Schors RC, Leurs R, Scheper RJ, Boorsma DM, Willemze R (1999) Human IP-9: a keratinocyte-derived high affinity CXC-chemokine ligand for the IP-10/Mig receptor (CXCR3). J Invest Dermatol 112(5):716–722. doi:10.1046/j.1523-1747.1999.00581.x
8. Loetscher M, Loetscher P, Brass N, Meese E, Moser B (1998) Lymphocyte-specific chemokine receptor CXCR3: regulation, chemokine binding and gene localization. Eur J Immunol 28 (11):3696–3705. doi:10.1002/(SICI)1521-4141(199811)28:11<3696::AID-IMMU3696>3.0.CO;2-W
9. Jenh CH, Cox MA, Hipkin W, Lu T, Pugliese-Sivo C, Gonsiorek W, Chou CC, Narula SK, Zavodny PJ (2001) Human B cell-attracting chemokine 1 (BCA-1; CXCL13) is an agonist for the human CXCR3 receptor. Cytokine 15(3):113–121. doi:10.1006/cyto.2001.0923
10. Mueller A, Meiser A, McDonagh EM, Fox JM, Petit SJ, Xanthou G, Williams TJ, Pease JE (2008) CXCL4-induced migration of activated T lymphocytes is mediated by the chemokine receptor CXCR3. J Leukocyte Biol 83:875–882. doi: 10.1189/jlb.1006645
11. Lasagni L, Francalanci M, Annunziato F, Lazzeri E, Giannini S, Cosmi L, Sagrinati C, Mazzinghi B, Orlando C, Maggi E, Marra F, Romagnani S, Serio M, Romagnani P (2003) An alternatively spliced variant of CXCR3 mediates the inhibition of endothelial cell growth induced by IP-10, Mig, and I-TAC, and acts as functional receptor for platelet factor 4. J Exp Med 197(11):1537–1549. doi: 10.1084/jem.20021897
12. Colvin RA, Campanella GSV, Sun JT, Luster AD (2004) Intracellular domains of CXCR3 that mediate CXCL9, CXCL10, and CXCL11 function. J Biol Chem 279(29):30219–30227

13. Smit MJ, Verdijk P, van der Raaij-Helmer EMH, Navis M, Hensbergen PJ, Leurs R, Tensen CP (2003) CXCR3-mediated chemotaxis of human T cells is regulated by a G(i)- and phospholipase C-dependent pathway and not via activation of MEK/p44/p42 MAPK nor Akt/PI-3 kinase. Blood 102(6):1959–1965. doi: 10.1182/blood-2002-12-3945
14. Scholten DJ, Canals M, Wijtmans M, de Munnik S, Nguyen P, Verzijl D, de Esch IJ, Vischer HF, Smit MJ, Leurs R (2012) Pharmacological characterization of a small-molecule agonist for the chemokine receptor CXCR3. Br J Pharmacol 166(3):898–911. doi:10.1111/j.1476-5381.2011.01648.x
15. Canals M, Scholten DJ, de Munnik S, Han MK, Smit MJ, Leurs R (2012) Ubiquitination of CXCR7 controls receptor trafficking. PLoS One 7(3):e34192. doi:10.1371/journal.pone.0034192
16. Dagan-Berger M, Feniger-Barish R, Avniel S, Wald H, Galun E, Grabovsky V, Alon R, Nagler A, Ben-Baruch A, Peled A (2006) Role of CXCR3 carboxyl terminus and third intracellular loop in receptor-mediated migration, adhesion and internalization in response to CXCL11. Blood 107(10):3821–3831. doi:10.1182/blood-2004-01-0214
17. Meiser A, Mueller A, Wise EL, McDonagh EM, Petit SJ, Saran N, Clark PC, Williams TJ, Pease JE (2008) The chemokine receptor CXCR3 is degraded following internalization and is replenished at the cell surface by de novo synthesis of receptor. J Immunol 180(10):6713–6724
18. Luttrell LM, Gesty-Palmer D (2010) Beyond desensitization: physiological relevance of arrestin-dependent signaling. Pharmacol Rev 62(2):305–330. doi:10.1124/pr.109.002436
19. Ferre S, Casado V, Devi LA, Filizola M, Jockers R, Lohse MJ, Milligan G, Pin JP, Guitart X (2014) G protein-coupled receptor oligomerization revisited: functional and pharmacological perspectives. Pharmacol Rev 66(2):413–434. doi:10.1124/pr.113.00805266/2/413
20. Scholten DJ, Canals M, Maussang D, Roumen L, Smit MJ, Wijtmans M, de Graaf C, Vischer HF, Leurs R (2012) Pharmacological modulation of chemokine receptor function. Br J Pharmacol 165(6):1617–1643. doi:10.1111/j.1476-5381.2011.01551.x
21. Vischer HF, Nijmeijer S, Smit MJ, Leurs R (2008) Viral hijacking of human receptors through heterodimerization. Biochem Biophys Res Commun 377 (1):93–97. doi:10.1016/j.bbrc.2008.09.082
22. Watts AO, van Lipzig MM, Jaeger WC, Seeber RM, van Zwam M, Vinet J, van der Lee MM, Siderius M, Zaman GJ, Boddeke HW, Smit MJ, Pfleger KD, Leurs R, Vischer HF (2013) Identification and profiling of CXCR3-CXCR4 chemokine receptor heteromer complexes. Br J Pharmacol 168(7):1662–1674. doi:10.1111/bph.12064
23. Cox MA, Jenh CH, Gonsiorek W, Fine J, Narula SK, Zavodny PJ, Hipkin RW (2001) Human interferon-inducible 10-kDa protein and human interferon-inducible T cell alpha chemoattractant are allotopic ligands for human CXCR3: differential binding to receptor states. Mol Pharmacol 59(4):707–715. doi: 10.1124/mol.59.4.707
24. Xanthou G, Williams TJ, Pease JE (2003) Molecular characterization of the chemokine receptor CXCR3: evidence for the involvement of distinct extracellular domains in a multi-step model of ligand binding and receptor activation. Eur J Immunol 33(10):2927–2936. doi:10.1002/eji.200324235
25 Watts AO, Scholten DJ, Heitman LH, Vischer HF, Leurs R (2012) Label-free impedance responses of endogenous and synthetic chemokine receptor CXCR3 agonists correlate with Gi-protein pathway activation. Biochem Biophys Res Commun 419(2):412–418. doi:10.1016/j.bbrc.2012.02.036
26. Kouroumalis A, Nibbs RJ, Aptel H, Wright KL, Kolios G, Ward SG (2005) The chemokines CXCL9, CXCL10, and CXCL11 differentially stimulate G alpha i-independent signaling and actin responses in human intestinal myofibroblasts. J Immunol 175(8):5403–5411. doi:10.4049/jimmunol.175.8.5403
27. Mantovani A (1999) The chemokine system: redundancy for robust outputs. Immunoly Today 20(6):254–257. doi:10.1016/S0167-5699(99)01469-3

28. Allen SJ, Crown SE, Handel TM (2007) Chemokine: receptor structure, interactions, and antagonism. Annu Rev Immunol 25:787–820. doi:10.1146/annurev.immunol.24.021605.090529
29. Clark-Lewis I, Mattioli I, Gong JH, Loetscher P (2003) Structure-function relationship between the human chemokine receptor CXCR3 and its ligands. J Biol Chem 278(1): 289–295. doi:10.1074/jbc.M209470200
30. Hasegawa H, Inoue A, Kohno M, Muraoka M, Miyazaki T, Terada M, Nakayama T, Yoshie O, Nose M, Yasukawa M (2006) Antagonist of interferon-inducible protein 10/CXCL10 ameliorates the progression of autoimmune sialadenitis in MRL/lpr mice. Arthrit Rheumat 54(4):1174–1183. doi: 10.1002/art.21745
31. Hensbergen PJ, van der Raaij-Helmer EMH, Dijkman R, van der Schors RC, Werner-Felmayer G, Boorsma DM, Scheper RJ, Willemze R, Tensen CP (2001) Processing of natural and recombinant CXCR3-targeting chemokines and implications for biological activity. Eur J Biochem 268(18):4992–4999. doi:10.1046/j.0014-2956.2001.02433.x
32. Proost P, Schutyser E, Menten P, Struyf S, Wuyts A, Opdenakker G, Detheux M, Parmentier M, Durinx C, Lambeir A-M, Neyts J, Liekens S, Maudgal PC, Billiau A, Van Damme J (2001) Amino-terminal truncation of CXCR3 agonists impairs receptor signaling and lymphocyte chemotaxis, while preserving antiangiogenic properties. Blood 98 (13):3554–3561. doi:10.1182/blood.V98.13.3554
33. Mellado M, Rodriguez-Frade JM, Vila-Coro AJ, Fernandez S, Martin de Ana A, Jones DR, Toran JL, Martinez AC (2001) Chemokine receptor homo- or heterodimerization activates distinct signaling pathways. EMBO J 20(10):2497–2507. doi:10.1093/emboj/20.10.2497
34. Schall TJ, Proudfoot AE (2011) Overcoming hurdles in developing successful drugs targeting chemokine receptors. Nat Rev Immunol 11(5):355–363. doi:10.1038/nri2972
35. Rosenblum JM, Zhang Q-W, Siu G, Collins TL, Sullivan T, Dairaghi DJ, Medina JC, Fairchild RL (2009) CXCR3 antagonism impairs the development of donor-reactive, IFN-γ-producing effectors and prolongs allograft survival. Transplantation 87(3):360–369. doi:10.1097/TP.0b013e31819574e9
36. Romagnani P, Crescioli C (2012) CXCL10: a candidate biomarker in transplantation. Clinica Chimica Acta 413(17–18):1364–1373. doi:10.1016/j.cca.2012.02.009
37. Hancock WW, Lu B, Gao W, Csizmadia V, Faia K, King JA, Smiley ST, Ling M, Gerard NP, Gerard C (2000) Requirement of the chemokine receptor CXCR3 for acute allograft rejection. J Exp Med 192(10):1515–1520. doi:10.1084/jem.192.10.1515
38. Mach F, Sauty A, Iarossi AS, Sukhova GK, Neote K, Libby P, Luster AD (1999) Differential expression of three T lymphocyte-activating CXC chemokines by human atheroma-associated cells. J Clin Invest 104(8):1041–1050. doi:10.1172/JCI6993
39. Mohan K, Issekutz TB (2007) Blockade of chemokine receptor CXCR3 inhibits T cell recruitment to inflamed joints and decreases the severity of adjuvant arthritis. J Immunol 179(12):8463–8469. doi:10.4049/jimmunol.179.12.8463
40. Saetta M, Mariani M, Panina-Bordignon P, Turato G, Buonsanti C, Baraldo S, Bellettato CM, Papi A, Corbetta L, Zuin R, Sinigaglia F, Fabbri LM (2002) Increased expression of the chemokine receptor CXCR3 and its ligand CXCL10 in peripheral airways of smokers with chronic obstructive pulmonary disease. Am J Respir Crit Care Med 165(10):1404–1409. doi: 10.1164/rccm.2107139
41. Sorensen TL, Tani M, Jensen J, Pierce V, Lucchinetti C, Folcik VA, Qin S, Rottman J, Sellebjerg F, Strieter RM, Frederiksen JL, Ransohoff RM (1999) Expression of specific chemokines and chemokine receptors in the central nervous system of multiple sclerosis patients. J Clin Invest 103(6):807–815. doi: 10.1172/JCI5150
42. Enghard P, Humrich JY, Rudolph B, Rosenberger S, Biesen R, Kuhn A, Manz R, Hiepe F, Radbruch A, Burmester GR, Riemekasten G (2009) CXCR3+CD4+ T cells are enriched in inflamed kidneys and urine and provide a new biomarker for acute nephritis flares in systemic lupus erythematosus patients. Arthrit Rheumat 60(1):199–206. doi:10.1002/art.24136

43. Melter M, Exeni A, Reinders ME, Fang JC, McMahon G, Ganz P, Hancock WW, Briscoe DM (2001) Expression of the chemokine receptor CXCR3 and its ligand IP-10 during human cardiac allograft rejection. Circulation 104(21):2558–2564. doi:10.1161/ hc4601.098010
44. Kao J, Kobashigawa J, Fishbein MC, MacLellan WR, Burdick MD, Belperio JA, Strieter RM (2003) Elevated serum levels of the CXCR3 chemokine ITAC are associated with the development of transplant coronary artery disease. Circulation 107(15):1958–1961. doi:10.1161/01.CIR.0000069270.16498.75
45. Bauer JW, Baechler EC, Petri M, Batliwalla FM, Crawford D, Ortmann WA, Espe KJ, Li W, Patel DD, Gregersen PK, Behrens TW (2006) Elevated serum levels of interferon-regulated chemokines are biomarkers for active human systemic lupus erythematosus. PLoS Med 3(12):e491. doi:10.1371/journal.pmed.0030491
46. Lit LC, Wong CK, Tam LS, Li EK, Lam CW (2006) Raised plasma concentration and ex vivo production of inflammatory chemokines in patients with systemic lupus erythematosus. Ann Rheumat Dis 65(2):209–215. doi:10.1136/ard.2005.038315
47. Kawada K, Sonoshita M, Sakashita H, Takabayashi A, Yamaoka Y, Manabe T, Inaba K, Minato N, Oshima M, Taketo MM (2004) Pivotal role of CXCR3 in melanoma cell metastasis to lymph nodes. Cancer Res 64(11):4010–4017. doi:10.1158/0008-5472.CAN-03-1757
48. Walser TC, Rifat S, Ma XR, Kundu N, Ward C, Goloubeva O, Johnson MG, Medina JC, Collins TL, Fulton AM (2006) Antagonism of CXCR3 inhibits lung metastasis in a murine model of metastatic breast cancer. Cancer Res 66(15):7701–7707. doi:10.1158/0008-5472. CAN-06-0709
49. Kawada K, Hosogi H, Sonoshita M, Sakashita H, Manabe T, Shimahara Y, Sakai Y, Takabayashi A, Oshima M, Taketo MM (2007) Chemokine receptor CXCR3 promotes colon cancer metastasis to lymph nodes. Oncogene 26(32):4679–4688. doi:10.1038/sj.onc. 1210267
50. Murakami T, Kawada K, Iwamoto M, Akagami M, Hida K, Nakanishi Y, Kanda K, Kawada M, Seno H, Taketo MM, Sakai Y (2013) The role of CXCR3 and CXCR4 in colorectal cancer metastasis. Int J Cancer 132(2):276–287. doi:10.1002/ijc.27670
51. Gao P, Zhou XY, Yashiro-Ohtani Y, Yang YF, Sugimoto N, Ono S, Nakanishi T, Obika S, Imanishi T, Egawa T, Nagasawa T, Fujiwara H, Hamaoka T (2003) The unique target specificity of a nonpeptide chemokine receptor antagonist: selective blockade of two Th1 chemokine receptors CCR5 and CXCR3. J Leukocyte Biol 73(2):273–280. doi:10.1189/jlb. 0602269
52. van Wanrooij EJ, De Jager SC, van Es T, de Vos P, Birch HL, Owen DA, Watson RJ, Biessen EA, Chapman GA, van Berkel TJ, Kuiper J (2008) CXCR3 antagonist NBI-74330 attenuates atherosclerotic plaque formation in LDL receptor–deficient mice. Arterioscler Thromb Vasc Biol 28:251–257. doi:10.1161/ ATVBAHA.107.147827
53. Liu C, Luo D, Reynolds BA, Meher G, Katritzky AR, Lu B, Gerard CJ, Bhadha CP, Harrison JK (2011) Chemokine receptor CXCR3 promotes growth of glioma. Carcinogenesis 32(2): 129–137. doi:10.1093/carcin/bgq224
54. Jenh CH, Cox MA, Cui L, Reich EP, Sullivan L, Chen SC, Kinsley D, Qian S, Kim SH, Rosenblum S, Kozlowski J, Fine JS, Zavodny PJ, Lundell D (2012) A selective and potent CXCR3 antagonist SCH 546738 attenuates the development of autoimmune diseases and delays graft rejection. BMC Immunol 13(1):2. doi: 10.1186/1471-2172-13-2
55. Kakuta Y, Okumi M, Miyagawa S, Tsutahara K, Abe T, Yazawa K, Matsunami K, Otsuka H, Takahara S, Nonomura N (2012) Blocking of CCR5 and CXCR3 suppresses the infiltration of macrophages in acute renal allograft rejection. Transplantation 93(1):24–31. doi:10.1097/TP. 0b013e31823aa585
56. Baker MS, Chen X, Rotramel AR, Nelson JJ, Lu B, Gerard C, Kanwar Y, Kaufman DB (2003) Genetic deletion of chemokine receptor CXCR3 or antibody blockade of its ligand IP-10 modulates posttransplantation graft-site lymphocytic infiltrates and prolongs functional graft survival in pancreatic islet allograft recipients. Surgery 134(2):126–133. doi:10.1067/ msy.2003.213

57. Zerwes HG, Li J, Kovarik J, Streiff M, Hofmann M, Roth L, Luyten M, Pally C, Loewe RP, Wieczorek G, Banteli R, Thoma G, Luckow B (2008) The chemokine receptor Cxcr3 is not essential for acute cardiac allograft rejection in mice and rats. Am J Transplant 8(8): 1604–1613. doi: 10.1111/j.1600-6143.2008.02309.x
58. Kwun J, Hazinedaroglu SM, Schadde E, Kayaoglu HA, Fechner J, Hu HZ, Roenneburg D, Torrealba J, Shiao L, Hong X, Peng R, Szewczyk JW, Sullivan KA, DeMartino J, Knechtle SJ (2008) Unaltered graft survival and intragraft lymphocytes infiltration in the cardiac allograft of Cxcr3−/− mouse recipients. Am J Transplant 8(8):1593–1603. doi:10.1111/j.1600-6143.2008.02250.x
59. Pradelli E, Karimdjee-Soilihi B, Michiels JF, Ricci JE, Millet MA, Vandenbos F, Sullivan TJ, Collins TL, Johnson MG, Medina JC, Kleinerman ES, Schmid-Alliana A, Schmid-Antomarchi H (2009) Antagonism of chemokine receptor CXCR3 inhibits osteosarcoma metastasis to lungs. Int J Cancer 125(11):2586–2594. doi:10.1002/ijc.24665
60. Goldberg-Bittman L, Sagi-Assif O, Meshel T, Nevo I, Levy-Nissenbaum O, Yron I, Witz IP, Ben-Baruch A (2005) Cellular characteristics of neuroblastoma cells: regulation by the ELR–CXC chemokine CXCL10 and expression of a CXCR3-like receptor. Cytokine 29(3): 105–117. doi:10.1016/j.cyto.2004.10.003
61. Cambien B, Karimdjee BF, Richard-Fiardo P, Bziouech H, Barthel R, Millet MA, Martini V, Birnbaum D, Scoazec JY, Abello J, Al Saati T, Johnson MG, Sullivan TJ, Medina JC, Collins TL, Schmid-Alliana A, Schmid-Antomarchi H (2009) Organ-specific inhibition of metastatic colon carcinoma by CXCR3 antagonism. Br J Cancer 100(11):1755–1764. doi:10.1038/sj.bjc.6605078
62. Winkler AE, Brotman JJ, Pittman ME, Judd NP, Lewis JS Jr, Schreiber RD, Uppaluri R (2011) CXCR3 enhances a T-cell-dependent epidermal proliferative response and promotes skin tumorigenesis. Cancer Res 71(17):5707–5716. doi:10.1158/0008-5472.CAN-11-0907
63. Giuliani N, Bonomini S, Romagnani P, Lazzaretti M, Morandi F, Colla S, Tagliaferri S, Lasagni L, Annunziato F, Crugnola M, Rizzoli V (2006) CXCR3 and its binding chemokines in myeloma cells: expression of isoforms and potential relationships with myeloma cell proliferation and survival. Haematologica 91(11):1489–1497
64. Wu Q, Dhir R, Wells A (2012) Altered CXCR3 isoform expression regulates prostate cancer cell migration and invasion. Mol Cancer 11:3. doi:10.1186/1476-4598-11-3
65. Walser TC, Ma X, Kundu N, Dorsey R, Goloubeva O, Fulton AM (2007) Immune-mediated modulation of breast cancer growth and metastasis by the chemokine Mig (CXCL9) in a murine model. J Immunother 30(5):490–498. doi:10.1097/CJI.0b013e318031b551
66. Andersson A, Yang SC, Huang M, Zhu L, Kar UK, Batra RK, Elashoff D, Strieter RM, Dubinett SM, Sharma S (2009) IL-7 promotes CXCR3 ligand-dependent T cell antitumor reactivity in lung cancer. J Immunol 182(11):6951–6958. doi:10.4049/jimmunol.0803340
67. Wendel M, Galani IE, Suri-Payer E, Cerwenka A (2008) Natural killer cell accumulation in tumors is dependent on IFN-gamma and CXCR3 ligands. Cancer Res 68(20):8437–8445. doi:10.1158/0008-5472.CAN-08-1440
68. Ehlert JE, Addison CA, Burdick MD, Kunkel SL, Strieter RM (2004) Identification and partial characterization of a variant of human CXCR3 generated by posttranscriptional exon skipping. J Immunol 173(10):6234–6240. doi:10.4049/jimmunol.173.10.6234
69. Yates-Binder CC, Rodgers M, Jaynes J, Wells A, Bodnar RJ, Turner T (2012) An IP-10 (CXCL10)-derived peptide inhibits angiogenesis. PLoS One 7(7):e40812. doi:10.1371/journal.pone.0040812
70. Datta D, Banerjee P, Gasser M, Waaga-Gasser AM, Pal S (2010) CXCR3-B can mediate growth-inhibitory signals in human renal cancer cells by down-regulating the expression of heme oxygenase-1. J Biol Chem 285(47):36842–36848. doi:10.1074/jbc.M110.170324
71. Lo BK, Yu M, Zloty D, Cowan B, Shapiro J, McElwee KJ (2010) CXCR3/ligands are significantly involved in the tumorigenesis of basal cell carcinomas. Am J Pathol 176(5): 2435–2446. doi:10.2353/ajpath.2010.081059

72. Yates CC, Krishna P, Whaley D, Bodnar R, Turner T, Wells A (2010) Lack of CXC chemokine receptor 3 signaling leads to hypertrophic and hypercellular scarring. Am J Pathol 176(4):1743–1755. doi:10.2353/ajpath.2010.090564
73. Yates CC, Whaley D, Kulasekeran P, Hancock WW, Lu B, Bodnar R, Newsome J, Hebda PA, Wells A (2007) Delayed and deficient dermal maturation in mice lacking the CXCR3 ELR-negative CXC chemokine receptor. Am J Pathol 171(2):484–495. doi: 10.2353/ajpath.2007.061092
74. Yates CC, Whaley D, Wells A (2012) Transplanted fibroblasts prevents dysfunctional repair in a murine CXCR3-deficient scarring model. Cell Transplant 21(5):919–931. doi:10.3727/096368911X623817
75. Yates CC, Whaley D, Y-Chen A, Kulesekaran P, Hebda PA, Wells A (2008) ELR-negative CXC chemokine CXCL11 (IP-9/I-TAC) facilitates dermal and epidermal maturation during wound repair. Am J Pathol 173(3):643–652. doi:10.2353/ajpath.2008.070990
76. Collins TL, Johnson MG, Medina JC (2007) In: Neote K, Letts GL, Moser B (eds) Chemokine biology-basic research and clinical application, vol 2. Birkhauser Verlag, Basel, Switzerland p 79
77. Wijtmans M, Verzijl D, Leurs R, de Esch IJ, Smit MJ (2008) Towards small-molecule CXCR3 ligands with clinical potential. ChemMedChem 3(6):861–872. doi:10.1002/cmdc.200700365
78. Wijtmans M, de Esch IJP, Leurs R (2011) In: Smit MJ, Lira SA, Leurs R (eds) Chemokine receptors as drug targets. Wiley-VCH, Weinheim, pp 301–315. doi:10.1002/9783527631995.ch13
79. Roumen L, Scholten DJ, de Kruijf P, de Esch IJP, Leurs R, de Graaf C (2012) C(X)CR in silico: Computer-aided prediction of chemokine receptor–ligand interactions. Drug Discov Today Technol 9(4):e281–e291. doi:10.1016/j.ddtec.2012.05.002
80. Scholten DJ, Roumen L, Wijtmans M, Verkade-Vreeker MC, Custers H, Lai M, de Hooge D, Canals M, de Esch IJ, Smit MJ, de Graaf C, Leurs R (2014) Identification of overlapping but differential binding sites for the high-affinity CXCR3 antagonists NBI-74330 and VUF11211. Mol Pharmacol 85(1):116–126. doi:10.1124/mol.113.088633
81. Katritch V, Cherezov V, Stevens RC (2013) Structure-function of the G protein-coupled receptor superfamily. Annu Rev Pharmacol Toxicol 53:531–556. doi:10.1146/annurev-pharmtox-032112-135923
82 Venkatakrishnan AJ, Deupi X, Lebon G, Tate CG, Schertler GF, Babu MM (2013) Molecular signatures of G-protein-coupled receptors. Nature 494(7436):185–194. doi:10.1038/nature11896
83. Wu B, Chien EY, Mol CD, Fenalti G, Liu W, Katritch V, Abagyan R, Brooun A, Wells P, Bi FC, Hamel DJ, Kuhn P, Handel TM, Cherezov V, Stevens RC (2010) Structures of the CXCR4 chemokine GPCR with small-molecule and cyclic peptide antagonists. Science 330(6007):1066–1071. doi:10.1126/science.1194396science.1194396
84. Tan Q, Zhu Y, Li J, Chen Z, Han GW, Kufareva I, Li T, Ma L, Fenalti G, Zhang W, Xie X, Yang H, Jiang H, Cherezov V, Liu H, Stevens RC, Zhao Q, Wu B (2013) Structure of the CCR5 chemokine receptor-HIV entry inhibitor maraviroc complex. Science 341(6152):1387–1390. doi:10.1126/science.1241475
85. Ondeyka JG, Herath KB, Jayasuriya H, Polishook JD, Bills GF, Dombrowski AW, Mojena M, Koch G, DiSalvo J, DeMartino J, Guan Z, Nanakorn W, Morenberg CM, Balick MJ, Stevenson DW, Slattery M, Borris RP, Singh SB (2005) Discovery of structurally diverse natural product antagonists of chemokine receptor CXCR3. Mol Diversity 9:123–129. doi:10.1007/s11030-005-1296-8
86. Schall TJ, Dairaghi DJ, McMaster BE (2001) Compounds and methods for modulating CXCR3 function. WO0116114
87. Johnson M, Li A-R, Liu J, Fu Z, Zhu L, Miao S, Wang X, Xu Q, Huang A, Marcus A, Xu F, Ebsworth K, Sablan E, Danao J, Kumer J, Dairaghi D, Lawrence C, Sullivan T, Tonn G, Schall T, Collins T, Medina J (2007) Discovery and optimization of a series of quinazolinone-

derived antagonists of CXCR3. Bioorg Med Chem Lett 17(12):3339–3343. doi:10.1016/j.bmcl.2007.03.106

88. Storelli S, Verdijk P, Verzijl D, Timmerman H, van de Stolpe AC, Tensen CP, Smit MJ, De Esch IJP, Leurs R (2005) Synthesis and structure-activity relationship of 3-phenyl-3H-quinazolin-4-one derivatives as CXCR3 chemokine receptor antagonists. Bioorg Med Chem Lett 15(11):2910–2913. doi: 10.1016/j.bmcl.2005.03.070
89. Heise CE, Pahuja A, Hudson SC, Mistry MS, Putnam AL, Gross MM, Gottlieb PA, Wade WS, Kiankarimi M, Schwarz D, Crowe P, Zlotnik A, Alleva DG (2005) Pharmacological characterization of CXC chemokine receptor 3 ligands and a small molecule antagonist. J Pharmacol Exp Ther 313(3):1263–1271. doi: 10.1124/jpet.105.083683
90. Medina JC, Johnson MG, Li A, Liu J, Huang AX, Zhu L, Marcus AP (2002) CXCR3 antagonists. WO02083143
91. Johnson MG (2006) Presented at the XIXth international symposium on medicinal chemistry, Istanbul, Turkey, Aug 29–Sep 2
92. Storelli S, Verzijl D, Al-Badie J, Elders N, Bosch L, Timmerman H, Smit MJ, De Esch IJP, Leurs R (2007) Synthesis and structure-activity relationships of 3H-quinazolin-4-ones and 3H-pyrido[2,3-d]pyrimidin-4-ones as CXCR3 receptor antagonists. Arch Pharm 340(6): 281–291. doi:10.1002/ardp.200700037
93. Floren LC (2003). Presented at inflammation 2003 – sixth world congress, Vancouver, Canada, Aug 2–6
94. Berry K, Friedrich M, Kersey K, Stempien M, Wagner F, van Lier J, Sabat R, Wolk K (2004) Evaluation of T0906487, a CXCR3 antagonist, in a phase 2a psoriasis trial. Inflammation Res Suppl 53:pS222
95. Tonn GR, Wong SG, Wong SC, Johnson MG, Ma J, Cho R, Floren LC, Kersey K, Berry K, Marcus AP, Wang X, Van Lengerich B, Medina JC, Pearson PG, Wong BK (2009) An inhibitory metabolite leads to dose- and time-dependent pharmacokinetics of AMG 487 in human subjects following multiple dosing. Drug Metab Disposit 37:502–513. doi: 10.1124/dmd.108.021931
96. Henne KR, Tran TB, VandenBrink BM, Rock DA, Aidasani DK, Subramanian R, Mason AK, Stresser DM, Teffera Y, Wong SG, Johnson MG, Chen X, Tonn GR, Wong BK (2012) Sequential metabolism of AMG 487, a novel CXCR3 antagonist, results in formation of quinone reactive metabolites that covalently modify CYP3A4 Cys239 and cause time-dependent inhibition of the enzyme. Drug Metab Disposit 40(7):1429–1440. doi: 10.1124/dmd.112.045708
97. Li A-R, Johnson MG, Liu J, Chen X, Du X, Mihalic JT, Deignan J, Gustin DJ, Duquette J, Fu Z, Zhu L, Marcus AP, Bergeron P, McGee LR, Danao J, Sullivan T, Ma J, Tang L, Tonn G, Collins T, Medina JC (2008) Optimisation of the heterocyclic core of the quinazolinone-derived CXCR3 antagonists. Bioorg Med Chem Lett 18:688–693. doi: 10.1016/j.bmcl.2007.11.060
98. Liu J, Fu Z, Li AR, Johnson M, Zhu L, Marcus A, Danao J, Sullivan T, Tonn G, Collins T, Medina J (2009) Optimization of a series of quinazolinone-derived antagonists of CXCR3. Bioorg Med Chem Lett 19(17):5114–5118. doi:10.1016/j.bmcl.2009.07.032
99. Bernat V, Heinrich MR, Baumeister P, Buschauer A, Tschammer N (2012) Synthesis and application of the first radioligand targeting the allosteric binding pocket of chemokine receptor CXCR3. ChemMedChem 7(8):1481–1489. doi:10.1002/cmdc.201200184
100. Chen X, Mihalic J, Deignan J, Gustin DJ, Duquette J, Du X, Chan J, Fu Z, Johnson M, Li AR, Henne K, Sullivan T, Lemon B, Ma J, Miao S, Tonn G, Collins T, Medina JC (2012) Discovery of potent and specific CXCR3 antagonists. Bioorg Med Chem Lett 22(1): 357–362. doi:10.1016/j.bmcl.2011.10.120
101. Chan J, Burke BJ, Baucom K, Hansen K, Bio MM, DiVirgilio E, Faul M, Murry J (2011) Practical syntheses of a CXCR3 antagonist. J Org Chem 76(6):1767–1774. doi:10.1021/jo102399a
102. Du X, Chen X, Mihalic J, Deignan J, Duquette J, Li A-R, Lemon B, Ma J, Miao S, Ebsworth K, Sullivan TJ, Tonn G, Collins T, Medina J (2008) Design and optimisation of

imidazole derivatives as potent CXCR3 antagonists. Bioorg Med Chem Lett 18:608–613. doi:10.1016/j.bmcl.2007.11.072
103. Du X, Gustin DJ, Chen X, Duquette J, McGee LR, Wang Z, Ebsworth K, Henne K, Lemon B, Ma J, Miao S, Sabalan E, Sullivan TJ, Tonn G, Collins TL, Medina JC (2009) Imidazo-pyrazine derivatives as potent CXCR3 antagonists. Bioorg Med Chem Lett 19(17): 5200–5204. doi:10.1016/j.bmcl.2009.07.021
104. Afantitis A, Melagraki G, Sarimveis H, Koutentis PA, Igglessi-Markopoulou O, Kollias G (2010) A combined LS-SVM & MLR QSAR workflow for predicting the inhibition of CXCR3 receptor by quinazolinone analogs. Mol Divers 14(2):225–235. doi:10.1007/s11030-009-9163-7
105. Allen DR, Bolt A, Chapman GA, Knight RL, Meissner JWG, Owen DA, Watson RJ (2007) Identification and structure-activity relationships of 1-aryl-3-piperidin-4-yl-urea derivatives as CXCR3 receptor antagonists. Bioorg Med Chem Lett 17(3):697–701. doi:10.1016/j.bmcl.2006.10.088
106. Watson RJ, Allen DR, Birch HL, Chapman GA, Hannah DR, Knight RL, Meissner JWG, Owen DA, Thomas EJ (2007) Development of CXCR3 antagonists. Part 2: Identification of 2-amino(4-piperidinyl)azoles as potent CXCR3 antagonists. Bioorg Med Chem Lett 17: 6806–6810. doi:10.1016/j.bmcl.2007.10.029
107. Watson RJ, Allen DR, Birch HL, Chapman GA, Galvin FC, Jopling LA, Knight RL, Meier D, Oliver K, Meissner JW, Owen DA, Thomas EJ, Tremayne N, Williams SC (2008) Development of CXCR3 antagonists. Part 3: tropenyl and homotropenyl-piperidine urea derivatives. Bioorg Med Chem Lett 18:147–151. doi:10.1016/j.bmcl.2007.10.109
108. Knight RL, Allen DR, Birch HL, Chapman GA, Galvin FC, Jopling LA, Lock CJ, Meissner JWG, Owen DA, Raphy G, Watson RJ, Williams SC (2008) Development of CXCR3 antagonists, Part 4: discovery of 2-amino-(4-tropinyl) quinolines. Bioorg Med Chem Lett 18:629–633. doi:10.1016/j.bmcl.2007.11.075
109. McGuinness BF, Carroll CD, Zawacki LG, Dong G, Yang C, Hobbs DW, Jacob-Samuel B, Hall JW 3rd, Jenh CH, Kozlowski JA, Anilkumar GN, Rosenblum SB (2009) Novel CXCR3 antagonists with a piperazinyl-piperidine core. Bioorg Med Chem Lett 19(17):5205–5208. doi:10.1016/j.bmcl.2009.07.020
110. Bongartz JP, Buntinx M, Coesemans E, Hermans B, Lommen GV, Wauwe JV (2008) Synthesis and structure-activity relationship of benzetimide derivatives as human CXCR3 antagonists. Bioorg Med Chem Lett 18:5819–5823. doi:10.1016/j.bmcl.2008.07.115
111. Shao Y, Anilkumar GN, Carroll CD, Dong G, Hall JW 3rd, Hobbs DW, Jiang Y, Jenh CH, Kim SH, Kozlowski JA, McGuinness BF, Rosenblum SB, Schulman I, Shih NY, Shu Y, Wong MK, Yu W, Zawacki LG, Zeng Q (2011) II. SAR studies of pyridyl-piperazinyl-piperidine derivatives as CXCR3 chemokine antagonists. Bioorg Med Chem Lett 21(5): 1527–1531. doi:10.1016/j.bmcl.2010.12.114
112. McGuinness BF, Rosenblum SF, Kozlowksi JA, Anilkumar GN, Kim SH, Shih N-Y, Jenh C-H, Zavodny PJ, Hobbs DW, Dong G, Shao Y, Zawacki LG, Yang C, Carroll CD (2006) Pyridyl and phenyl substituted piperazine-piperidines with CXCR3 antagonist activity. WO2006088919
113. Wijtmans M, Verzijl D, van Dam CM, Bosch L, Smit MJ, Leurs R, de Esch IJ (2009) Exploring a pocket for polycycloaliphatic groups in the CXCR3 receptor with the aid of a modular synthetic strategy. Bioorg Med Chem Lett 19(8):2252–2257. doi:10.1016/j.bmcl.2009.02.093
114. Kim SH, Anilkumar GN, Zawacki LG, Zeng Q, Yang DY, Shao Y, Dong G, Xu X, Yu W, Jiang Y, Jenh CH, Hall JW 3rd, Carroll CD, Hobbs DW, Baldwin JJ, McGuinness BF, Rosenblum SB, Kozlowski JA, Shankar BB, Shih NY (2011) III. Identification of novel CXCR3 chemokine receptor antagonists with a pyrazinyl-piperazinyl-piperidine scaffold. Bioorg Med Chem Lett 21(23):6982–6986. doi:10.1016/j.bmcl.2011.09.120
115. Nair AG, Wong MKC, Shu Y, Jiang Y, Jenh C-H, Kim SH, Yang D-Y, Zeng Q, Shao Y, Zawacki LG, Duo J, McGuinness BF, Carroll CD, Hobbs DW, Shih N-Y, Rosenblum SB,

Kozlowski JA (2014) IV. Discovery of CXCR3 antagonists substituted with heterocycles as amide surrogates: Improved PK, hERG and metabolic profiles. Bioorg Med Chem Lett 24(4): 1085–1088. doi:10.1016/j.bmcl.2014.01.009
116. Thoma G, Baenteli R, Lewis I, Wagner T, Oberer L, Blum W, Glickman F, Streiff MB, Zerwes HG (2009) Special ergolines are highly selective, potent antagonists of the chemokine receptor CXCR3: discovery, characterization and preliminary SAR of a promising lead. Bioorg Med Chem Lett 19(21):6185–6188. doi:10.1016/j.bmcl.2009.09.002
117. Thoma G, Baenteli R, Lewis I, Jones D, Kovarik J, Streiff MB, Zerwes HG (2011) Special ergolines efficiently inhibit the chemokine receptor CXCR3 in blood. Bioorg Med Chem Lett 21(16):4745–4749. doi:10.1016/j.bmcl.2011.06.070
118. Christen S, Holdener M, Beerli C, Thoma G, Bayer M, Pfeilschifter JM, Hintermann E, Zerwes HG, Christen U (2011) Small molecule CXCR3 antagonist NIBR2130 has only a limited impact on type 1 diabetes in a virus-induced mouse model. Clin Exp Immunol 165(3): 318–328. doi:10.1111/j.1365-2249.2011.04426.x
119. Hayes ME, Wallace GA, Grongsaard P, Bischoff A, George DM, Miao W, McPherson MJ, Stoffel RH, Green DW, Roth GP (2008) Discovery of small molecule benzimidazole antagonists of the chemokine receptor CXCR3. Bioorg Med Chem Lett 18(5):1573–1576. doi:10.1016/j.bmcl.2008.01.074
120. Hayes ME, Breinlinger EC, Wallace GA, Grongsaard P, Miao W, McPherson MJ, Stoffel RH, Green DW, Roth GP (2008) Lead identification of 2-iminobenzimidazole antagonists of the chemokine receptor CXCR3. Bioorg Med Chem Lett 18(7):2414–2419. doi:10.1016/j.bmcl.2008.02.049
121. Wang Y, Busch-Petersen J, Wang F, Kiesow TJ, Graybill TL, Jin J, Yang Z, Foley JJ, Hunsberger GE, Schmidt DB, Sarau HM, Capper-Spudich EA, Wu Z, Fisher LS, McQueney MS, Rivero RA, Widdowson KL (2009) Camphor sulfonamide derivatives as novel, potent and selective CXCR3 antagonists. Bioorg Med Chem Lett 19:114–118. doi:10.1016/j.bmcl.2008.11.008
122. Wijtmans M, Verzijl D, Bergmans S, Lai M, Bosch L, Smit MJ, de Esch IJ, Leurs R (2011) CXCR3 antagonists: quaternary ammonium salts equipped with biphenyl- and polycycloaliphatic-anchors. Bioorg Med Chem 19(11):3384–3393. doi:10.1016/j.bmc.2011.04.035
123. Crosignani S, Missotten M, Cleva C, Dondi R, Ratinaud Y, Humbert Y, Mandal AB, Bombrun A, Power C, Chollet A, Proudfoot A (2010) Discovery of a novel series of CXCR3 antagonists. Bioorg Med Chem Lett 20(12):3614–3617. doi:10.1016/j.bmcl.2010.04.113
124. Cole AG, Stroke IL, Brescia MR, Simhadri S, Zhang JJ, Hussain Z, Snider M, Haskell C, Ribeiro S, Appell KC, Henderson I, Webb ML (2006) Identification and initial evaluation of 4-N-aryl-[1,4]diazepane ureas as potent CXCR3 antagonists. Bioorg Med Chem Lett 16(1): 200–203. doi:10.1016/j.bmcl.2005.09.020
125. Stroke IL, Cole AG, Simhadri S, Brescia MR, Desai M, Zhang JJ, Merritt JR, Appell KC, Henderson I, Webb ML (2006) Identification of CXCR3 receptor agonists in combinatorial small-molecule libraries. Biochem Biophys Res Commun 349(1):221–228. doi:10.1016/j.bbrc.2006.08.019
126. Afantitis A, Melagraki G, Sarimveis H, Igglessi-Markopoulou O, Kollias G (2009) A novel QSAR model for predicting the inhibition of CXCR3 receptor by 4-N-aryl-[1,4] diazepane ureas. Eur J Med Chem 44(2):877–884. doi:10.1016/j.ejmech.2008.05.028
127. Vummidi BR, Noreen F, Alzeer J, Moelling K, Luedtke NW (2013) Photodynamic agents with anti-metastatic activities. ACS Chem Biol 8(8):1737–1746. doi:10.1021/cb400008t
128. Wise E, Pease JE (2007) Unravelling the mechanisms underpinning chemokine receptor activation and blockade by small molecules: a fine line between agonism and antagonism? Biochem Soc Trans 35:755–759. doi:10.1042/BST0350755
129. Nedjai B, Li H, Stroke IL, Wise EL, Webb ML, Merritt JR, Henderson I, Klon AE, Cole AG, Horuk R, Vaidehi N, Pease JE (2012) Small molecule chemokine mimetics suggest a molecular basis for the observation that CXCL10 and CXCL11 are allosteric ligands of CXCR3. Br J Pharmacol 166(3):912–923. doi:10.1111/j.1476-5381.2011.01660.x

130. O'Boyle G, Fox CR, Walden HR, Willet JD, Mavin ER, Hine DW, Palmer JM, Barker CE, Lamb CA, Ali S, Kirby JA (2012) Chemokine receptor CXCR3 agonist prevents human T-cell migration in a humanized model of arthritic inflammation. Proc Natl Acad Sci U S A 109(12):4598–4603. doi:10.1073/pnas.1118104109
131. O'Boyle G, Mellor P, Kirby JA, Ali S (2009) Anti-inflammatory therapy by intravenous delivery of non-heparan sulfate-binding CXCL12. FASEB J 23(11):3906–3916. doi:10.1096/fj.09-134643
132. Wijtmans M, Scholten DJ, Roumen L, Canals M, Custers H, Glas M, Vreeker MC, de Kanter FJ, de Graaf C, Smit MJ, de Esch IJ, Leurs R (2012) Chemical subtleties in small-molecule modulation of peptide receptor function: the case of CXCR3 biaryl-type ligands. J Med Chem 55(23):10572–10583. doi:10.1021/jm301240t
133. Ballesteros JA, Weinstein H (1995) Integrated methods for the construction of three dimensional models and computational probing of structure-function relations in G-protein coupled receptors. Methods Neurosci 25:366–428. doi:10.1016/S1043-9471(05)80049-7
134. Hollenstein K, de Graaf C, Bortolato A, Wang MW, Marshall FH, Stevens RC (2014) Insights into the structure of class B GPCRs. Trends Pharmacol Sci 35(1):12–22. doi:10.1016/j.tips.2013.11.001
135. Wu H, Wang C, Gregory KJ, Han GW, Cho HP, Xia Y, Niswender CM, Katritch V, Meiler J, Cherezov V, Conn PJ, Stevens RC (2014) Structure of a class C GPCR metabotropic glutamate receptor 1 bound to an allosteric modulator. Science 344(6179):58–64. doi:10.1126/science.1249489
136. Wang C, Wu H, Katritch V, Han GW, Huang XP, Liu W, Siu FY, Roth BL, Cherezov V, Stevens RC (2013) Structure of the human smoothened receptor bound to an antitumour agent. Nature 497(7449):338–343. doi:10.1038/nature12167
137. Kooistra AJ, Roumen L, Leurs R, de Esch IJ, de Graaf C (2013) From heptahelical bundle to hits from the Haystack: structure-based virtual screening for GPCR ligands. Methods Enzymol 522:279–336. doi:10.1016/B978-0-12-407865-9.00015-7
138. Michino M, Abola E, participants GD, Brooks CL 3rd, Dixon JS, Moult J, Stevens RC (2009) Community-wide assessment of GPCR structure modelling and ligand docking: GPCR Dock 2008. Nat Rev Drug Discov 8(6):455–463. doi:10.1038/nrd2877
139. Kufareva I, Rueda M, Katritch V, Stevens RC, Abagyan R, participants GD (2011) Status of GPCR modeling and docking as reflected by community-wide GPCR Dock 2010 assessment. Structure 19(8):1108–1126. doi:10.1016/j.str.2011.05.012
140. Manglik A, Kruse AC, Kobilka TS, Thian FS, Mathiesen JM, Sunahara RK, Pardo L, Weis WI, Kobilka BK, Granier S (2012) Crystal structure of the micro-opioid receptor bound to a morphinan antagonist. Nature 485(7398):321–326. doi:10.1038/nature10954nature10954
141. Roumen L, Sanders MP, Vroling B, de Esch IJ, de Vlieg J, Leurs R, Klomp JP, Nabuurs SB, de Graaf C (2011) In Silico Veritas: the pitfalls and challenges of predicting GPCR-ligand interactions. Pharmaceuticals 4(9):1196–1215. doi:10.1021/ci200088d
142. Brelot A, Heveker N, Montes M, Alizon M (2000) Identification of residues of CXCR4 critical for human immunodeficiency virus coreceptor and chemokine receptor activities. J Biol Chem 275(31):23736–23744. doi:10.1074/jbc.M000776200
143. Kofuku Y, Yoshiura C, Ueda T, Terasawa H, Hirai T, Tominaga S, Hirose M, Maeda Y, Takahashi H, Terashima Y, Matsushima K, Shimada I (2009) Structural basis of the interaction between chemokine stromal cell-derived factor-1/CXCL12 and its G-protein-coupled receptor CXCR4. J Biol Chem 284(50):35240–35250. doi:10.1074/jbc.M109.024851
144. Colvin RA, Campanella GS, Manice LA, Luster AD (2006) CXCR3 requires tyrosine sulfation for ligand binding and a second extracellular loop arginine residue for ligand-induced chemotaxis. Mol Cell Biol 26(15):5838–5849. doi:10.1128/MCB.00556-06
145. Anghelescu AV, DeLisle RK, Lowrie JF, Klon AE, Xie X, Diller DJ (2008) Technique for generating three-dimensional alignments of multiple ligands from one-dimensional alignments. J Chem Inform Model 48(5):1041–1054. doi:10.1021/ci700395f

146. Rosenkilde MM, Andersen MB, Nygaard R, Frimurer TM, Schwartz TW (2007) Activation of the CXCR3 chemokine receptor through anchoring of a small molecule chelator ligand between TM-III, −IV, and −VI. Mol Pharmacol 71(3):930–941. doi: 10.1124/mol.106.030031
147. Scholten DJ (2012) Chemokine receptors CXCR3 and CXCR7: allosteric ligand binding, biased signaling, and receptor regulation. VU University Amsterdam, Amsterdam
148. Zhang J, Chen P, Yuan B, Ji W, Cheng Z, Qiu X (2013) Real-space identification of intermolecular bonding with atomic force microscopy. Science 342(6158):611–614. doi:10.1126/science.1242603science.1242603
149. Pierce AC, Sandretto KL, Bemis GW (2002) Kinase inhibitors and the case for CH…O hydrogen bonds in protein-ligand binding. Proteins 49(4):567–576. doi:10.1002/prot.10259
150. van Linden OP, Kooistra AJ, Leurs R, de Esch IJ, de Graaf C (2014) KLIFS: a knowledge-based structural database to navigate kinase-ligand interaction space. J Med Chem 57(2): 249–277. doi:10.1021/jm400378w
151. Becker OM, Marantz Y, Shacham S, Inbal B, Heifetz A, Kalid O, Bar-Haim S, Warshaviak D, Fichman M, Noiman S (2004) G protein-coupled receptors: in silico drug discovery in 3D. Proc Natl Acad SciUSA 101(31):11304–11309. doi:10.1073/pnas.0401862101
152. Bayry J, Tchilian EZ, Davies MN, Forbes EK, Draper SJ, Kaveri SV, Hill AV, Kazatchkine MD, Beverley PC, Flower DR, Tough DF (2008) In silico identified CCR4 antagonists target regulatory T cells and exert adjuvant activity in vaccination. Proc Natl Acad Sci USA 105(29):10221–10226. doi:10.1073/pnas.0803453105
153. Kellenberger E, Springael JY, Parmentier M, Hachet-Haas M, Galzi JL, Rognan D (2007) Identification of nonpeptide CCR5 receptor agonists by structure-based virtual screening. J Med Chem 50(6):1294–1303. doi:10.1021/jm061389p
154. Kim J, Yip ML, Shen X, Li H, Hsin LY, Labarge S, Heinrich EL, Lee W, Lu J, Vaidehi N (2012) Identification of anti-malarial compounds as novel antagonists to chemokine receptor CXCR4 in pancreatic cancer cells. PLoS One 7(2):e31004. doi:10.1371/journal.pone.0031004
155. Huang D, Gu Q, Ge H, Ye J, Salam NK, Hagler A, Chen H, Xu J (2012) On the value of homology models for virtual screening: discovering hCXCR3 antagonists by pharmacophore-based and structure-based approaches. J Chem Inform Model 52(5): 1356–1366. doi:10.1021/ci300067q
156. Mysinger MM, Weiss DR, Ziarek JJ, Gravel S, Doak AK, Karpiak J, Heveker N, Shoichet BK, Volkman BF (2012) Structure-based ligand discovery for the protein-protein interface of chemokine receptor CXCR4. Proc Natl Acad Sci USA 109(14):5517–5522. doi:10.1073/pnas.1120431109
157. Vitale RM, Gatti M, Carbone M, Barbieri F, Felicita V, Gavagnin M, Florio T, Amodeo P (2013) Minimalist hybrid ligand/receptor-based pharmacophore model for CXCR4 applied to a small-library of marine natural products led to the identification of phidianidine a as a new CXCR4 ligand exhibiting antagonist activity. ACS Chem Biol 8(12):2762–2770. doi:10.1021/cb400521b
158. Yoshikawa Y, Oishi S, Kubo T, Tanahara N, Fujii N, Furuya T (2013) Optimized method of G-protein-coupled receptor homology modeling: its application to the discovery of novel CXCR7 ligands. J Med Chem 56(11):4236–4251. doi:10.1021/jm400307y
159. Evers A, Klabunde T (2005) Structure-based drug discovery using GPCR homology modeling: successful virtual screening for antagonists of the alpha1A adrenergic receptor. J Med Chem 48(4):1088–1097. doi:10.1021/jm0491804
160. Kolb P, Rosenbaum DM, Irwin JJ, Fung JJ, Kobilka BK, Shoichet BK (2009) Structure-based discovery of beta2-adrenergic receptor ligands. Proc Natl Acad Sci USA 106(16):6843–6848. doi:10.1073/pnas.0812657106
161. Varady J, Wu X, Fang X, Min J, Hu Z, Levant B, Wang S (2003) Molecular modeling of the three-dimensional structure of dopamine 3 (D3) subtype receptor: discovery of novel and

potent D3 ligands through a hybrid pharmacophore- and structure-based database searching approach. J Med Chem 46(21):4377–4392. doi:10.1021/jm030085p

162. Carlsson J, Coleman RG, Setola V, Irwin JJ, Fan H, Schlessinger A, Sali A, Roth BL, Shoichet BK (2011) Ligand discovery from a dopamine D3 receptor homology model and crystal structure. Nat Chem Biol 7(11):769–778. doi:10.1038/nchembio.662
163. de Graaf C, Kooistra AJ, Vischer HF, Katritch V, Kuijer M, Shiroishi M, Iwata S, Shimamura T, Stevens RC, de Esch IJ, Leurs R (2011) Crystal structure-based virtual screening for fragment-like ligands of the human histamine H(1) receptor. J Med Chem 54(23):8195–8206. doi:10.1021/jm2011589
164. Kooistra AJ, Kuhne S, de Esch IJ, Leurs R, de Graaf C (2013) A structural chemogenomics analysis of aminergic GPCRs: lessons for histamine receptor ligand design. Br J Pharmacol 170(1):101–126. doi:10.1111/bph.12248
165. Mason JS, Bortolato A, Congreve M, Marshall FH (2012) New insights from structural biology into the druggability of G protein-coupled receptors. Trends Pharmacol Sci 33(5): 249–260. doi:10.1016/j.tips.2012.02.005
166. Nicholls DJ, Tomkinson NP, Wiley KE, Brammall A, Bowers L, Grahames C, Gaw A, Meghani P, Shelton P, Wright TJ, Mallinder PR (2008) Identification of a putative intracellular allosteric antagonist binding-site in the CXC chemokine receptors 1 and 2. Mol Pharmacol 74(5):1193–1202. doi:10.1124/mol.107.044610
167. Salchow K, Bond ME, Evans SC, Press NJ, Charlton SJ, Hunt PA, Bradley ME (2010) A common intracellular allosteric binding site for antagonists of the CXCR2 receptor. Br J Pharmacol 159(7):1429–1439. doi:10.1111/j.1476-5381.2009.00623.x
168. de Kruijf P, Lim HD, Roumen L, Renjaan VA, Zhao J, Webb ML, Auld DS, Wijkmans JC, Zaman GJ, Smit MJ, de Graaf C, Leurs R (2011) Identification of a novel allosteric binding site in the CXCR2 chemokine receptor. Mol Pharmacol 80(6):1108–1118. doi:10.1124/mol.111.073825

Top Med Chem (2015) 14: 187–242
DOI: 10.1007/7355_2014_40

Published online: 5 March 2014

# Selective and Dual Targeting of CCR2 and CCR5 Receptors: A Current Overview

**Anna Junker, Artur Kamil Kokornaczyk, Ann Kathrin Strunz, and Bernhard Wünsch**

**Abstract** The chemokine receptor 2 (CCR2) and chemokine receptor 5 (CCR5) are important mediators of leukocyte trafficking in inflammatory processes. The emerging evidence for a role of CCR2 and CCR5 receptors in human inflammatory diseases led to a growing interest in CCR2- and CCR5-selective antagonists. In this review, we focus on the recent development of selective CCR2/CCR5 receptor ligands and dual antagonists. Several compounds targeting CCR2, e.g., INCB8761 and MK0812, were developed as promising candidates for clinical trials, but failed to show clinical efficacy as presumed from preclinical models. The role of CCR5 receptors as the second co-receptor for the HIV-host cell fusion led to the development of various CCR5-selective ligands. Maraviroc is the first CCR5-targeting drug for the treatment of HIV-1 infections on the market. The role of CCR5 receptors in the progression of inflammatory processes fueled the use of CCR5 antagonists for the treatment of rheumatoid arthritis. Unfortunately, the use of maraviroc for the treatment of rheumatoid arthritis failed due to its inefficacy. Some of the ligands, e.g., TAK-779 and TAK-652, were also found to be dual antagonists of CCR2 and CCR5 receptors. The fact that CCR2 and CCR5 receptor antagonists contribute to the treatment of inflammatory diseases renders the development of dual antagonists as promising novel therapeutic strategy.

**Keywords** Atherosclerosis, AZD5672, AZD5672, CCR2, CCR5, Chemokine receptors, GSK163929, hERG, HIV-1, INCB10820, INCB3284, INCB3344, INCB8761, Inflammation JNJ17166864, Maraviroc, MCP-1, MIP-1, MK0483, MK0812, PF-232798, PF-4136309, PF-4254196, RANTES, RS504393, SKB3380732, TAK-220, TAK-652, TAK-779, UK-107,543, UK-347,503, UK-427,857

---

A. Junker, A.K. Kokornaczyk, A.K. Strunz, and B. Wünsch (✉)
Westfälische Wilhelms-Universität Münster, PharmaCampus, Institut für Pharmazeutische und Medizinische Chemie, Corrensstraße 48, 48149 Münster, Germany
e-mail: wuensch@uni-muenster.de

## Contents



# Abbreviations

| | |
|---|---|
| 5-HT | 5-Hydroxytryptamine |
| AT | Angiotensin receptors |
| CHO | Chinese hamster ovary |
| CYP450 | Cytochrome P450 enzyme |
| EMEA | European Medicines Evaluation Agency |
| ET | Endothelin receptor |
| FDA | Food and Drug Administration |
| GPCR | G-protein-coupled receptor |
| hERG | Human ether-a-go-go-Related Gene |
| HIV | Human immunodeficiency virus |
| $IC_{50}$ | Half maximal inhibitory concentration |
| MS | Multiple sclerosis |
| PET | Positron emission tomography |
| RA | Rheumatoid arthritis |
| SAR | Structure affinity relationship |
| SP | Spiropiperidine |
| TdP | Torsades de pointes |
| TM | Transmembrane domain |

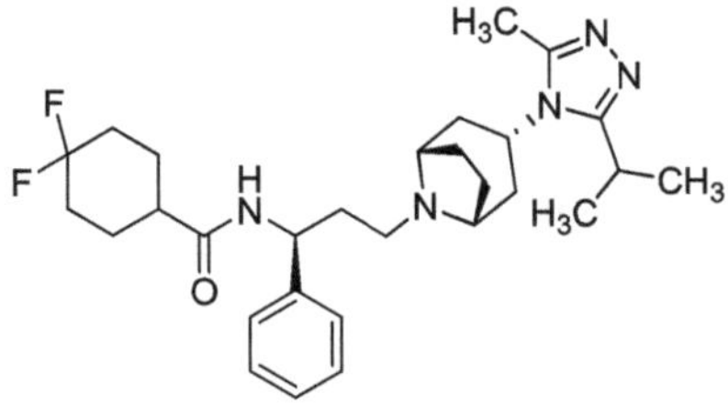

**Fig. 1** **1** (UK-427,827, maraviroc, Celsentri®)

# 1 Introduction

In the last 25 years, chemokines and their receptors have become promising targets in many fields of research. Because the chemokine receptors 2 (CCR2) and 5 (CCR5) represent highly interesting candidates of the chemokine receptor family, much of investigations have recently been carried out in the development of ligands for these receptors. Maraviroc (**1**) is the first and to date the only FDA- and EMEA-approved drug on the market, targeting the CCR5 receptor (see Fig. 1) [1, 2]. The intention of this review is to elucidate the structure activity relationships of various small-molecule CCR2 and CCR5 ligands. The focus will be on the receptor binding affinity, antiviral activity for the treatment of HIV, and chemotactic activity for the treatment of atherosclerosis. In addition to the receptor binding affinity, many further aspects, which play a crucial role in drug development, will be discussed, e.g., physicochemical properties, lipophilicity, and the affinity to the human Ether-à-go-go-Related Gene (hERG), a $K^+$-channel which would lead to severe side effects induced by the compounds.

The CCR2 receptor has become a promising target in the therapy of atherosclerosis. The concerted action of the chemokine ligand CCL2 and the CCR2 receptor plays an important role in the recruitment of monocytes from the bone marrow into the arterial wall, which is known to be an early key step in atherosclerotic plaque formation. Lesions of the arterial endothelium are caused by mechanic injury or toxins and lead to migration of monocytes into the subendothelium that is mediated by adhesion molecules and chemokine receptors. In the artery wall, monocytes differentiate into macrophages, which develop to foam cells by taking up blood lipids [3–5]. Advanced plaques become unstable and can suddenly rupture. They expose their content to the blood, leading to platelet aggregation and occlusion of the blood vessel. Thrombosis, stroke, and myocardial infarction result as serious long-term complications. With regard to the increasing number of patients, the use of small-molecule CCR2 antagonists in atherosclerosis has attracted significant attention in the last years [6]. The CCR2 receptor is linked also to progression and development of other inflammatory diseases like multiple sclerosis (MS) and rheumatoid arthritis (RA) [7].

The CCR5 and the CXCR4 receptors are mainly known as co-receptors required for the development of the HIV-1 infection [8]. The binding of gp120 of the virus to CD4 receptors on T-lymphocytes and macrophages leads to a conformational change of gp120 and enables the interaction with the CCR5 co-receptor. This triggers conformational changes in gp41, which leads to the fusion of the virus with the host cell [9]. At the beginning of an infection, the vast majority of the transmitted virus strains is M- or R5-tropic and uses the CCR5 receptor as a co-receptor. The T- or X4-tropic virus strains, which use the CXCR4 receptor as a co-receptor, are associated with an advanced disease progression [10]. A genetic polymorphism of the CCR5 receptor, characterized by a deletion of 32 bp in the gene segment encoding the receptor (CCR5Δ32 bp), results in a lack of function of the CCR5 receptor in homozygote individuals. These CCR5Δ32 bp mutations are found in 1–5% among uninfected Caucasian population and is exceedingly rare in infected patients (<0.1%), which indicates that CCR5Δ32 bp homozygotes are strongly resistant against HIV-1 infection [11–13]. The CCR5Δ32 bp mutant has also been linked to reduced susceptibility to coronary artery diseases and myocardial infarction [14, 15]. CCR5Δ32 bp polymorphism, as well as the function of the CCR5 receptor as a co-receptor for the HIV-host cell fusion, stimulated the beginning of several drug development programs by different pharmaceutical companies since the early 2000s.

In this review, we will focus on selective CCR2 (Sect. 2) and CCR5 (Sect. 3) receptor ligands, which can be used in the treatment of several immunological diseases including arthritis [16], asthma [17], multiple sclerosis [18], vascular diseases [19, 20], and HIV-1 infection [7, 21]. The fact that the antagonism of both CCR2 and CCR5 receptors may contribute to the treatment of inflammatory diseases makes the development of dual antagonists attractive. Dual CCR2 and CCR5 antagonists will be discussed in Sect. 4.

## 2 CCR2 Receptor Ligands

The CCR2 receptor plays an important role in an inflammatory response and is involved in several diseases of the immune system including atherosclerosis. The interaction of the chemokine CCL2 with the CCR2 receptor is responsible for the recruitment of blood monocytes to the site of inflammation and is also an early key step in the pathogenesis of atherosclerosis. The CCR2 receptor represents a promising therapeutic target for the treatment of atherosclerosis and is discussed as potential PET (positron emission tomography) target for diagnostic use [7]. Current atherosclerosis treatments are restricted to manipulation of indirect mechanisms, e.g., the modulation of cholesterol or triglyceride concentration, control of homoeostasis, or reduction of other risk factors associated with the metabolic syndrome. With regard to millions of patients (and numbers rising) who suffer under this chronic inflammation, CCR2 antagonists have attracted substantial attention during the past years [20].

Early developments of CCR2 antagonists in the late 1990s have been already reviewed elsewhere [6, 22–24]. This chapter will analyze structural features from different classes of CCR2 antagonists that were published until August 2013.

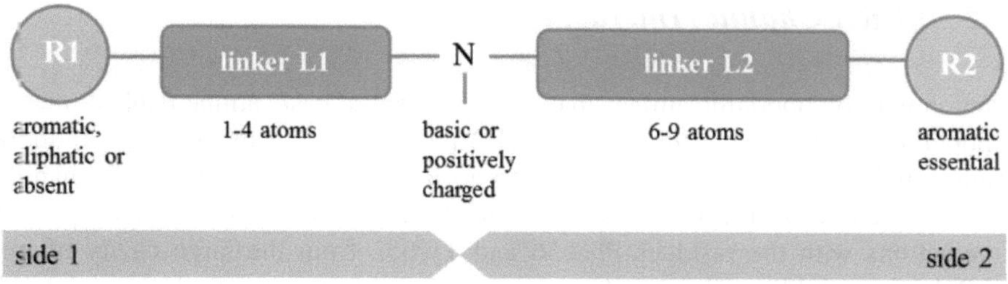


**Fig. 2** General pharmacophore model for CCR2 antagonists, modified according to ref. [27]

We will compare them with regard to their structure activity relationships (SARs), explain strategies that led to increasing CCR2 affinity and selectivity, and elucidate the influence on the CYP system and hERG inhibition as well. The affinity of ligands to the CCR2 receptor is usually determined in the radioligand displacement assays, where the chemokine [$^{125}$I] CCL2 is used as the radioligand [25]. Chemotactic assays are generally used to determine the compounds' ability to inhibit the CCL2-stimulated chemotaxis in human peripheral blood monocytes [26].

## 2.1 Pharmacophore Model

Several series of CCR2 antagonists from different structural classes have been described in patents and publications. The majority of known CCR2 antagonists consist of a basic center flanked by two lipophilic residues as aromatic rings or one aromatic and one aliphatic moiety as demonstrated in the pharmacophore model in Fig. 2. The basic amine or quaternary ammonium ion of the ligand essentially anchors a small-molecule ligand to Glu291 (in the transmembrane domain 7 (TM7)) of the CCR2 receptor by a salt-bridge formation [28].

The aryl or heteroaryl motif R2 on side 2 is another essential feature of this pharmacophore model. The substitution pattern of this aromatic system greatly influences the CCR2 receptor affinity. The linker L2 of 6–9 atoms with lipophilic or polar, peptidic or saturated, and unsaturated or aromatic structural elements is well tolerated by the receptor and determines the specificity of binding.

The left part of the molecules tolerates more variations: the moiety R1 can be an aliphatic or aromatic ring. The linker L1 can be short and aliphatic (1–4 atoms), can be part of a ring system, or it can represent an aliphatic system in case of absence of a ring system [27]. Compounds with an aromatic ring in the R1 position show interactions with the hERG channel [29].

Most of the CCR2 ligands, discussed in this chapter, correspond to the described pharmacophore model. The report of the crystal structure of the CXCR4 receptor [30], which has a high sequence homology with the CCR2 receptor, initiated various structure-based approaches in the CCR2 ligand design and the investigations of the ligand-receptor interactions.

## 2.2 hERG Channel Interaction

Predominantly, lipophilic amines that contain a central basic amine flanked by two lipophilic moieties according to the CCR2 pharmacophore model (Fig. 2) show high affinity for the human Ether-a-go-go-Related Gene (hERG) $K^+$-channel [29, 31]. The aromatic moieties of a drug form π–π-stacking and hydrophobic interactions with the residues Phe656 and Tyr652 from the large cavity of the hERG channel [32]. The hERG gene encodes the α-subunit of the inwardly rectifying $K^+$-channel, which is highly expressed in the human heart. This channel is involved in the repolarization of the cell. The mutation of the hERG gene or a channel blockade by drugs can lead to prolongation of the QT interval in the electrocardiogram in severe arrhythmia.

High sensitivity of the hERG channel to a blockage by many drugs and with the resulting cardiovascular adverse events like torsades de pointes (TdP), which can degenerate into ventricular fibrillation, led to the requirement of regulatory agencies that the effect of novel drugs on the hERG channel has to be investigated and reported [33]. Hence, considering hERG channel blockade is essential to improve cardiovascular safety of novel CCR2 antagonists. The increase of overall polarity of the drugs by introduction of hydrophilic ring systems and substituents on side 2 has been a successful strategy to eliminate hERG inhibition [31]. Further, successful approaches are the attenuation of the $pK_a$ value of the basic amine, modification of its steric environment, and the formation of zwitterions [32, 34].

## 2.3 Pyrrolidine-Based CCR2 Ligands

Incyte's INCB3344 (**2**), one of the first potent pyrrolidine derivatives with a 3,4-methylenedioxyphenyl residue (Table 1), was well investigated in receptor binding and chemotaxis assays with human (hCCR2) and murine (mCCR2) receptors [35]. Despite the fact that human and murine CCR2 receptors show high sequence homology, binding affinities of ligands differ considerably among species [36]. INCB3344 (**2**) showed the $IC_{50}$ values of 5.1 nM in hCCR2-binding and 3.8 nM in chemotaxis assay. **2** has been used as a tool in rodent in vivo efficacy models for multiple sclerosis, arthritis, and obesity and was effective in lowering macrophage levels in the targeted tissue [25, 26, 37]. Despite a high selectivity over other chemokine receptors, INCB3344 (**2**) was not a suitable clinical candidate due to its moderate hERG binding activity ($IC_{50} = 13$ μM) and inhibition of CYP 3A4 [6, 35]. This data led to further structural modifications of **2**. The removal of the ethoxy group at the quaternary carbon at position 3 of the pyrrolidine ring led to a loss of the mCCR2 affinity but retained the hCCR2 affinity. Previous SAR studies proved that the trifluoromethylphenyl residue on side 2 was crucial for the CCR2 binding affinity. The (*R*)-configuration at the position 3 of the pyrrolidine ring was also known to be important for the CCR2 affinity [38, 39]. The optimization

**Table 1** CCR2 antagonists with pyrrolidine structure, inhibitory effects on CCL2 binding to human CCR2 receptor

| Compounds | | hCCR2 $IC_{50}$ (nM) |
|---|---|---|
| **2** (INCB3344) | 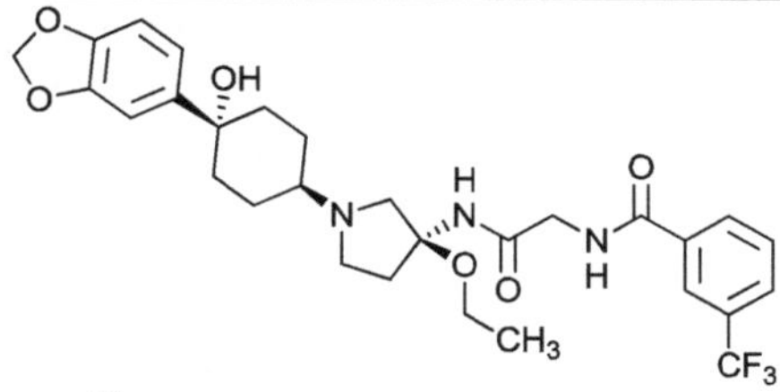 | 5.1 |
| **3** (INCB3284) | 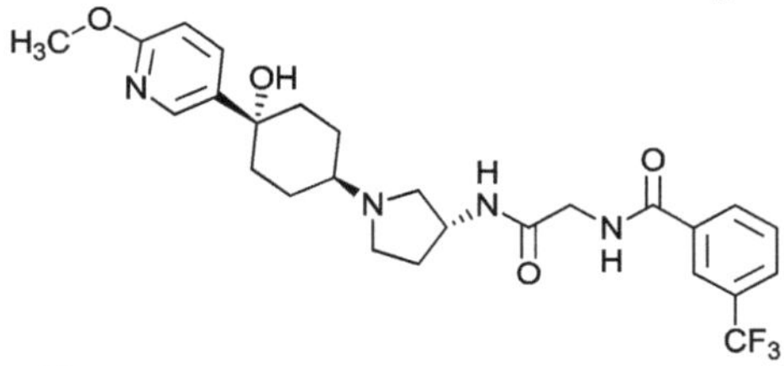 | 3.7 |
| **4** (INCB8761/PF-4136309) | 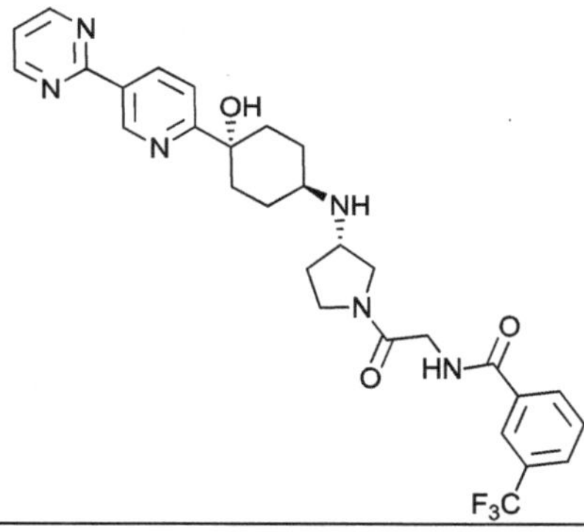 | 5.2 |

involved the replacement of one phenyl ring by a heteroaromatic ring to reduce hydrophobicity (logP) as in INCB3284 (**3**). **3** includes a 6-methoxy-3-pyridyl moiety on side 1 and is a selective and potent CCR2 antagonist with $IC_{50}$ values of 3.7 nM in hCCR2 binding and 4.7 nM in chemotaxis assay. In contrast to INCB3344 (**2**), INCB3284 (**3**) was a substrate for CYP 3A4 and CYP 2D6, but had no inhibitory or inducing effects on the CYP system. The inhibition of the hERG-associated potassium channel was rather low ($IC_{50} = 84$ μM) [35]. The balanced profile and safety data made INCB3284 (**3**) a promising candidate for phase I and phase II clinical trials, which were unfortunately terminated ahead of schedule [40].

A further clinical candidate from this series, which was tested in osteoarthritis and liver fibrosis, was INCB8761 (a.k.a. PF-4136309) (**4**), which belongs to the group of "inverse" pyrrolidines. Molecular modeling led to a new series of compounds, in which the contacts to the (R)-3-aminopyrrolidine as the main functional group on side 1 and 3-trifluoromethylbenzoyl aminoacetyl moiety on side 2 in the INCN3284 (**3**) series were reversed [39].

By analogy to the INCB3344 and INCB3284 series, the stereochemistry at the cyclohexyl ring required to be cis. Because previous studies demonstrated that the hydroxyl group and heteroaryl moiety at position 4 of the cyclohexyl ring led to weak hERG blockade and low intrinsic clearance, both structural elements were retained. The 6-methoxy-3-pyridyl moiety, previously present in the lead compound **3**, was replaced by a (pyrimidin-2-yl)-pyridin-2-yl residue in the potent analog INCB8761 (**4**) ($IC_{50} = 5.2$ nM). Compound **4** demonstrated no significant inhibition of other chemokine receptors or any influence on the CYP system. In contrast to the INCB3284 series, (*S*)-configuration in position 3 of the pyrrolidine ring is preferred over the (*R*)-enantiomer [27, 39].

The length of both linkers in compounds **2**, **3**, and **4** corresponds well to the pharmacophore model, where the linker L1 contains 4 carbon atoms and the linker L2 7 carbon atoms.

## 2.4 Piperidine-Based CCR2 Ligands

Merck has disclosed a variety of CCR2 antagonists, which contain a piperidine ring and a cyclopentanecarboxamide substructure. They identified a series of pyrido-annulated piperidines like MK0812 (**5**) that has a tetrahydro-3-trifluoromethyl-1,6-naphthyridine substructure (Table 2) [41]. This compound contains four chiral centers and is a potent CCR2 antagonist with an $IC_{50}$ value of 5.0 nM and inhibits the chemotaxis with an $IC_{50}$ value of 0.2 nM. MK0812 (**5**) became a clinical candidate in arthritis and multiple sclerosis, but failed in the phase II due to lack of efficacy and was therefore discontinued from the company's pipeline [37, 40].

MK 0483 (**6**) is another potent clinical candidate derived by Merck with an $IC_{50}$ value of 4.0 nM in the CCR2 binding [27]. **6** contains a piperidine and a 1,3-phenoxazine system instead of the tetrahydro-1,6-naphthyridine moiety as in **5** [24]. Further structural variations led to compound **7** not only with a binding affinity of 1.3 nM but also with a potent hERG inhibition ($IC_{50} = 54$ nM). To minimize effects on the hERG channel, the 4-fluorophenyl substituent in the R1 moiety was replaced by diverse more polar aryl and heteroaryl residues. The lowest inhibition of the hERG channel was observed after the introduction of a carboxylic acid in position 3 of the phenyl substituent, unfortunately the CCR2 binding also decreased in similar range [42]. A benzylamide incorporated in **7** is also a promising common structural element of Merck's spirocyclic CCR2 antagonists **13** and **14** (see Sect. 2.6).

Further development of piperidine-based CCR2 ligands led to a series exemplified by compound **8**. The potent CCR2 antagonist **8** includes a heteroaromatic system of an indole, representing R1, a central core with a cyclohexyl and a piperidine ring and a *trans*-configured cinnamide instead of the benzamide as in 7. In respect to

**Table 2** CCR2 antagonists with piperidine and cyclopentancarboxamide substructure, inhibitory effects on the CCL2 binding to the CCR2 receptor

| Compounds | | hCCR2 $IC_{50}$ (nM) |
|---|---|---|
| **5** (MK0812) | 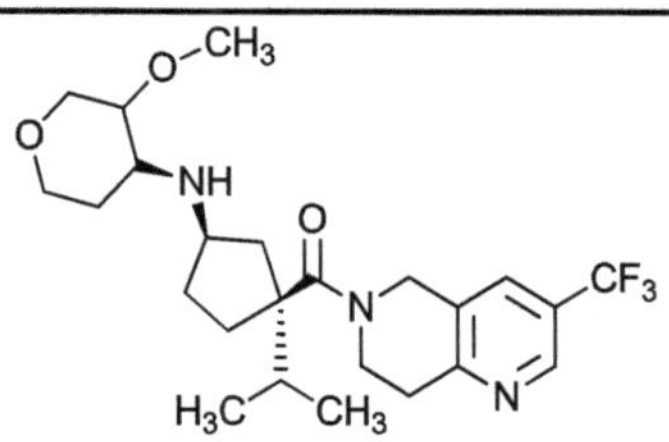 | 5.0 |
| **6** (MK0483) | 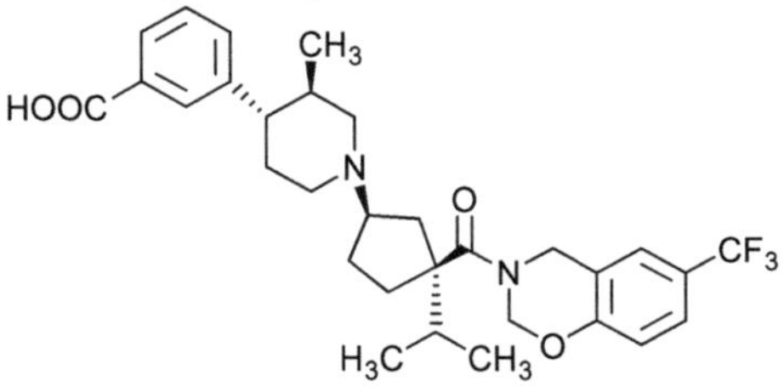 | 4.0 |
| **7** | 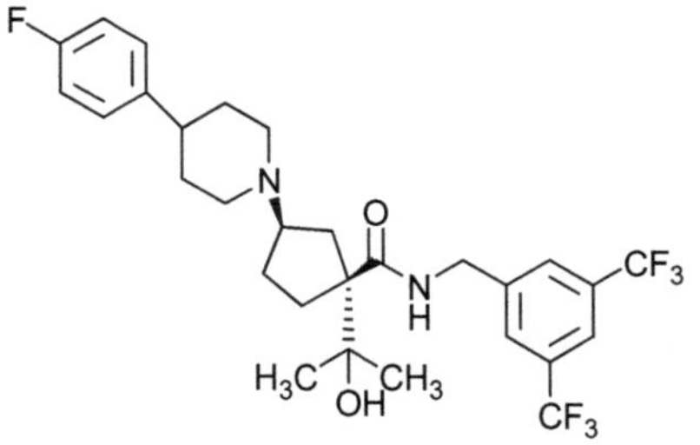 | 1.3 |
| **8** | 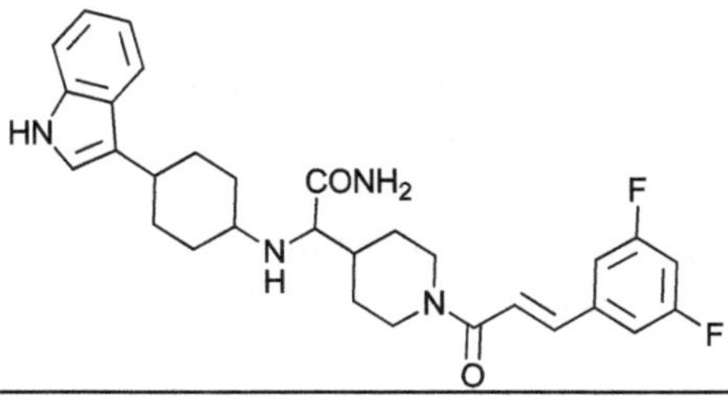 | 12 (*trans*)<br>740 (*cis*) |

cyclohexane stereochemistry, the *trans*-substituted compounds were more active in the CCR2 binding ($IC_{50}$ (hCCR2) = 12 nM) than the cis-configured derivatives ($IC_{50}$ (hCCR2) = 240 nM). It was shown that two substituents in *meta* or *para* position provided the highest CCR2 affinity. Although the in vitro hERG binding of **8** was rather high ($IC_{50}$ = 8 μM), an influence on hemodynamic parameters in a guinea pig model was not observed. Compound **8** also reached animal studies in an inflammation model (thioglycollate-induced peritonitis) [43].

**Table 3** CCR2 antagonists with piperazine structure, inhibitory effects on CCL2 binding to human CCR2 receptor

| Compounds | | hCCR2 $IC_{50}$ (nM) |
|---|---|---|
| **9** (PF-4254196) | 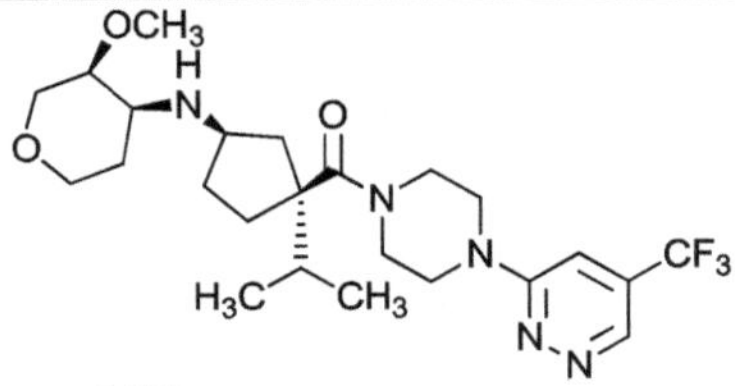 | 8.1 |
| **10** | 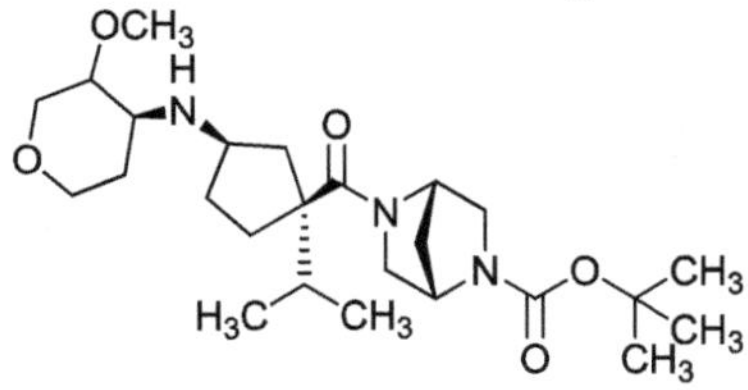 | 2.9 |
| **11** | 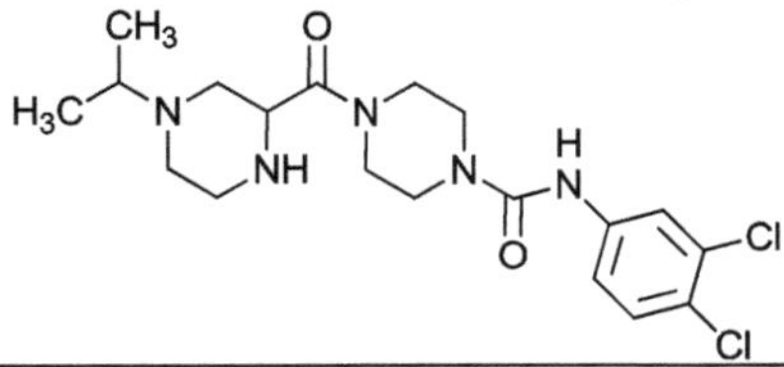 | 3.5 |

## *2.5 Piperazine-Based CCR2 Ligands*

The replacement of the trifluoromethyl naphthyridine group in MK0812 (**5**) by a (trifluoromethyl pyridazinyl)piperazine moiety led to a new series of piperazine-based CCR2 antagonists. PF-4254196 (**9**) is a potent ligand of the CCR2 receptor ($IC_{50}$ = 8.1 nM) without any cardiovascular liabilities ($IC_{50}$ (hERG) = 31.3 μM) (Table 3) [34]. Similar to Merck's piperidines MK0812 (**5**) and MK0483 (**6**), piperazines **9** and **10** also include a cyclopentane core with an amino substituent in position 3 and a carboxamide and isopropyl substituent in position 1. The development of PF-4254196 (**9**) started with modifications of the spacer length between the cyclopentane carboxamide and the trifluoromethyl containing aryl residues in existing series of CCR2 ligands. Prior compounds included a trifluoromethyl-substituted pyridine but showed a significant hERG inhibition. To eliminate the cardiovascular risk modifications of both, the side 1 tetrahydropyran ring and side 2 heterocycle were explored. Based on former SAR studies, a substitution of side 2 with more polar and/or potential π–π-stacking residues was expected to be well tolerated [34].

The introduction of a pyridazine ring led to PF-4254196 (**9**), which displayed a significantly better CCR2/hERG index and selectivity than the corresponding pyridine-containing compound. Pyridine-based compounds were also reported to be dual CCR2 and CCR5 antagonists [38].

Further modifications led to a methano-bridged piperazine derivative **10** (2,5-diazobicyclo[2.2.1]heptane). Although **10** contains a Boc (*tert*-butoxycarbonyl) moiety instead of an aryl or heteroaryl ring, it showed improved CCR2 affinity ($IC_{50} = 2.9$ nM) [44].

A further series of potent CCR2 ligands contains an aromatic ring on side 1 and a second terminal piperazine ring on side 2 connected via a carbonyl linker. The prototype of this series **11** demonstrated a high binding affinity to the CCR2 receptor ($IC_{50} = 3.5$ nM) and a significant reduction of hERG activity compared to methylene-linked subseries and other heterocycles at the side 2. In the hERG assay, a 10,000-fold selectivity for the CCR2 receptor over hERG was observed. Compound **11** did not interact with other chemokine receptors except with the CCR5 receptor ($IC_{50} = 22$ nM) and can therefore be considered as a dual antagonist with the CCR2 receptor preference [45–47].

## 2.6 *Spiropiperidine-Based CCR2 Ligands*

A prominent example of the spiropiperidine class of CCR2 inhibitors, RS504393 (**12**), was reported by Roche/Iconix. The central structural element of this class included a tertiary amine in a benzannulated piperidine ring system and an orthogonal relationship between the 3,1-benzoxazin-2-one and the piperidine ring, caused by the spirocyclic connection of the rings (Table 4). Another important aspect is the hydrogen binding potential of the urethane moiety and the restriction to small substituents at the benzoxazine heterocycle. SAR studies led to RS504393 (**12**) ($IC_{50} = 89$ nM) as the most affine compound of this benzoxazine class. The SARs of spiropiperidines were extensively investigated. In these spirobenzoxazine systems, the position 4 of the piperidine is disubstituted by a spiro-phenyl urethane system. Only piperidines or linear alkyl chains were accepted by the CCR2 receptor, other substituents were inactive or revealed reduced affinity. This group of compounds is highly selective for the CCR2 receptor [48].

Site-directed mutagenesis of acidic residues Glu291 and Asp284 in the CCR2 receptor to Ala, Asn, or Gln showed the importance of both Glu291 and Asp284 for ligand interactions via hydrogen bonding towards the tertiary amine of the piperidine ring. For this reason, spiropiperidines show an affinity to receptors in which a glutamic acid residue is in a similar position as in the CCR2 receptor [49]. It was also known that spiropiperidines prevent CCL2 binding by occupying the same region in the inter-helical bundle on the extracellular side [48]. In contrast, the CCR2 receptor binding of CCR2 antagonists without a basic amino moiety was not affected by Glu291 mutations [22].

Merck's compound **13** was claimed as a CCR2 antagonist for the potential treatment of inflammatory and rheumatic diseases [49]. This spiro[indene-piperidine] **13** showed $IC_{50}$ values of 1.3 and 0.45 nM in the CCR2 binding and the chemotactic assay, respectively. **13** possessed a high selectivity over other chemokine receptors including the CCR5 receptor ($\approx$500-fold).

**Table 4** CCR2 antagonists with spiropiperidine structure, inhibitory effects on CCL2 binding to human CCR2 receptor

| Compounds | | hCCR2 $IC_{50}$ (nM) |
|---|---|---|
| **12** (RS504393) | 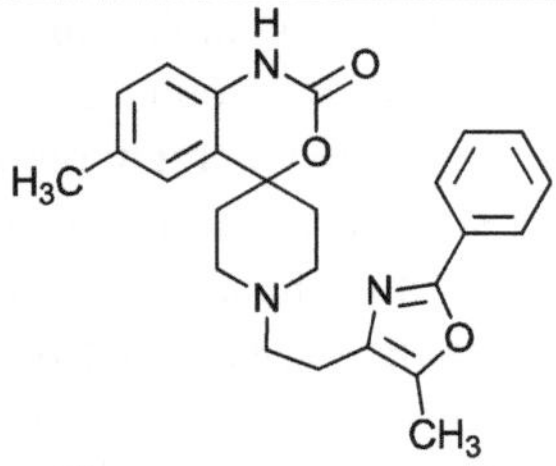 | 89 |
| **13** | 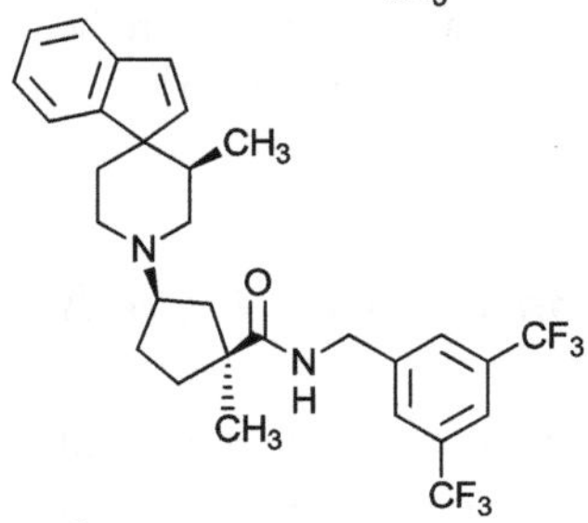 | 1.3 |
| **14** | 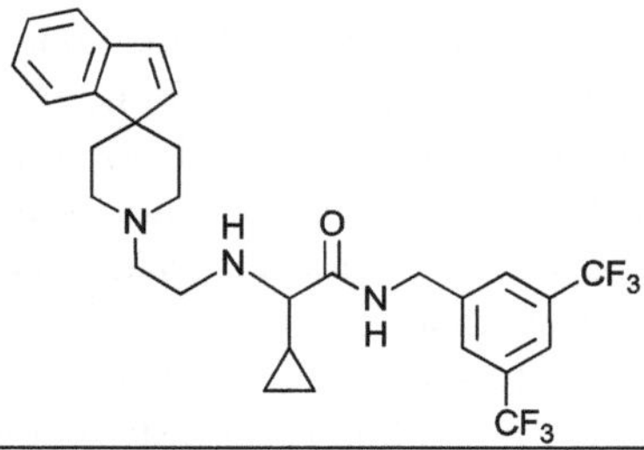 | 4 |

The tested isomer **13**, shown in Table 4, was found to be the only active stereoisomer of the four possible stereoisomers. The series of spiropiperidines similar to **13** demonstrated that the presence of methyl groups in position 3 of the 1,3-disubstituted cyclopentane ring and in the piperidine ring is crucial for the high affinity at the human and mouse CCR2 receptors. The introduction of an additional methyl group in position 4 of cyclopentane instead of the methyl group in position 3 led to a total loss of the CCR2 activity, as the removal of the methyl group in position 3 of the piperidine ring eliminated CCR2 activity in the assay with both human and murine CCR2 receptors [6, 50]. Compared to analogs without a methyl group in position 1 of the cyclopentane ring, the CCR2 binding affinity increased twofold [51]. Compound **13** also belongs to the series of cyclopentyl and cyclobutyl constrained analogs, in which a quaternary carbon substitution of the central cyclopentane ring was preferred for CCR2 binding [37].

The spiropiperidine **14** from Merck contains a tertiary carbon in side 2. The side 1 is represented by a spiro[indene-piperidine] in compounds **13** and **14**, but the cyclopentyl ring of **13** is replaced by an open chain, which includes an additional

**Table 5** CCR2 antagonists with quaternary ammonium salt structure, inhibitory effects on the CCL2 binding to the CCR2 receptor

| Compounds | | hCCR2 $IC_{50}$ (nM) |
|---|---|---|
| **15** (TAK-779) | 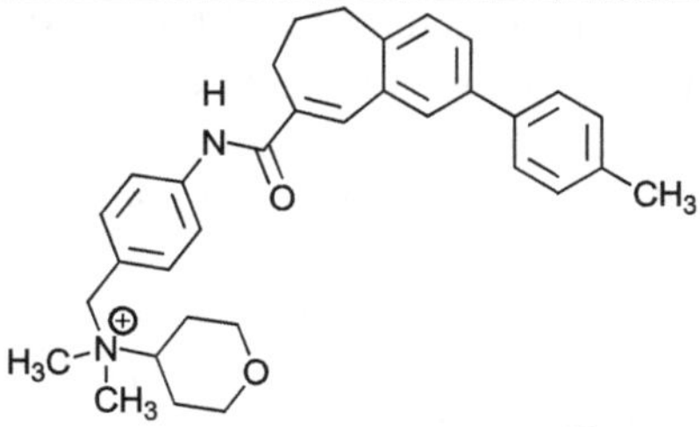 | 27 |
| **16** (JNJ17166864) | 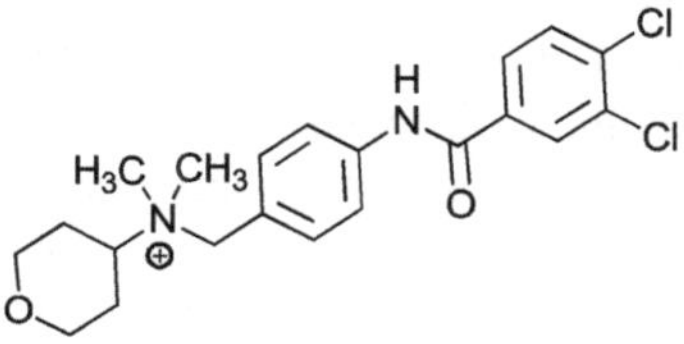 | 20 |
| **17** | 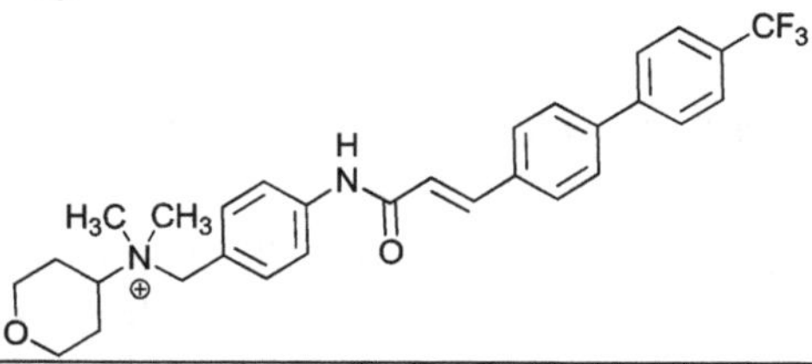 | 10 |

secondary amine and a cyclopropylmethyl moiety leading to a tertiary carbon atom. The substituent of the tertiary carbon atom was also varied. A *p*-fluorophenyl substituent at that position led to Merck's "compound 26", a dual CCR2 and CCR5 receptor antagonist, which is, apart from this substituent, identical to the cyclopropylmethyl derivative **14**. In contrast to the *p*-fluorophenyl derivative, the cyclopropyl derivative **14** was selective for the CCR2 receptor, showing an $IC_{50}$ value of 4 nM and promising pharmacokinetic properties [22, 23]. In both **13** and **14**, the aromatic residue R1 is represented by an indene and linked to the central tertiary amine via a 2-carbon linker. The size of linker L2 is broadly based, including 2 atoms in **12,** 6 in **13**, and 7 in **14**.

## 2.7 *Quaternary Ammonium Salts*

Potent CCR2 receptor antagonists that contain a quaternary ammonium salt have also been reported (Table 5). The quaternary ammonium moiety of these compounds is expected to form an ionic interaction with Glu291 in the binding pocket of the CCR2 receptor [27]. Developments of these types of CCR2 antagonists started from TAK-779 (**15**, see also Sect. 3.1), first developed as a CCR5 antagonist and later found to have also significant CCR2 affinity ($IC_{50} = 27$ nM) [40, 52].

Compounds **16** and **17**, both showing promising binding affinities in a $^{125}$I-labeled CCL2 assay using a THP-1 cell line, resulted from the systematic modifications of side 2 in **15** [53]. JNJ171668 (**16**), developed by Johnson&Johnson, is a potent candidate with a binding affinity of 20 nM. In line with its quaternary ammonium structure, JNJ171668 (**16**) had poor oral bioavailability, but entered clinical trials for the treatment of allergic rhinitis as a nasal application [40]. The 3,4-dichloro phenyl ring of **16** led to higher binding affinity at the CCR2 receptor than its analogs with other substitution patterns. From a series of different biphenyl-containing compounds, it was evident that the presence of a chloro or bromo group leads to improved binding affinities compared to electron-donating (OMe, Me) and electron-withdrawing groups (CN, $CF_3$). Structural modifications on the side 2 were also found to be responsible for selectivity related to the interactions with the CCR2 and CCR5 receptor. A different modification on side 2 led to compound **17** with a linker L2 consisting of 9 carbon atoms. **17** contains a biphenyl moiety attached to the 3-position of acrylamide displaying the binding affinity of 10 nM [53]. In comparison with the pharmacophore model, the linker L1 is missing in **15**, **16**, and **17**, whereas R1 is represented by a tetrahydropyran moiety.

## *2.8 Latest Structural Developments in CCR2 Ligands*

In recent years, new structural classes of CCR2 ligands appeared in the literature, mostly including only small series of compounds. Here, we want to highlight three promising classes, each exemplified by a typical ligand: (1) sulfonamides are represented by **18**, (2) azetidines by **19**, and (3) bicyclic compounds by **20**.

ChemoCentryx and GlaxoSmithKline started the development of sulfonamides as CCR2 ligands [6]. The efforts resulted in the triazolyl-substituted compound as a promising example. Modifications of the substitution pattern led to the trichloro-substituted *N*-phenylbenzenesulfonamide **18** as a potent CCR2 ligand (the GTPγS accumulation assay, $IC_{50} = 10$ nM). **18** inhibited monocyte recruitment but also showed inhibitory effects on the CYP2C19 and CYP2C9 activity (Table 6). The investigation in a thioglycollate-induced peritonitis model for inflammation in a mouse strain with the human CCR2 receptor knocked-in (hCCR2KI mouse) verified the dose-dependent and strain-specific inhibition of monocyte recruitment by **18** [54].

In an analogy to piperidines (see Sect. 2.4) and pyrrolidines (see Sect. 2.3), azetidine-based CCR2 ligands were described. In order to eliminate the zwitterionic character of a piperidine-based amino acid derivative developed by Johnson&Johnson and to increase solubility, a cyclohexylazetidine system was prepared first [31, 55]. Different six- and five-membered heterocycles were introduced in the 4-position of the cyclohexane ring, which reduced the hERG channel activity. Finally, the thiazole derivative **19** was identified as the most potent candidate. A cis orientation of the thiazolyl and the azetidinyl substituents on the cyclohexane ring was found to be essential for the high CCR2 binding affinity [55]. Compound **19** revealed an $IC_{50}$ value of 37 nM in the CCR2 binding assay and did not interact with the hERG channel ($IC_{50} > 50$ μM). A promising cardiovascular safety profile was confirmed in an anesthetized dog safety study. A high selectivity against other chemokine receptors

**Table 6** Latest CCR2 antagonists form different structural classes, inhibitory effects on CCL2 binding to the CCR2 receptor

| Compounds | | hCCR2 $IC_{50}$ (nM) |
|---|---|---|
| **18** | 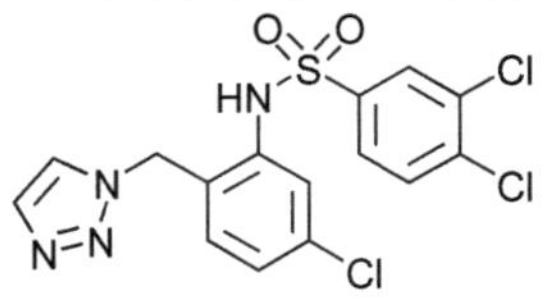 | [a] |
| **19** | 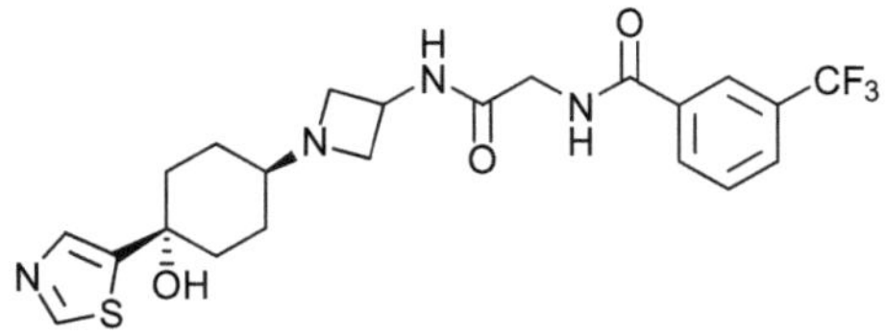 | 37 |
| **20** | 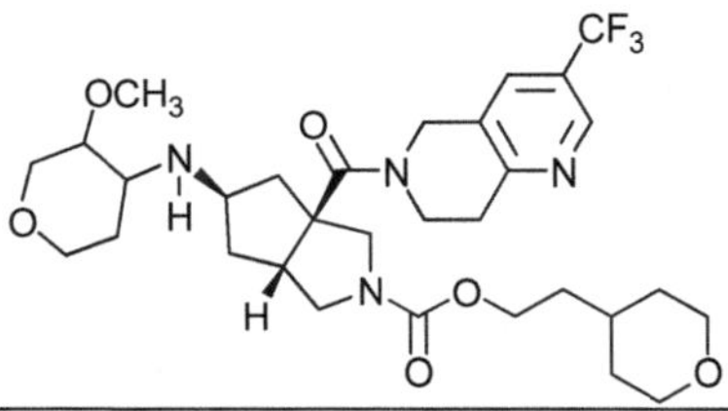 | 31 |

[a]$K_i = 10$ nM

was found, but the pharmacological profile remains to be reported [56]. In 2013, further variations of the heteroaromatic substituents at the *N*-acylglycine moiety were published, which started from **19** as a lead compound. Although various compounds with the high CCR2 affinity, promising functional activity and low hERG channel affinity were identified, none of these compounds displayed a promising pharmacokinetic profile [57].

In 2013, Cai et al. published a novel series of CCR2 antagonists with a bicyclo [3.3.0]octane or bicyclo[4.3.0]nonane scaffold. This class of ligands was designed according to the CCR2 pharmacophore model mentioned above. The first generation was based on a 7-aminobicyclo[3.3.0]octane system [58]. Replacement of the methylene moiety in 3-position by an amino moiety led to the closely related class of 7-amino-3-azabicyclo[3.3.0]octanes. A systematic evaluation of the substituents on the exocyclic amino moiety resulted in **20**, the most promising compound of this series. **20** displayed the high CCR2 affinity ($IC_{50} = 31$ nM) and low cardiovascular risk (hERG $IC_{50} > 50$ μM). The clinical potential of this new candidate will be evaluated after investigation of its in vivo properties [59].

## 2.9 *Binding Poses of Ligands in the CCR2 Receptor*

As predicted by computational-based homology modeling studies that included few potent CCR2 ligands, different amino acids were postulated to be essential for the ligand binding to the CCR2 receptor. The most recent studies were based on the

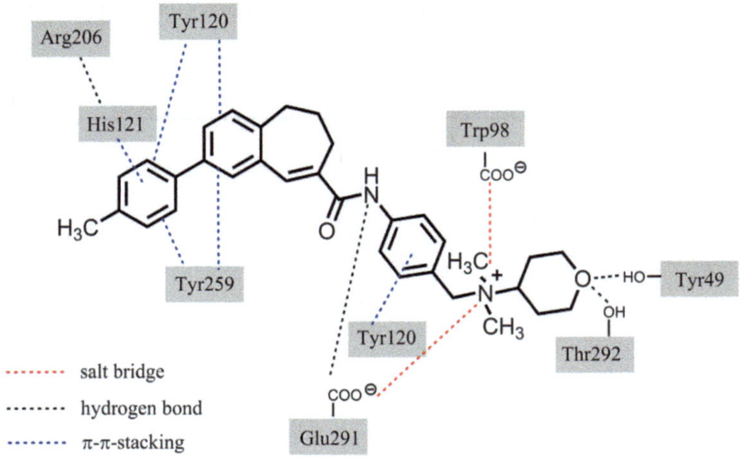


**Fig. 3** Schematic presentation of interactions between the CCR2 receptor and TAK-779 (**15**), modified according to [28, 60, 62]

homology modeling, where in 2010 published crystal structure of the closely related CXCR4 receptor served as a template. The CXCR4 structure shows a higher sequence homology as well as a larger binding pocket than the structure of bovine rhodopsin. The binding pocket of the CCR2 receptor is formed by transmembrane domains TM2, TM3, TM5, TM6, and TM7 [28]. Glu291, located on the transmembrane region 7 (TM7), was proposed to be an important anchor residue of various CCR2 ligands, including spiropiperidines [48, 60, 61]. The central basic amine, present in most CCR2 receptor ligands, forms a salt bridge to this conserved acidic residue.

Mutagenesis studies also implied that Tyr120 and His121 might be crucial because of their ability to form hydrogen bonds with endogenous ligands or synthetic molecules. Hydrophobic interactions of the ligands were observed with aromatic residues Tyr49 and Trp98 [28]. The predicted binding site of TAK-779 (**15**) was studied best. The residues of the CCR2 receptor that strongly interact with TAK-779 are shown in Fig. 3.

The most important interaction is the electrostatic interaction of the carboxylate of Glu291 with the quaternary ammonium group of TAK-779. The tetrahydropyran oxygen forms hydrogen bonds with both Tyr49 in TM1 and Thr292 in TM7. The biaryl system on the other site is fixed between the hydroxyphenyl moieties of Tyr120 and Tyr259 and to a lesser extent by His121 via π–π-stacking interactions. His121 also interacts with Arg206 (TM 5), which results in a weaker ligand-histidine interaction. In the CCR5 receptor, His121 is replaced by phenylalanine (see Fig. 5), which cannot form interaction with arginine and therefore adopts an alternative rotameric conformation increasing the π-π-interactions. In case of TAK779, the residues Tyr49, Trp98, Tyr120, and His121 are discussed to form an aromatic cluster contributing to the CCR2 receptor binding [60, 62]. For the binding of the spirocyclic antagonist RS504393 (**12**), Glu291 and Asp284 were identified as hydrogen bond partners for the tertiary amine within the piperidine ring [48, 49]. Expectedly, the binding of the CCR2 antagonists devoid of basic amine was not affected by Glu291 mutations [22, 23].

### *2.10 Conclusion*

Altogether diverse compounds with high affinity to the CCR2 receptor have been identified. A few CCR2 antagonists were investigated in clinical trials. Up to now, CCR2 antagonists have not shown promising clinical efficacy in inflammatory diseases as presumed from preclinical models. Whether this failure is a result of wrong target selection, off-target effects or poor drug-like properties of the small-molecule antagonist remains to be elucidated [27, 37, 41].

## 3 CCR5 Ligands

The unique opportunity to study the impact of CCR5 receptor antagonists by exploiting the CCR5Δ32 bp polymorphism as well as its function as a co-receptor for the HIV-host cell fusion brought the CCR5 receptor into focus of many pharmaceutical companies. There are various CCR5 antagonists reported in the literature so far. The CCR5 receptor ligand maraviroc (**1**, Celsentri®, UK-427,827) (Fig. 1) developed by Pfizer is the only CCR5 ligand approved for the treatment of confirmed R5-tropic HIV-1 infection by the FDA and EMEA [2]. On account of this, many further investigations have already been undertaken, starting from maraviroc (**1**) as a lead compound.

The intention of this chapter is to summarize the SARs of various CCR5 antagonists focusing on CCR5 antagonists derived from TAK compounds (Takeda), maraviroc, and related tropane-based CCR5 ligands. Several different aspects such as antiviral activity, CYP inhibition, leading to several drug-drug interactions and severe adverse effects will be discussed [65, 66]. Inhibition of the hERG $K^+$-channel is a common challenge in developing CCR5 selective ligands due to the basic amine, which is required for the interaction with Glu283 of CCR5 receptor [29, 65, 66]. Therefore, the affinity to the hERG $K^+$-channel [67] and the oral bioavailability, which significantly influence the CCR5 ligand development, will be discussed.

### *3.1 CCR5 Ligands Developed by Takeda Inc*

#### 3.1.1 Quaternary Ammonium Salts and Tertiary Amine-Based CCR5 Antagonists

Takeda Pharmaceutical Company has set themselves the task of creating a new class of antihuman immunodeficiency virus 1 (HIV-1) entry inhibitors. One way to

**Fig. 4** TAK-779 (**15**)

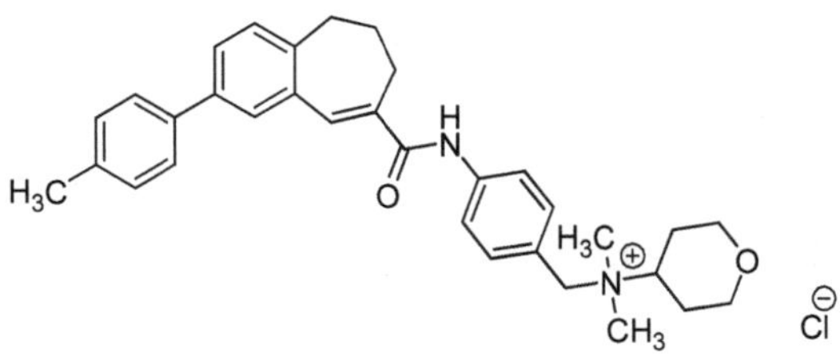

inhibit HIV-1 replication is to prevent the viral entry into the target cell. The potential of this approach is shown by T20, a peptide that prevents the conformational change in the viral glycoprotein gp41 that drives membrane fusion [68]. Therefore, Takeda has designed several compounds which were based on hits of a high-throughput screening (HTS). The most promising compound resulting from these hits was TAK-779 (**15**) (Fig. 4).

TAK-779 (**15**) antagonizes the binding of CCL5 to CCR5-expressing Chinese hamster ovary (CHO) cells completely at a concentration of 100 nM and showed an $IC_{50}$ value of 1.4 nM. Moreover, **15** was shown to block membrane fusion of HIV-1 at nanomolar concentrations. The binding of CCL3 and CCL4 to the CCR5-expressing cells was also blocked with $IC_{50}$ values around 1.0 nM. Although TAK-779 inhibited the binding of [$^{125}$I]-CCL2 to CCR2 in CHO/CCR2 cells, its $IC_{50}$ value for CCR2 receptor ($IC_{50}$ = 25 nM) was approximately 20-fold higher than that for CCR5 receptor [52]. The sequence homology between CCR5 and CCR2 receptors is 76% [69], which might explain the dual antagonistic character.

Molecular modeling and mutagenesis studies have shown that the active site of the CCR5 receptor is very hydrophobic with multiple aromatic residues forming a tight binding pocket [70]. The benzene ring of the benzo[7]annulene moiety of **15** was observed to interact with aromatic side chains of Tyr108 and Trp248 via a T-shaped π–π-stacking. Additionally Tyr108 forms a hydrogen-bond interaction between the phenolic OH group and the carbonyl moiety of **15**. Strong hydrophobic interaction between the *p*-tolyl group and Ile198 on TM5 accompanied by some weaker interactions of **15** with Thr195, Ile198, Phe109, Trp248, and Tyr251 were also found. The limited ionic interaction between the quaternary ammonium moiety of **15** and Glu283 was caused by the steric shielding of the positively charged center (Fig. 5) [63].

Due to the fact that **15** inhibits the CCR5 receptor, an anti-R5 HIV-1 assay was performed. The measured effective concentrations in the anti-fusion assay were 1.2 nM ($EC_{50}$) and 5.7 nM ($EC_{90}$) [71]. It was additionally shown that TAK-779 did not interact with CCR1, CCR3, or CCR4 receptors. The quaternary ammonium moiety of TAK-779 led to a good binding affinity, but poor oral bioavailability, which required further optimization.

In order to develop an active CCR5 antagonist with a reasonable oral bioavailability, derivatives of tertiary amines were investigated. The tertiary amine **21**, derived from TAK-779 by removing one $CH_3$ group, resulted in decrease of CCR5 affinity but increased oral bioavailability. In order to enhance the CCR5, affinity modifications of the [7]annulene ring were undertaken [72]. The exchange

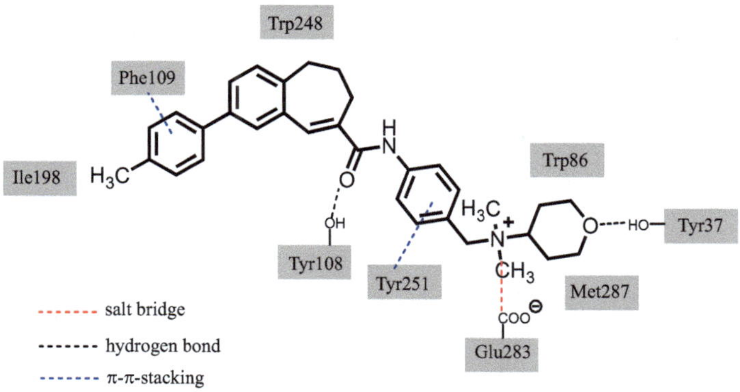


**Fig. 5** Schematic presentation of interactions between CCR5 receptor and TAK-779 (**15**), modified according to [63, 64]

**Table 7** CCR5 affinities of quaternary ammonium salts **15** and **25** and the tertiary amines **21–24**, **26**, and **27** [72, 73, 75] and inhibitory effects on [$^{125}$I]-CCL5 binding to CCR5-expressing CHO cells

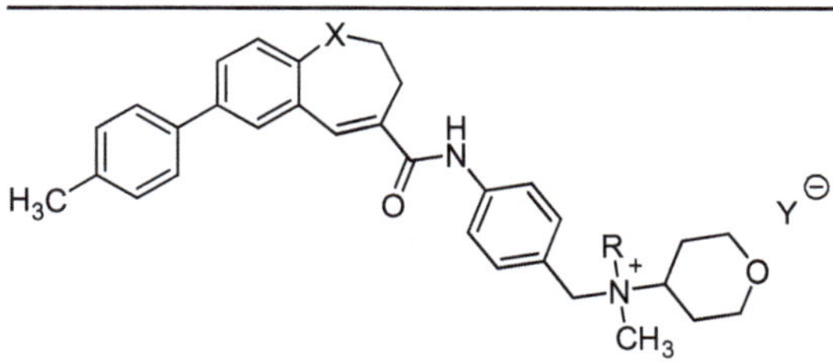

| Compounds | *X* | *R* | *Y* | CCR5 $IC_{50}$ (nM) |
|---|---|---|---|---|
| **15** | $CH_2$ | $CH_3$ | Cl | 1.4 |
| **21** | $CH_2$ | – | | 950 |
| **22** | S | – | | 800 |
| **23** | SO | – | | 300 |
| **24** | $SO_2$ | – | | 200 |
| **25** | O | $CH_3$ | Cl | 1.4 |
| **26** | O | – | | 530 |
| **27** | $NCH_3$ | – | | 130 |

at the 5-$CH_2$-moiety of the benzo[7]annulene of TAK-779 (**15**) by a S-atom (**22**) did not significantly increase CCR5 receptor affinity [73]. The introduction of a sulfoxide (**23**) or a sulfone (**24**) led to a slightly increased affinity [74]. The exchange of the 5-$CH_2$-group by an O-atom resulted in the benzoxepine **25** with high CCR5 affinity, which could not be retained in the tertiary amine **26**. The highest affinity of the tertiary amines was found for the 1-methyl-1-benzazepine **27** ($IC_{50}$ = 130 nM) [72]. Because the compounds **24** and **27** possess high oral bioavailability (>50%) in rats [73], further variations of the benzothiepine-1,1-dioxide and the 1-benzazepine cores were envisaged (Table 7).

In order to enhance the CCR5 receptor affinity, modifications of the *p*-methyl group of **24** were investigated (Table 8). Replacement of the methyl group by an ethyl substituent (**28**) led to a 3-fold increased CCR5 receptor affinity, which was

**Table 8** Variation of the *p*-$CH_3$ group [75], inhibitory effects on the [$^{125}$I]-CCL5 binding to CCR5-expressing CHO cells

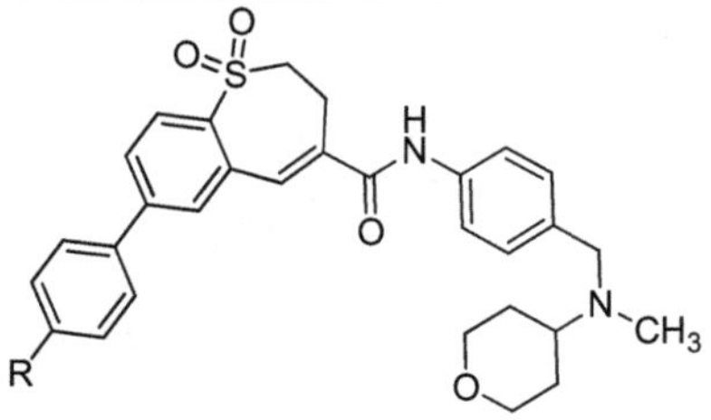


| Compounds | *R* | $IC_{50}$ (nM) |
|---|---|---|
| **24** | $H_3C$– | 200 |
| **28** | Et– | 60 |
| **29** | PrO– | 35 |
| **30** | $BuO(CH_2)_2O$– | 27 |

**Table 9** Variation of the N-substituent

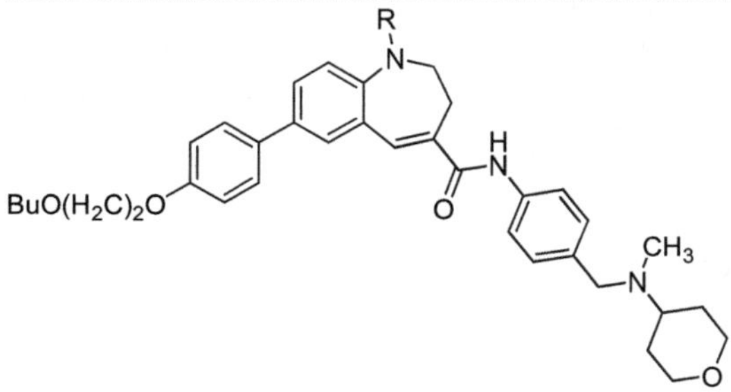


| Compounds | *R* | CCR5 $IC_{50}$ (nM)[a] | Membrane fusion $IC_{50}$ (nM)[b] |
|---|---|---|---|
| **31** | Et | 5.6 | 1,000 |
| **32** | Pr | 3.5 | 54 |
| **33** | *i*-Bu | 3.6 | 1.7 |
| **34** | 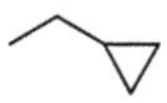 | 4.5 | 150 |
| **35** | Bn | 5.3 | 2.3 |
| **36** | 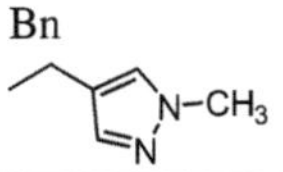  | 2.7 | 1.2 |

[a]Inhibitory effects on [$^{125}$I]-CCL5 binding to CCR5-expressing CHO cells
[b]Inhibition of membrane fusion [75]

further increased by introduction of alkoxy groups (compounds **29** and **30**) [75]. The butoxyethoxy derivative **30** was chosen for further optimization.

The combination of the butoxyethoxy group with a 1-benzazepine scaffold led to the next series of CCR5 antagonists. Homologation of the N-methyl group to an N ethyl group (**31**) resulted in an increased CCR5 affinity, but low inhibition of HIV-1 envelope-mediated membrane fusion (Table 9). Introduction of propyl (**32**), isobutyl (**33**), benzyl (**35**), and methylpyrazolyl (**36**) residues increased the inhibitory activity, whereas the cyclopropylmethyl group (**34**) resulted in a remarkable drop of inhibitory activity [75].

**Table 10** Introduction of a sulfoxide, inhibitory effects on the binding of [$^{125}$I]-CCL5 to CCR5-expressing CHO cells

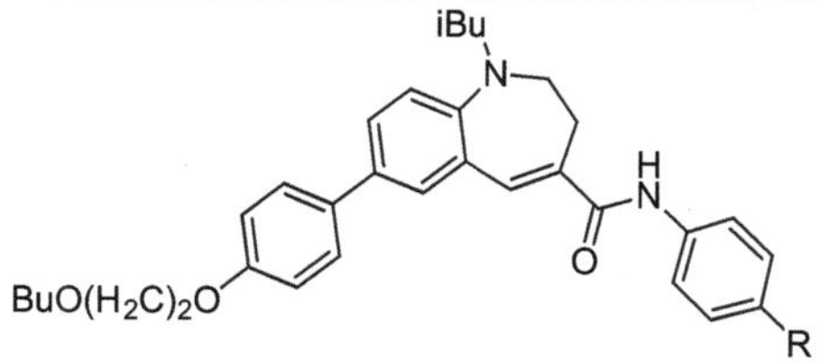


| Compounds | *R* | CCR5 $IC_{50}$ (nM) |
|---|---|---|
| **37** | 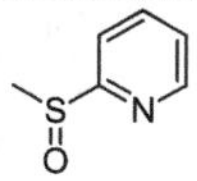 | 450 |
| **38** | 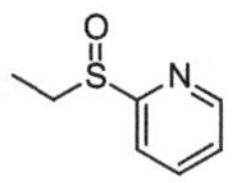 | 250 |
| **39** | 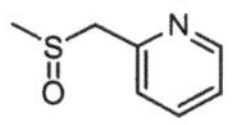 | 67 |
| **40** | 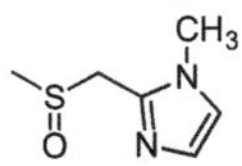 | 14 |
| **41** | 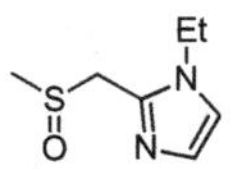 | 6.6 |
| **42** | 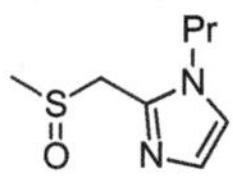 | 4.3 |
| (*S*)-**42** | 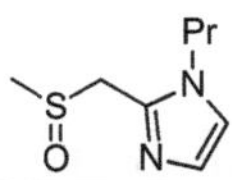 | 1.9 |

### 3.1.2 1-Benzazepine and 1-Benzazocine-Based CCR5 Ligands

Compounds **37–42** were synthesized to examine the effect of various sulfoxides on the CCR5 affinity (Table 10). The methylimidazolyl-sulfinyl derivative **40** led to increased CCR5 receptor affinity. Elongation of the alkyl substituent to an ethyl (**41**) or propyl (**42**) moiety led to increased CCR5 affinity. The enantiomer (*S*)-**42** was found to be more potent than the (*R*)-enantiomer (*R*)-**42** [72].

Expansion of the seven-membered azepine ring of (*S*)-**42** to an azocine ((*S*)-**43**), azonine ((*S*)-**44**), and azecine ring ((*S*)-**45**) led to a series of potent compounds (Table 11) [72].

Due to the high CCR5 affinity, virus fusion inhibition, and oral bioavailability, TAK-652 ((*S*)-**43**) became a promising HIV-1 entry inhibitor for clinical studies. It was shown that TAK-652 inhibited the binding of CCL5 ($IC_{50}$ = 3.1 nM), CCL3,

**Table 11** Variation of ring size

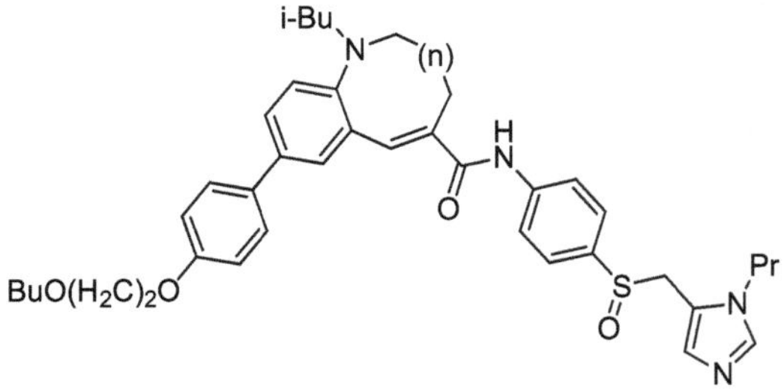


| Compounds | *n* | CCR5 $IC_{50}$ (nM)[a] | Membrane fusion $IC_{50}$ (nM)[b] |
|---|---|---|---|
| (*S*)-**42** | 0 | 1.9 | 1.0 |
| (*S*)-**43** (TAK-652) | 1 | 3.1 | 0.1 |
| (*S*)-**44** | 2 | 6.8 | 0.48 |
| (*S*)-**45** | 3 | 67 | – |

[a]Inhibitory effects on the binding of [$^{125}$I]-CCL5 to CCR5-expressing CHO cells
[b]Inhibitory effects on the binding of HIV-1 envelope-mediated membrane fusion

and CCL4 ($IC_{50}$ = 2.3 nM) to the CCR5 receptor and also blocked CCL2 binding ($IC_{50}$ = 5.9 nM) to the CCR2 receptor. In further tests, the inhibitory effect on the fusion between the HIV-1 envelop protein and the cell membrane was investigated with TAK-652 ($IC_{50}$ = 0.1 nM). The replication of all HIV-1 isolates in peripheral blood mononuclear cells (PBMCs) was inhibited by TAK-652 with $EC_{50}$ and $EC_{90}$ values of 0.061 and 0.25 nM, respectively [76].

### 3.1.3 Propandiamine-Based CCR5 Ligands

Compounds **46–48** represent CCR5 antagonists with entirely different structures. The core structure of **46–48** is characterized by an *N*-(3-piperidinopropyl) carboxamide. The propandiamine substructure has become an important pharmacophore element for the development of CCR5 antagonists (Table 12).

The HTS of the Takeda's compound library led to the discovery of *N*-(piperidinopropyl)carboxamide **46** with low micromolar CCR5 binding affinity. Subsequent optimization resulted in a series of piperidine-4-carboxamides, exemplified by **47**, which had low nanomolar affinity for CCR5 receptors and exhibited high anti-HIV-1 activity [77]. The fast metabolism of **47** stimulated further optimization, which led to the most promising derivative **48** (TAK-220) of this new series (Table 12). It demonstrated high inhibition of the [$^{125}$I]-CCL5 binding to CCR5-expressing CHO cells ($IC_{50}$ = 3.5 nM), high inhibition for HIV-1 membrane fusion ($IC_{50}$ = 0.42 nM), and also high metabolic stability upon incubation with human hepatic microsomes [78]. A comparison of binding affinities and antiviral activity of TAK-779 (**15**), TAK-652 (**43**), and TAK-220 (**48**) is summarized in Table 13.

**Table 12** Inhibitory effects on the binding of [$^{125}$I]-CCL5 to CCR5-expressing CHO cells

| Compounds | | CCR5 $IC_{50}$ (nM) |
|---|---|---|
| **45** | 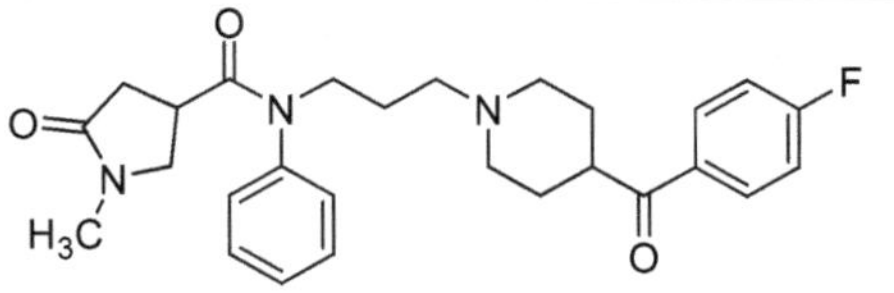 | 1,900 |
| **47** | 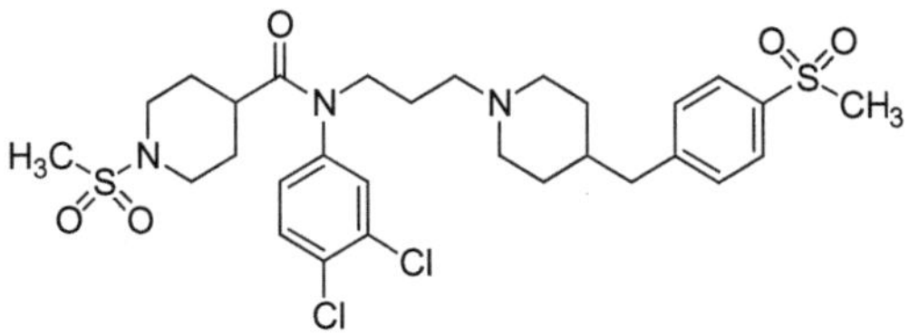 | 2.3 |
| **48** (TAK-220) | 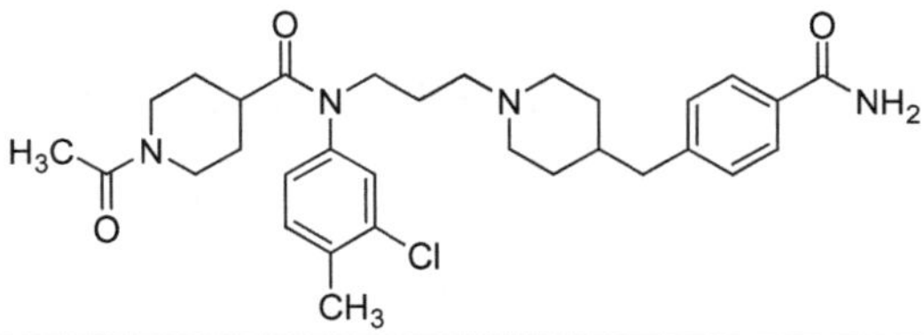 | 3.5 |

**Table 13** CCR5 and CCR2 binding affinities of compounds **15** (TAK-779), **43** (TAK-652) and **48** (TAK-220)

| Compounds | CCR5 $IC_{50}$ (nM)[a] | CCR2 $IC_{50}$ (nM)[b] | Membrane fusion $IC_{50}$ (nM)[c] |
|---|---|---|---|
| **15** (TAK-779) | 1.4 | 27 | 15 |
| **43** (TAK-652) | 3.1 | 5.9 | 0.1 |
| **48** (TAK-220) | 3.5 | – | 0.42 |

[a]Inhibitory effects ($IC_{50}$) on the binding of [$^{125}$I]-CCL5 to CCR5-expressing CHO cells
[b]Inhibitory effects ($IC_{50}$) on the binding of [$^{125}$I]-MIC-1 to CCR2b-expressing CHO cells
[c]Inhibition ($IC_{50}$) of HIV-1 envelope-mediated membrane fusion

The conformational flexibility of TAK-220 is higher than that of TAK-779 due to the higher number of rotatable bonds. Docking into a 3-D homology model of the CCR5 receptor showed that TAK-220 forms a strong salt bridge with Glu283. Mutagenesis studies indicated that the residues Trp86, Tyr108, Trp248, Tyr251, and Met287 (see Fig. 5) are important for TAK-779 binding, but have little effects on TAK-220 binding. However the hydrophobic interaction of TAK-220 with Ile198 is as strong as in case of TAK-779 binding (Fig. 6). The 3-chloro-4-methylphenyl group of TAK-220 is placed in the similar region within the helical bundle between Phe109, Trp248, and Tyr251 as the phenyl group of maraviroc (**1**) [63, 79].

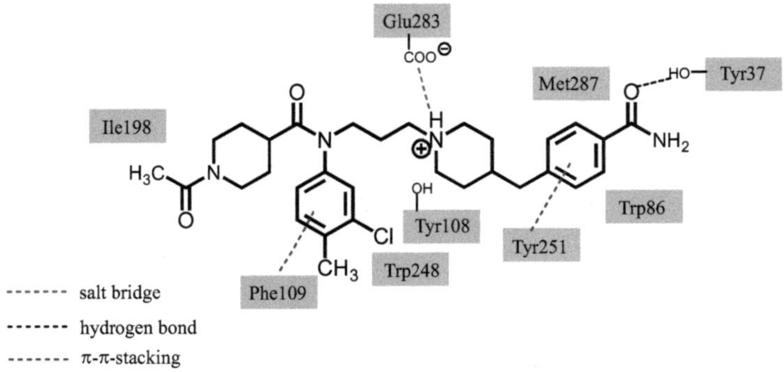


**Fig. 6** Schematic presentation of interactions between the CCR5 receptor and TAK-220 (**48**), modified according to [79]

## *3.2 The Development of Maraviroc and Related Tropane-Based CCR5 Ligands*

### 3.2.1 1-(3,3-Diphenylpropyl)-Piperidinyl and 1-(3-Amido-3-Phenylpropyl)-Piperidinyl-Based CCR5 Ligands

The first 1-(3,3-diphenylpropyl)piperidine-based CCR5 antagonists were found by AstraZeneca and Pfizer by an HTS of their compound libraries [45, 80]. AstraZeneca's screen resulted in two closely related hits, **49** and **50,** and Pfizer's screen in the hits **51** (UK-107,543) and **52** (Table 14).

AstraZeneca's approach focused on the development of CCR5-selective antagonists for the treatment of chronic inflammatory diseases, such as rheumatoid arthritis [81] and inflammatory bowel disease [82]. Compounds **49** and **50** demonstrated similar CCR5 binding affinities in low micromolar range, indicating no advantage of the cyclized N-substituent of **49**. The SAR investigations, in which the substituent $R^2$ of the acyl group was varied, revealed that neither (hetero)aromatic nor aliphatic groups increased the CCR5 receptor affinity (Table 15). The phenylacetyl derivative **53** was the only compound with potency in the high sub-micromolar range. Introduction of substituents at the *o*- and *m*-position of the phenylacetyl group did not significantly affect the binding affinity. However, an increase in affinity was observed after introduction of polar electron-withdrawing substituents in *p*-position. In particular, sulfamoyl (**55**), *N,N*-dimethylsulfamoyl (**56, 57**), and methylsulfonyl (**58**) groups showed nanomolar CCR5 affinities. Replacing the methyl (**56**) with the ethyl group (**57**) at the amide N-atom slightly increased the CCR5 affinity (Table 15). Replacement of the amide substructure by a sulfonamide moiety was shown to be detrimental, whereas the introduction of a urea moiety instead of the amide retained the CCR5 affinity. Compounds **55, 56,**

**Table 14** HTS hits **49**, **50** (AstraZeneca), **51** (UK-107,543), and **52** (Pfizer)

| Compounds | | CCR5 $IC_{50}$ (nM) |
|---|---|---|
| **49** | 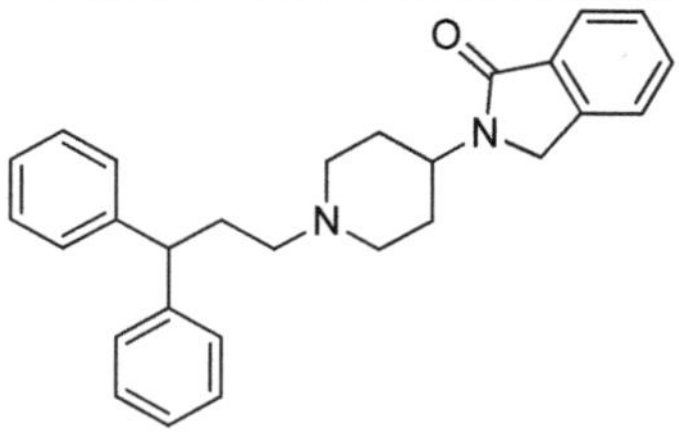 | 1,900[a] |
| **50** | 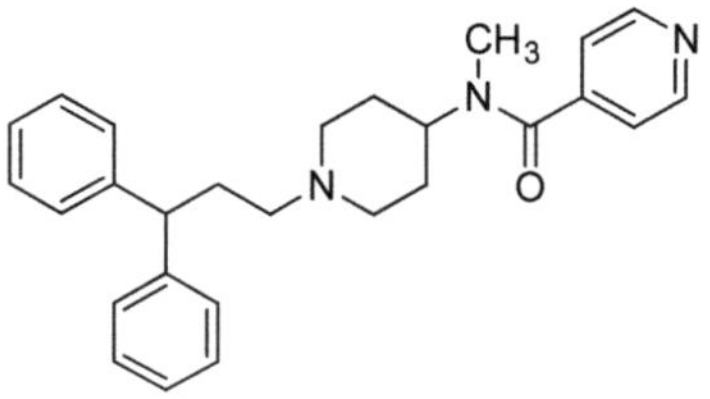 | 2,300[a] |
| **51** (UK-107,543) | 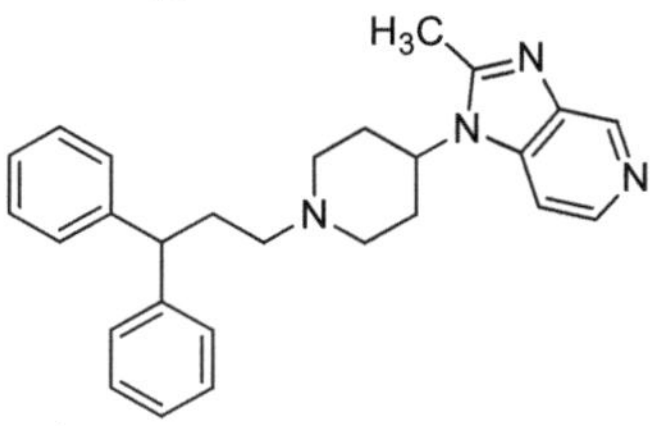 | 400[b] |
| **52** | 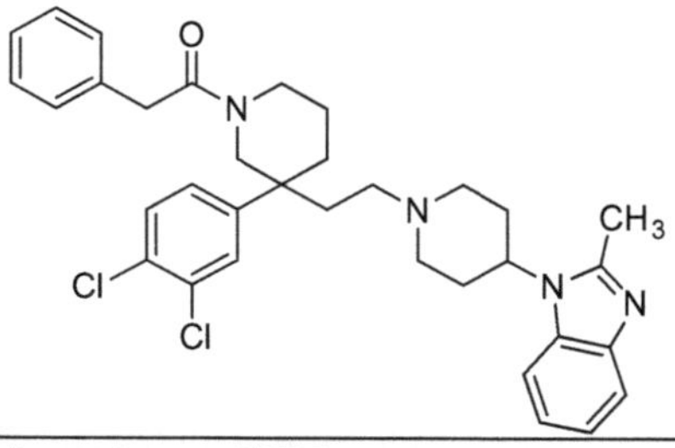 | 1,100[b] |

[a]Inhibition of [$^{125}$I]-CCL5 binding to human CCR5 receptors
[b]Inhibition of CCL4 binding to the human CCR5 receptor in stably expressed HEK-293 cells

and **57** showed no affinity towards CCR1, CCR2b, CCR3, CXCR1, and CXCR2 receptors. The *N*,*N*-dimethylsulfonamide **57** displayed micromolar affinity to muscarinic and serotonergic receptors as well as the hERG channel [45]. Therefore, the methyl sulfone **58** was chosen as the new lead compound for further ligand development.

Next SAR studies around the benzhydryl structure were undertaken. Introduction of one **60** or two **59** fluorine atoms into the *p*-position of the phenyl rings resulted in decreased CCR5 affinity. In contrast, one chlorine atom in *p*-position (**61**) was highly beneficial. Therefore, further substituents in *p*-position of one

**Table 15** CCR5 receptor affinities of compounds **49, 50, 53–57**. Inhibitory effects of [$^{125}$I]-CCL5 to human CCR5 receptors

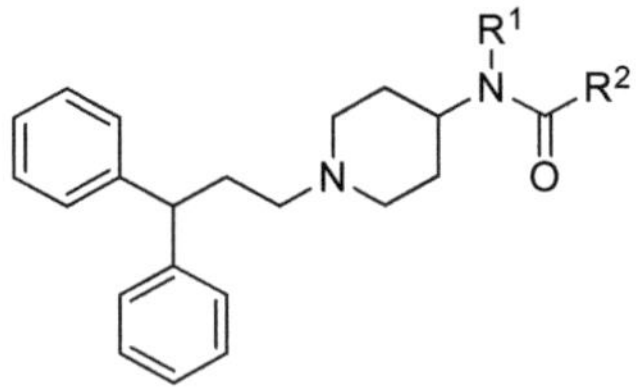


| Compounds | $R^1$ | $R^2$ | CCR5 $IC_{50}$ (nM) |
|---|---|---|---|
| **49** | $H_3C$– | – | 4,100 |
| **50** | $H_3C$– | 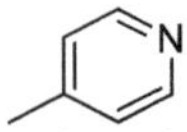  | 6,100 |
| **53** | $H_3C$– | 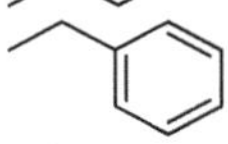 | 810 |
| **54** | $H_3C$– | 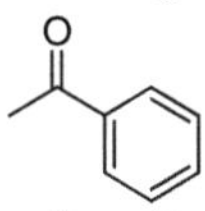  | 5,900 |
| **55** | $H_3C$– | 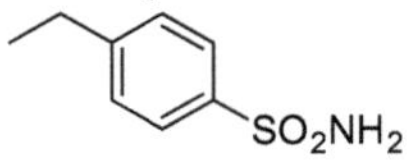  | 91 |
| **56** | $H_3C$– | 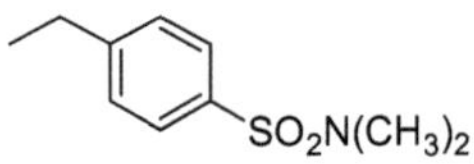  | 46 |
| **57** | Ethyl– | 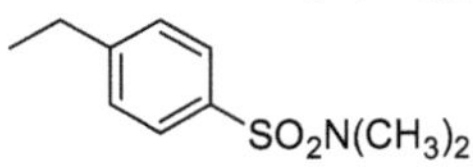  | <10 |
| **58** | Ethyl– | 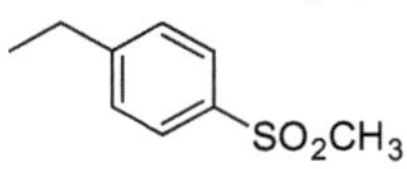  | 76 |

phenyl ring, retaining the other phenyl ring unsubstituted, were investigated. Compounds with strongly electron-withdrawing substituents such as trifluoromethyl (**62**), cyano (**64**), and methylsulfonyl (**66**) were highly potent, but also the methoxy derivative **65** demonstrated high CCR5 affinity (Table 16). Electronic effects alone are not sufficient to explain the SAR. The weakly potent difluoro (**59**) and 4-fluoro (**60**) compounds as well as the methylsulfonyl (**66**) derivative demonstrated sufficient metabolic stability and a good pharmacokinetic (PK) profile. The potential of the fluorine atoms to reduce oxidative metabolism on the phenyl rings of the diphenylpropyl moiety indicated a possible approach to improve oral bioavailability in this series of CCR5 ligands [47].

The (*S*)-enantiomer of the methylsulfonyl derivative (*S*)-**66** was found to be twice as potent as the racemate **66.** Moreover the enantioselective synthesis of (*S*)-**66** was used to prepare several analogs with various substituents at the second phenyl ring.

**Table 16** $IC_{50}$ values of 1-(3,3-diphenylpropyl)-piperidine derivatives **58–66** with various substituents at the phenyl residues. Inhibition of CCL3 binding to the human CCR5 receptor

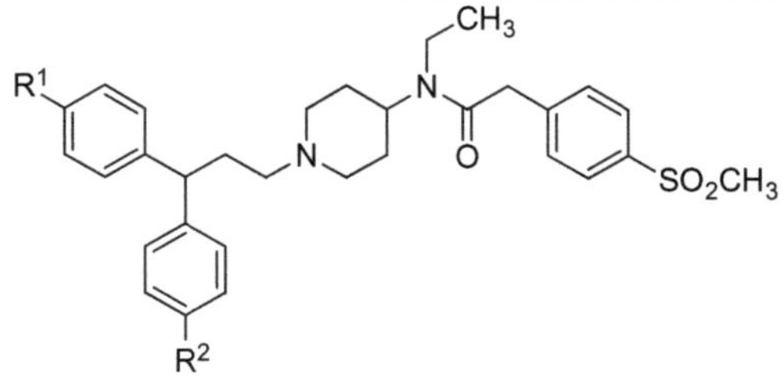


| Compounds | $R^1$ | $R^2$ | CCR5 $IC_{50}$ (nM) |
|---|---|---|---|
| **58** | H– | H– | 18 |
| **59** | F– | F– | 780 |
| **60** | F– | H– | 310 |
| **61** | Cl– | H– | 8.5 |
| **62** | $F_3C$– | H– | 2.3 |
| **63** | $H_3CO_2C$– | H– | 7.1 |
| **64** | CN– | H– | <1.0 |
| **65** | $H_3CO$– | H– | 6.3 |
| **66** | $H_3CSO_2$– | H– | 1.7 |

Since fluorine atom in *p*-position of the second phenyl moiety led to dramatic loss of the CCR5 affinity, the *m*-position was addressed for the introduction of the fluoro substituent (**67**). **67** displayed increased CCR5 affinity, but also fast clearance and short half-life. This could be improved by introduction of halogen atoms at both *m*-positions (Table 17). The 3,5-*di*fluoro (**69**), 3-fluoro (**67**), and 5-chloro (**68**) analogs showed favorable pharmacokinetic profiles. The 3,5-difluoro derivative **69** displayed no affinity towards CCR1, CCR2b, CCR3, CXCR1, and CXCR2 receptors and other human G-protein-coupled receptors (human $M_1$, $M_2$, and 5-$HT_{2A}$ receptor). Unfortunately, an inhibition of CYP 2D6 (1.6 μM) and hERG ion channel binding (7.3 μM) were detected [46].

In order to reduce cardiotoxicity, the benzhydryl part of the molecule was further modified. SAR investigation clearly indicated the requirement of one phenyl substituent, whereas the replacement of the second phenyl ring by other substituents was tolerated. Introduction of a piperazine and C-atom-linked piperidine ring led to compounds with reduced lipophilicity, promising CCR5 affinity and decreased hERG ion channel binding (Table 18) [83]. The piperazine derivative **71** showed only moderate bioavailability in dogs and very fast plasma clearance in rats, whereas the C-linked piperidine with the methylsulfonyl substituent at the N-atom demonstrated good bioavailability in both species with high selectivity over CYP 1A1, 2C9, 2C19, 2D6, and 3A4 enzymes. Therefore **72** (AZD5672) was selected as drug candidate for the treatment of rheumatoid arthritis (RA). The development of AZD5672 (**72**) was terminated in a phase IIb study with RA

**Table 17** CCR5 affinity of compounds **66–70** with various substituents in the *m*-position of the second phenyl ring. Inhibition of [$^{125}$I]-CCL3 binding to the human CCR5 receptor

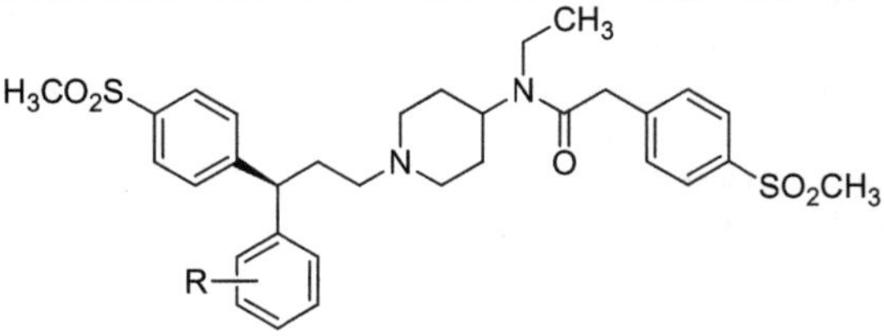


| Compounds | *R* | CCR5 $IC_{50}$ (nM) |
|---|---|---|
| (*S*)-**66** | H– | 0.76 |
| (*R*)-**67** | 3-F– | 0.22 |
| (*R*)-**68** | 3-Cl– | 1.0 |
| (*R*)-**69** | 3,5-*di*-F– | 0.32 |
| (*R*)-**70** | 3-F–, 5-Cl– | 1.1 |

**Table 18** Variations (**71**, **72**) of the second phenyl ring

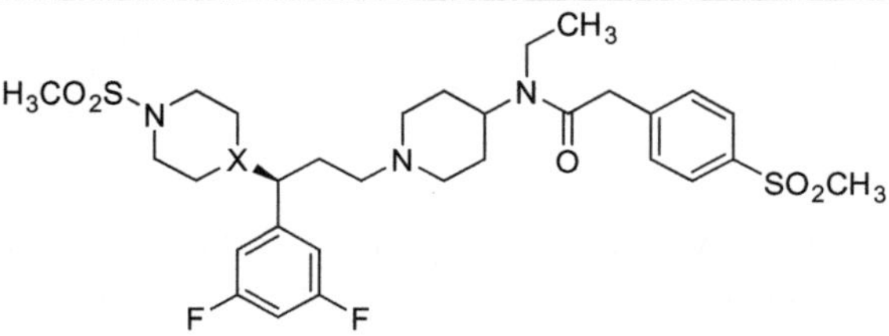


| Compounds | *X* | CCR5 $IC_{50}$ (nM)[a] | hERG inhibition $IC_{50}$ (nM)[b] |
|---|---|---|---|
| **71** | N | 3.7 | >32,000 |
| **72** (AZD5672) | CH | 0.26 | 24,000 |

[a]Inhibition of [$^{125}$I]-CCL3 binding to the human CCR5 receptor
[b]The concentration required to inhibit binding of [$^{3}$H]dofetilide binding to hERG stably expressed on HEK-293 cells

patients, due to absence of statistically significant effects of AZD5672 on symptoms of RA [83].

Pfizer's approach focused on the development of CCR5 selective ligands for the treatment of HIV-1 infection [84, 85]. Compounds **51** and **52** showed weak CCR5 binding affinity, antiviral activity could not be detected, and, moreover, high affinity to the CYP 2D6 enzyme was found (Tables 14 and 19) [80].

The replacement of the imidazopyridine structure, responsible for the interaction with CYP 2D6 by coordination of the pyridine N-atom to the heme iron, led to benzimidazole **73**. Compound **73** showed potent inhibition of CCL4 binding and much weaker CYP 2D6 inhibition, but still no antiviral activity (Table 19). In order

**Table 19** The CCR5 receptor affinity and antiviral activity of compounds **51, 52,** and **73–77**

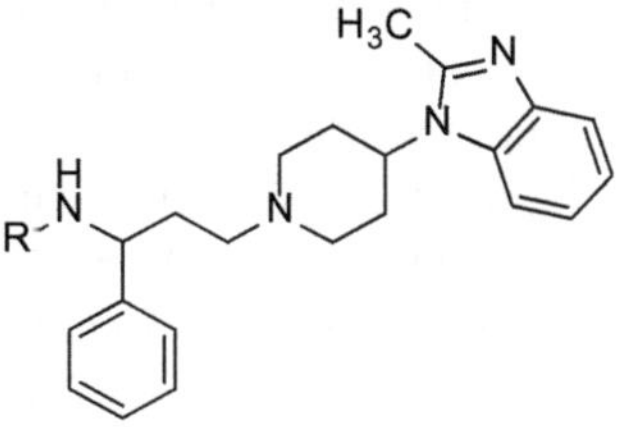


| Compounds | *R* | CCR5 $IC_{50}$ (nM)[a] | Antiviral activity $IC_{50}$ (nM)[b] |
|---|---|---|---|
| **51** | – | 400 | – |
| **52** | – | 1,100 | – |
| **73** | 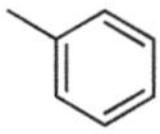 | 4 | – |
| **74** | 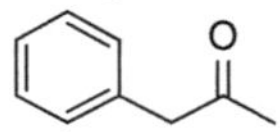 | 100 | 740 |
| **75** | 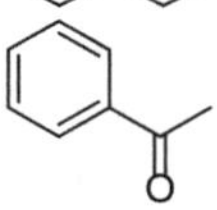 | 45 | 210 |
| (*S*)-**75** | 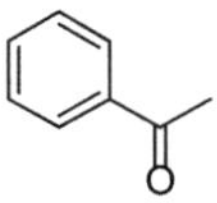 | 13 | 190 |
| **76** | 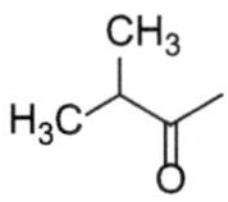 | 50 | 700 |
| **77** | 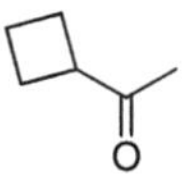 | 40 | 75 |
| (*S*)-**77** (UK-347,503) | 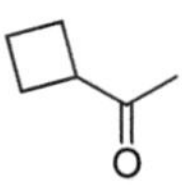 | 20 | 73 |

[a]Inhibition of the [$^{125}$I]-CCL4 binding to the human CCR5 receptor
[b]Antiviral activity determined against HIV-Bal in PM-1 cells [80]

to increase the polarity, one of the phenyl groups of the diphenylmethyl moiety was replaced by an amide bearing substructure already found in compound **52**. Amides **74–77** inhibited the CCL4 binding but, more interestingly, moderate levels of antiviral activity determined against HIV-Bal in PM-1 cells [80, 86] were found (Table 19).

The benzamide **75**, the isobutyramide **76**, and the cyclobutanecarboxamide **77** were found to be the most active antiviral compounds in this series. The data of CCL4 inhibition and antiviral activity do not correlate, indicating that the binding domains of HIV gp120 and CCL4 are distinct and separate. In order to determine the eutomers within the amide series, the benzamide **75** and the cyclobutanecarboxamide **77** were synthesized stereoselectively. The (*S*)-enantiomers (*S*)-**75** and (*S*)-**77** had higher CCR5 affinity and antiviral activity than the (*R*)-enantiomers. Compound UK-347,503 ((*S*)-**77**) also showed decreased affinity to the CYP 2D6 enzyme and was chosen as a lead compound for further CCR5 antagonist development [80].

### 3.2.2 Tropane-Based CCR5 Ligands

Compounds with affinity to the CYP 2D6 enzyme have a basic amino group in 5–7 Å distance to a possible site of oxidation. The basic amine is interacting with Asp301 of the enzyme [87, 88]. In order to avoid CYP 2D6 affinity a series of analogs of **77** with a modified piperidine ring was designed.

The benzimidazole derivatives *exo*-**78** and *endo*-**78** demonstrated increased inhibition of viral replication (Table 20) combined with reduced CYP 2D6 affinity. The similar activity of the *exo*- and the *endo*-isomers is caused by different orientations of the bridged piperidine rings. The benzimidazole forces the isomer *endo*-**78** into a boat conformation, well overlapping with the chair conformation of *exo*-**78** [1]. All compounds **78–80** demonstrated potent antiviral activity against clinically relevant CCR5-tropic viruses, but failed in safety screenings due to high inhibition of the hERG ion channel [89]. Therefore the next aim was to obtain selectivity against the hERG channel.

The first approach to overcome hERG affinity was driven by the exploration of the prodrug concept of compound **81**. Compound **81** demonstrated high bioavailability and high hERG binding but was rapidly oxidized to the highly selective primary metabolites tetrahydropyran S-oxide **82** and S,S-dioxide **83, 82** and **83** displayed high inhibition of cell-cell fusion without any binding to the hERG channel at 10 μM in in vitro assays (Table 21). However, the bioavailability of metabolites **82** and **83** after *p.o.* administration of **81** to rats was lower than 10%. Gut wall metabolism and excretion by the liver were suggested to be responsible to the failure of **81** as oxidizable prodrug [74].

Because the first strategy to overcome hERG affinity by a prodrug concept failed, the second strategy focused on the modification of the basicity and steric environment of the central amino moiety and alteration of the orientation and substitution patterns of the aromatic rings in lead compound **78**. In the oxagranatane *exo*-**80** (Table 20), the basicity of the central amino group is reduced to p$K_a$ 6.0 compared to p$K_a$ 7.8 of *exo*-**78**. However, the hERG channel affinity was not

**Table 20** CCL4 inhibitory activity and antiviral activity of piperidine analogs [1]

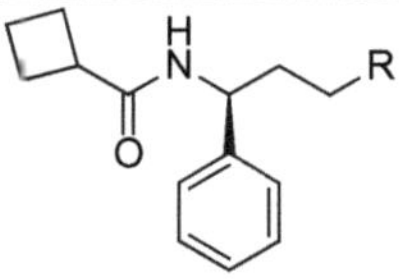

| Compounds | *R* | CCR5 $IC_{50}$ (nM)[a] | Antiviral activity $IC_{90}$ (nM)[b] | hERG channel inhibition[c] |
|---|---|---|---|---|
| *exo*-**78** | 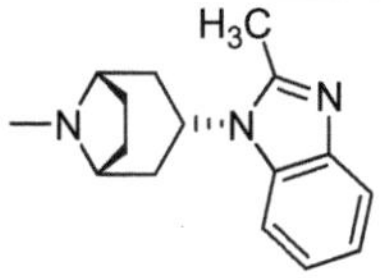 | 2 | 13 | 80% at 300 nM |
| *endo*-**78** | 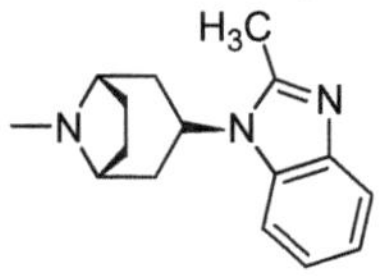 | 6 | 3 | 99% at 300 nM |
| **79** | 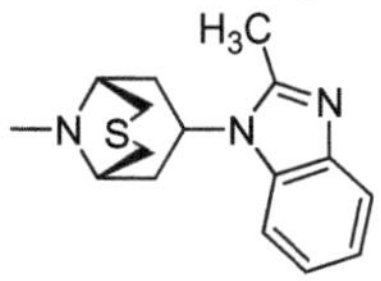 | 21.5 | – | – |
| *exo*-**80** | 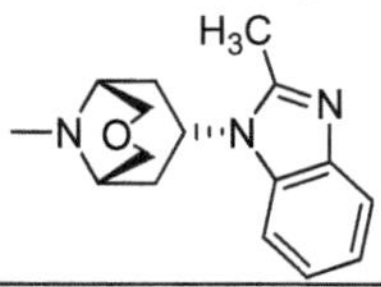 | 9.0 | – | 70% at 300 nM |

[a]Inhibition of the [$^{125}$I]-CCL4 binding to the human CCR5 receptor
[b]Antiviral activity determined against HIV-Bal in PM-1 cells
[c]Inhibition (%) of [$^{3}$H]dofetilide binding to hERG stably expressed on HEK-293 cells

reduced, which suggested that the basic center itself is not essential for hERG binding [90]. Docking of **78** into a hERG channel model indicated a lipophilic interaction of the phenyl ring of the benzimidazole moiety with the hERG channel residues. In order to inhibit this overlap, a triazole moiety (**84**) instead of the benzimidazole ring (**78**) was introduced, which dramatically decreased the hERG affinity. The introduction of an isopropyl side chain at the triazole motif (**85**) increased the antiviral activity, but the added lipophilicity increased the affinity to the hERG ion channel as well. The cyclobutyl group of the amide **78** was shown to overlap nicely with the lipophilic binding pocket of the hERG channel. In order to interrupt this interaction, polar fluorinated groups were introduced (**86**, **1**). The 4,4-difluorocyclohexyl derivative **1** does not show any binding to the hERG channel, even at a concentration of 1,000 nM. Combined with low nanomolar antiviral potency (Table 22) [89], a broad anti-R5 HIV-1 spectrum and no inhibition of CYP 1A2, 2C9, 2C19, 3A4, and 2D6 enzymes compound **1** were characterized

**Table 21** Compounds **82** and **83** as active metabolites of **81**

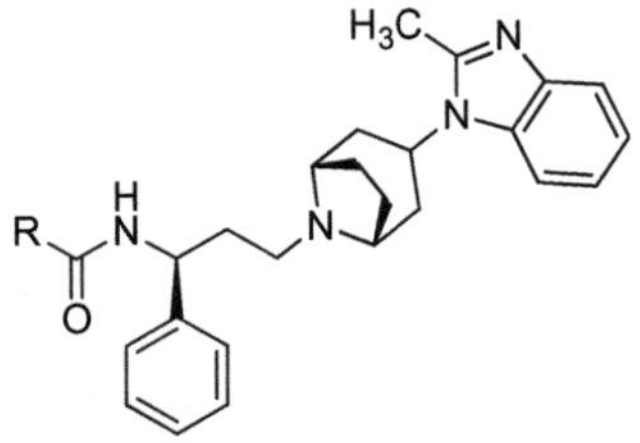


| Compounds | *R* | Membrane fusion $IC_{50}$ (nM)[a] | hERG inhibition $IC_{90}$ (nM)[b] |
|---|---|---|---|
| **81** | 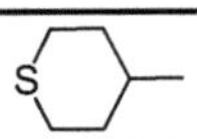 | 0.2 | 740 |
| **82** | 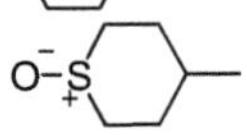 | 0.2 | >10,000 |
| **83** | 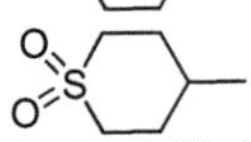 | 0.2 | >10,000 |

[a]Inhibition gp160 fusion
[b]Inhibition of [$^3$H]dofetilide binding to hERG stably expressed on HEK-293 cells

as an inverse agonist of CCR5 receptors, stabilizing the receptor in the inactive conformation [91].

Maraviroc (**1**, UK-427,857) was the product of a long optimization process leading to the first CCR5 ligand approved for the treatment of confirmed R5-tropic HIV-1 infection on the market [2]. The use of maraviroc for the HIV-1 therapy is currently complicated by the growing number of maraviroc-resistant HI-virus strains ($MVC^{RES}$) [92–94], which makes the ligands with an improved resistance profile desirable.

The crystal structure of the CCR5-maraviroc complex, reported in 2013, shows the binding of the ligand at the bottom of a pocket formed by residues from helices TM1, 2, 3, 5, 6, and 7. The tropane N-atom is protonated and forms a salt bridge with Glu283. The NH moiety of the amide forms a hydrogen bond with the phenolic group of Tyr251 (Fig. 7). The length of the propyl chain between the two N-atoms correlates with the positions of Glu283 and Tyr251 in the receptor. The fluorine atoms in the cyclohexane ring form two hydrogen bonds with Thr195 and Thr245. The phenyl group interacts with five aromatic residues, Tyr108, Phe109, Phe112, Trp248, and Tyr 251 in the binding pocket. The interaction of the benzene ring with Trp248 is believed to prevent the activation-related motion of the receptor, which underlines the inverse agonist character of maraviroc (see Fig. 7) [95]. Compared to the CXCR4/IT1t structure [30], the binding site of maraviroc (**1**) was found to be deeper, without any contact to extracellular loops. The availability of the X-ray crystal structure of the CCR5 receptor will help to promote the development of novel potent CCR5 ligands with optimized properties.

**Table 22** Binding data of compounds **1**, **84–86**

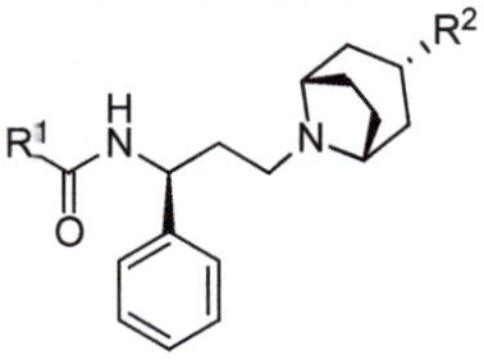

| Compounds | $R^1$ | $R^2$ | Antiviral activity $IC_{50}$ (nM)[a] | hERG inhibition[b] |
|---|---|---|---|---|
| **84** |  | 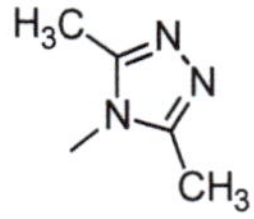 | 13 | 10% at 100 nM |
| **85** | 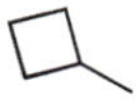 | 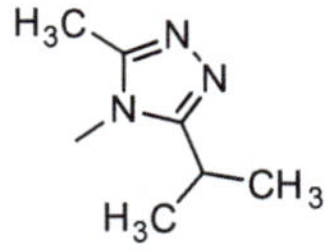 | 8 | 14% at 100 nM<br>30% at 300 nM |
| **86** |  | 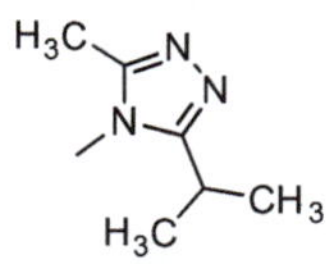 | 14 | 14% at 300 nM |
| **1** (UK-427,857, maraviroc) | 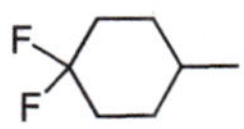 | 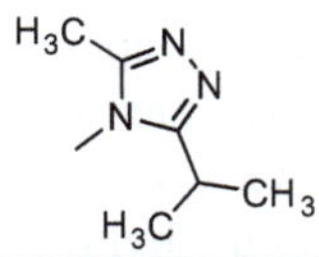 | 2 | 0% at 300 nM<br>0% at 1,000 nM |

[a]Inhibitory effect on HIV-Bal virus replication of in PM-1 cells
[b]Percentage inhibition of [$^3$H]dofetilide binding to hERG stably expressed on HEK-293 cells

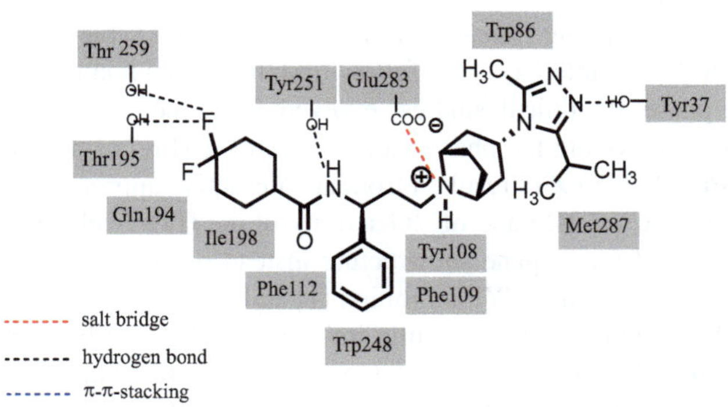


**Fig. 7** Schematic representation of interactions between the CCR5 receptor and maraviroc (**1**), modified according to [95]

**Table 23** CCR5 ligands **87** and **88** developed by lead deconstruction strategy. Inhibition of CCL5-stimulated [$^{35}$S]-GTPγS accumulation to CCR5-expressing CHO cell membranes

| Compounds | | CCR5 $IC_{50}$ (nM) |
|---|---|---|
| **87** TD0444 | 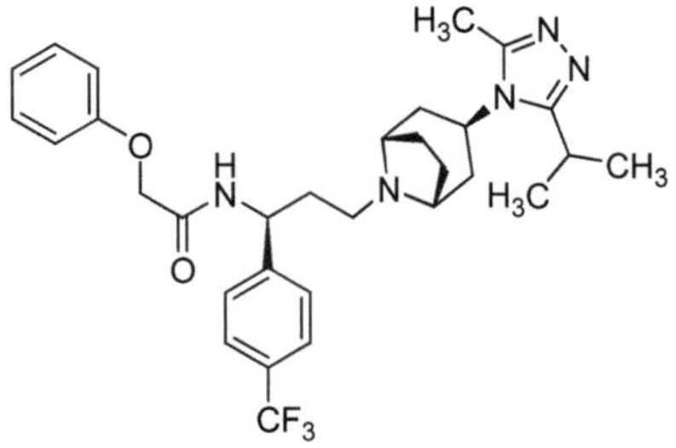 | 253 |
| *exo*-**88** | 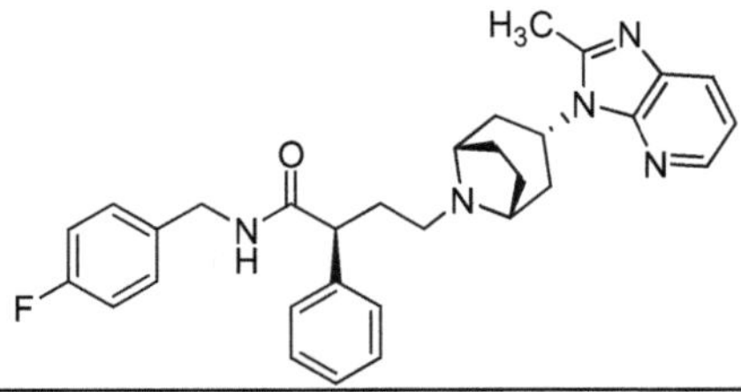 | 14 |

*Long* and coworkers from the Shanghai Institute of Materia Medica developed the lead compound **1** in more detail by applying lead deconstruction strategy. This approach combines privileged structures of a lead compound with new motifs. Replacement of the difluorocyclohexyl moiety of maraviroc by a phenoxy group and the introduction of the trifluoromethyl group at the *p*-position of the phenyl ring resulted in the moderate CCR5 ligand **87** (TD0444, Table 23). Further improvement of the CCR5 affinity was achieved by introduction of an *exo*-oriented 2-methyl-3*H*-imidazo[4,5-*b*]pyridine-3-yl residue instead of the triazolyl moiety and inversion of the amide substructure, which led to the potent CCR5 ligand **88**, whereas the corresponding *endo*-isomer of **88** is inactive (Table 23) [96, 97].

PF-232798 (**90c**, Table 24) is the follow-up clinical candidate of maraviroc (**1**), currently in phase II clinical studies, evolved from the efforts to increase the absorption and improving the pharmacokinetic profile (PK) of maraviroc (**1**). The structure of PF-232798 (**90c**) resulted from an alternative approach which intended to circumvent the CYP 2D6 and hERG activity of the HTS lead UK-107,543 (**51**). The introduction of the tropane substructure instead of the piperidine moiety was previously proven to reduce CYP inhibition [1] and was therefore incorporated into the new lead compound. The lipophilic imidazopyridine and benzimidazole substructures of **51** and **78** were shown to be responsible for the inhibition of CYP 2D6 and high hERG binding [80]. In order to prevent lipophilic interactions with the hERG ion channel, the imidazopyridine substructure was replaced by more polar 1,4,6,7-tetrahydro-imidazo[4,5-c]pyridine, which led to the 3-substituted (**89a–c**) and 1-substituted (**90a–d**) series of compounds. The methyl carbamates **89a** and **89b** demonstrated high hERG inhibition. Reducing the size of the amide substituent

**Table 24** 3-substituted (**89a–c**) and 1-substituted (**90a–d**) 1,4,6,7-tetrahydro-imidazo[4,5-c] pyridines

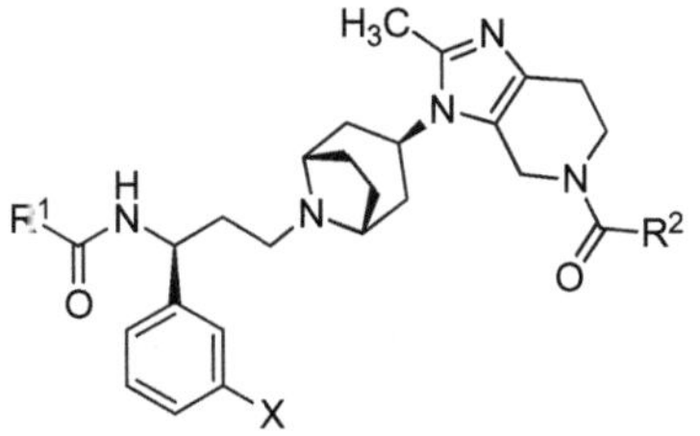

**89**

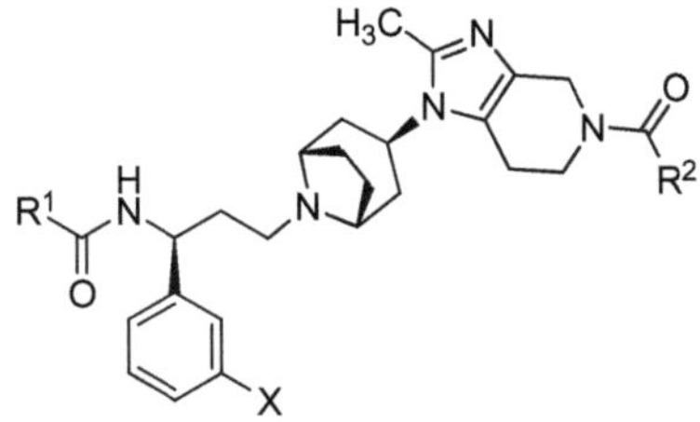

**90**

| Compounds | *X* | *R*[1] | *R*[2] | Membrane fusion $IC_{50}$ (nM)[a] | hERG channel inhibition[b] |
|---|---|---|---|---|---|
| **89a** | H | $H_3CO$– | $H_3C$– | 0.6 | 57% |
| **89b** | F | $H_3CO$– | $H_3C$– | 0.2 | 28% |
| **89c** | F | $H_3C$– | $H_3CO$– | 0.1 | 0% |
| **90a** | F | $H_3C$– | $H_3CO$– | <0.1 | 2 μM |
| **90b** | F | $H_3C$– | EtO– | <0.1 | 5 μM |
| **90c** (PF-232798) | F | $H_3C$– | $^i$PrO– | <0.1 | 12 μM |
| **90d** | F | $H_3C$– | $^t$BuO– | <0.1 | 6 μM |

[a]Inhibitory effect on gp160 fusion
[b]Inhibition of [$^3$H]dofetilide binding to hERG stably expressed on HEK-293 cells

to an acetyl group (**89c**) significantly increased the selectivity for hERG ion channel within the series **89.** Also the hERG affinity was reduced by incorporation of a *m*-fluoro substituent into the phenyl ring (**89b**). Switching the substitution position of the 1,4,6,7-tetrahydro-imidazo[4,5-c]pyridine slightly improved the gp160 inhibition from **89c** to **90a**, but was detrimental in terms of hERG binding. The lowest hERG inhibition ($IC_{50}$ = 12 μM) could be achieved by introduction of an isopropoxycarbonyl substituent **90c** (Table 24). Compound **90c** demonstrated complete oral absorption in rat and dog that was accompanied by improved metabolic stability compared to maraviroc (**1**) and other compounds in this series. Moreover, PF-232798 (**90c**) displayed antiviral activity against maraviroc-resistant viruses and was therefore chosen as the follow-up clinical candidate of maraviroc (**1**) [98].

### 3.2.3 1-Amido-1-Phenyl-3-Piperidinylbutane-Based CCR5 Ligands

The growing number of reports on maraviroc-resistant HI-viruses [93, 99, 100] underlines the need for development of a next generation of ligands with different resistance profile. The tropane-core represents the central structural motif of all previously successfully developed CCR5 ligands. The key features of the tropane moiety are an increased steric hindrance around the basic amino group and restricted conformational flexibility of the molecule. In order to retain the steric

**Table 25** Heterocycle-substituted piperidines **91–94** bearing an α-methyl moiety

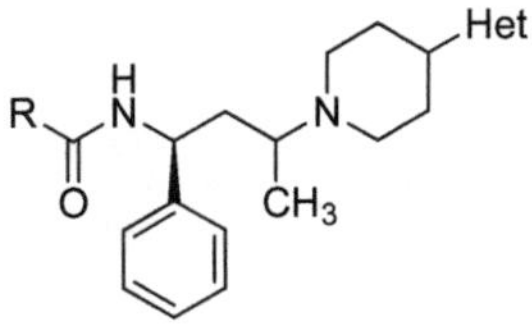


| Compounds | *R* | Het | Membrane fusion $IC_{50}$ (nM)[a] | hERG inhibition $IC_{50}$ (μM)[b] |
|---|---|---|---|---|
| **91** | 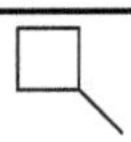 | 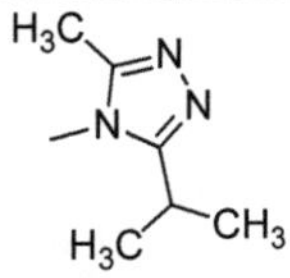 | 4.3 | 2.1 |
| **92** | 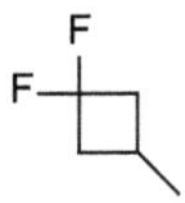 | 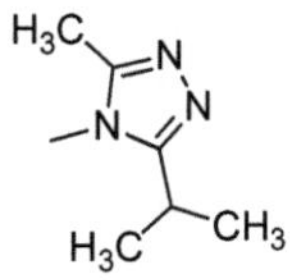 | 1.3 | 2.3 |
| (*R*)-**93** | 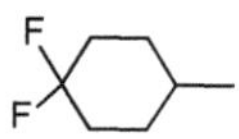 | 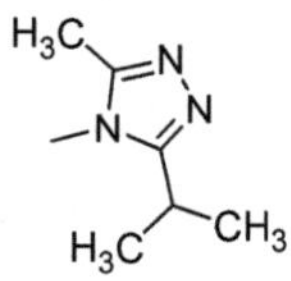 | 0.48 | >10 |
| (*S*)-**93** | 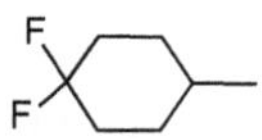 | 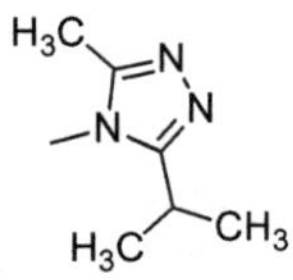 | 50 | >10 |
| (*R*)-**94** | 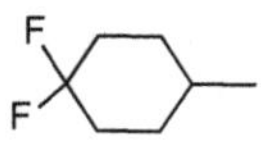 | 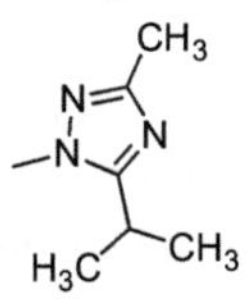 | 3.0 | >10 |

[a]Inhibitory effect on gp120-sCD4 complex binding
[b]Inhibition of [$^3$H]dofetilide binding to hERG stably expressed on HEK-293 cells

hindrance, the introduction of an additional methyl moiety into the propyl chain was envisaged. Several heterocycles as well as different amido substituents were screened for their antiviral activity, oral bioavailability, and low propensity towards hERG ion channel inhibition and interaction with a range of CYP enzymes. A large series of piperidines **91–94**, substituted with a *N*-heterocycle, bearing an α-methyl moiety were prepared (Table 25) [101, 102].

The cyclobutyl derivate **91** reveals high antiviral activity combined with the low hERG inhibition. Introduction of two additional fluorine atoms at the alkyl substituent (**92**) was shown to be beneficial in terms of antiviral activity, as it was already observed during the optimization of maraviroc [89]. Increasing the size of the alkyl substituent to a cyclohexyl group (**93**) inhibits the interactions with the hERG

**Table 26** First, potent 4,4-disubstituted piperidine-based CCR5 ligands **95** and **96**

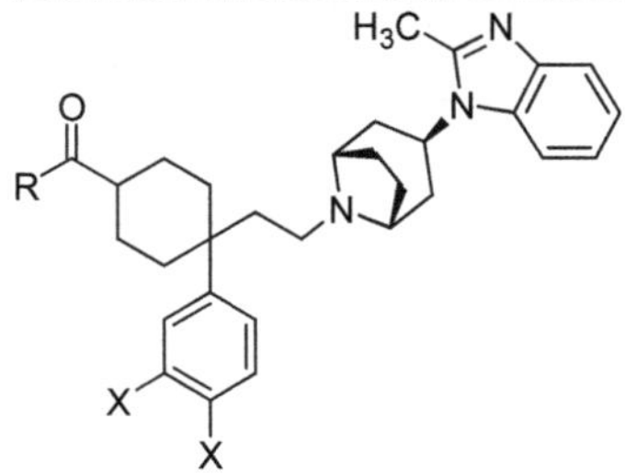


| Compounds | *R* | *X,Y* | CCR5 $IC_{50}$ (nM)[a] | Antiviral activity $IC_{50}$ (nM)[b] |
|---|---|---|---|---|
| **95** | 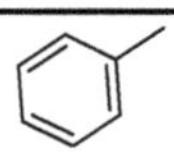 | H, H | 16.8 | – |
| **96** | 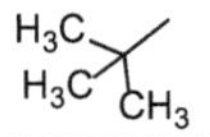  | Cl, H | 1.0 | 16.8 |

[a]Inhibitory potency ($IC_{50}$) of the [$^{125}$I]-CCL4 binding to the human CCR5 receptor
[b]Inhibitory potency of replication of HIV-Bal in PM-1 cells

channel leading to an affinity higher than 10 μM. Enantioselective synthesis of **93** allowed the determination of (*R*)-**93** as eutomer, with a 100-fold higher antiviral activity than the (*S*)-enantiomer. Bioisosteric replacement of the 1,3,4-triazolyl residue of (*R*)-**93** by a 1,2,4-triazol-1-yl ring in (*R*)-**94** resulted in a highly metabolically stable and potent CCR5 ligand. The 1,2,4-triazole (*R*)-**94** displays excellent whole-cell antiviral activity ($IC_{90} = 2.6$ nM), complete oral absorption in rat, and does not inhibit human CYP 1A2, 2C9, and 2D6 enzymes. Moderate affinity towards CYP 3A4 ($IC_{50} = 3.4$ μM) was also found [101, 102]. Therefore, compound (*R*)-**94** represents the new lead compound for the development of novel CCR5 ligands with new resistance profile.

### 3.2.4 4,4-Disubsituted Piperidine-Based CCR5 Ligands

GlaxoSmithKline's approach combining in-house HTS with computer-assisted drug design resulted in identification of two 4,4-disubstituted piperidines **95** and **96** (Table 26). Compounds **95** and **96** show high CCR5 affinity and high antiviral activity combined with promising pharmacokinetic profile in rodents. Unfortunately, **95** and **96** showed also moderate hERG affinity of $IC_{50}$ of 2 μM and 10 nM, respectively, which required further optimization [103].

The optimization focused on the substitution pattern of the phenyl rings, which led to the discovery of several potent CCR5 ligands with decreased inhibition of hERG. Introduction of a sulfonamide and two halogen substituents at the phenyl ring (**97**) turned out to decrease hERG affinity and retained antiviral activity. The exchange of the positions of the fluoro and chloro atoms from **97** to **98** decreased

**Table 27** Potent 4,4-disubsituted piperidine-based CCR5 ligands **97–100** with decreased inhibition of hERG

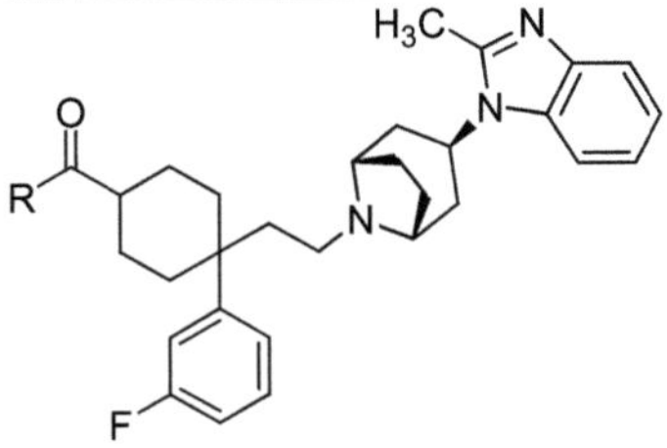

| Compounds | *R* | Antiviral activity $IC_{50}$ (nM)[a] | hERG inhibition $IC_{50}$ (μM)[b] |
|---|---|---|---|
| **97** | 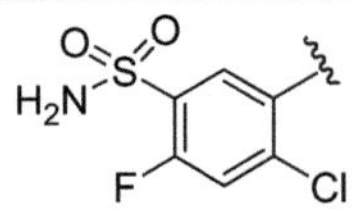 | 4.4 | >32 |
| **98** | 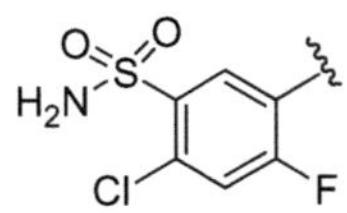 | 43 | >57 |
| **99** (GSK163929) | 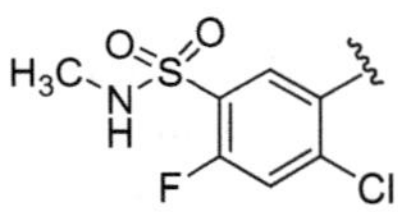 | 4.3 | 19 |
| **100** | 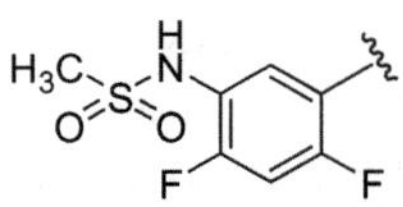 | 6.2 | 100 |

[a]Inhibitory potency of replication of HIV-Bal in PM-1 cells
[b]hERG patch-clamp assay

the antiviral activity. The reverse sulfonamide **100** is less hydrophobic than the forward sulfonamide **99** and was found to be very potent in anti-HIV assays (Table 27) [104]. However, compound **99** (GSK163929) was favored for further investigation as a clinical candidate, due to the potential aniline metabolite formation from reverse sulfonamide **100**.

GSK163929 (**99**) revealed high oral bioavailability and good pharmacokinetic profile in rats and dogs. Seven-day safety studies in both species did not show any adverse effects at therapeutic doses. The high antiviral activity, favorable pharmacokinetic profile, and safety data support further development of **99** in phase I clinical studies [104].

### 3.2.5 2-Phenylbutane-1,4-Diamine-Based CCR5 Ligands

Investigations of the GlaxoSmithKline research laboratories revealed a new class of CCR5 ligands with a 2-phenylbutane-1,4-diamine core structure. The sulfonamide

**Table 28** Antiviral activity of 2-phenyl-1,4-butanediamine-based CCR5 ligands **101–105**. Inhibitory effect on replication of HIV-Bal in PM-1 cells

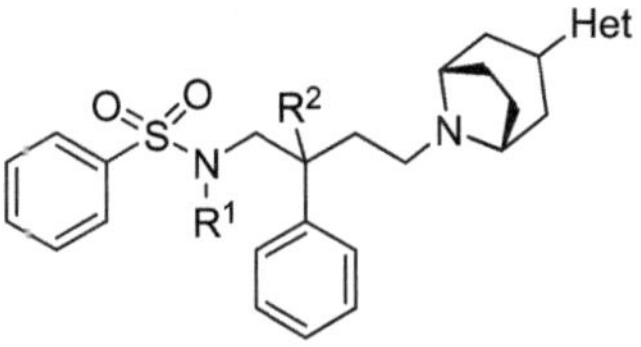


| Compounds | $R^1$ | $R^2$ | Het | Antiviral activity $IC_{50}$ (nM) |
|---|---|---|---|---|
| **101** | $H_3C$– | H– | 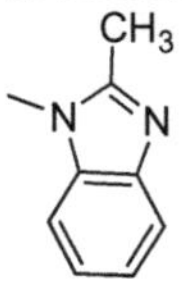 | 8 |
| **102** | H– | H– | 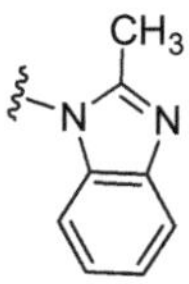 | 65 |
| **103** | $H_3C$– | $H_3C$– | 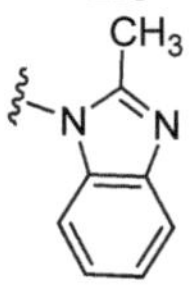 | 9 |
| **104** | H– | H– | 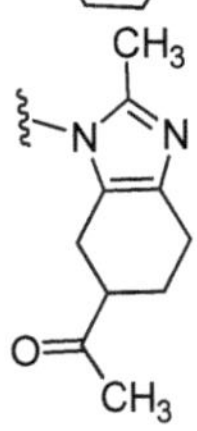 | 3 |
| **105** | H– | H– | 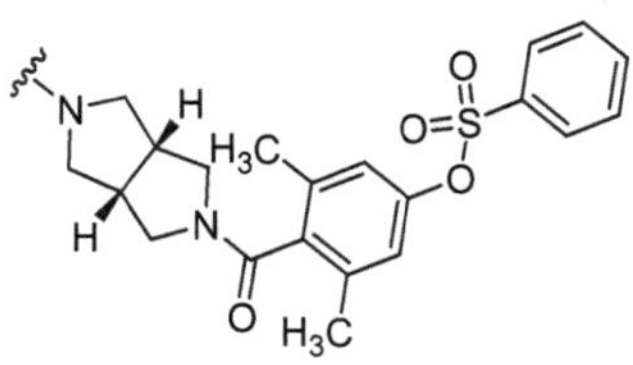 | 8 |

substituent turned out to produce superior antiviral activity over all other amides tested. Removal of the methyl substituent at the amide N-atom of the benzimidazole derivative **101** led to an 8-fold loss in the antiviral activity (**102**). Introduction of a second methyl moiety at position 3 of the butanediamine linker (**103**) increased the activity (Table 28). Compounds bearing other heterocycles than benzimidazole were less sensitive to the effect of *N*-methyl substitution [105].

Compounds **102** to **104** exhibit moderate to fast clearance, only **101** and **105** showed measurable bioavailability. All derivatives were rapidly metabolized and therefore further optimization is required in order to increase metabolic stability and improve the pharmacokinetic profile [105].

### *3.3 Conclusion*

The development of TAK-779 (**15**), maraviroc (**1**), and their follow-up compounds exemplifies the SARs of CCR5 receptor ligands and the existing hurdles. The CCR5 receptor antagonist maraviroc (**1**) developed by Pfizer has already been approved for the treatment of confirmed R5-tropic HIV-1 infection [2]. However, the increasing number of maraviroc-resistant HI-virus strains makes the development of ligands with a distinct resistance profile highly desirable [93]. The role of CCR5 receptors in the development and progression of inflammatory diseases led to an increased interest of using CCR5 antagonists for the treatment of rheumatoid arthritis [106]. Unfortunately, the use of **1** for the treatment of rheumatoid arthritis failed due to low efficacy [107]. The reasons for this failure remain to be elucidated.

## 4 Dual CCR2/CCR5 Ligands

According to the involvement of both CCR2 and CCR5 receptors in the pathogenesis of inflammatory diseases [108], these receptors have become attractive targets for the pharmaceutical industry. The chemokine system is very complex, and the CCR2 receptor binds multiple endogenous ligands including CCL2, which binds exclusively to the CCR2 receptor, as well as CCL8, CCL7, and CCL13 [109] which are rather unselective. The CCR2 receptor is abundantly expressed on blood monocytes and regulates their migration from the bone marrow into inflamed tissue, whereas the CCR5 receptor is expressed on macrophages. The in vivo function of the CCR5 receptor is less well defined than that of the CCR2 receptor, but has been shown to be related to the activation, survival, and retention of macrophages in the core of inflammation and is associated with Th1 cell recruitment and activation. The CCR5 receptor also binds various ligands, including CCL3, CCL4, CCL5, CCL8, and CCL3L1. CCL4 and CCL3 are known to be selective for the CCR5 receptor. During the differentiation of monocytes, a reciprocal pattern of expression and function of the CCR2 and CCR5 receptor was observed, showing a downregulation of CCR2 and an upregulation of the CCR5 receptor expression [110, 111]. CCR2 and CCR5 receptors are expressed on different cells but in a complementary manner. Both receptors are important in the mediation of leukocyte trafficking in case of inflammation, e.g., during the pathogenesis of cardiovascular (atherosclerosis) and immunological diseases (rheumatoid arthritis, Crohn's disease, transplant rejection). Thus targeting both receptors with dual antagonists appears to have therapeutic potential [112].

### 4.1 Benefits of Dual CCR2/CCR5 Ligands

The clinical efficacy of compounds acting selectively with a particular chemokine receptor remains to be shown. Clinical trials with various chemokine ligands failed because a benefit at a critical endpoint was not shown. The complex pharmacology of chemokines and their receptors most likely contributed to the failures. More than 50 different chemokines have been identified that interact with more than 20 classical and atypical chemokine receptors. A few chemokines show a one-to-one specificity as CCL2 is specific for the CCR2 receptor, while other chemokines are promiscuous and bind to different receptors [109, 113, 114]. Also, various chemokine receptors have more than one chemokine ligand, which usually leads to differential functional response mediated by the same chemokine receptor.

Due to the fact that multiple chemokine receptors are involved in the pathophysiology of a disease, dual antagonists that target and inhibit two most prominent receptors could be a possibility to enhance the therapeutic effect. A successful example for a dual antagonism in the field of GPCRs is PS433540, which inhibits the $AT_1$ receptor and the $ET_A$ receptor. PS433540 appears to be successful as antihypertensive in rats and reached phase IIa clinical trials. Therefore, promiscuous compounds that target several receptors are suggested to be particularly effective for the treatment of complex diseases, for example, multiple sclerosis or rheumatoid arthritis, in which both CCR2 and CCR5 receptors are involved [41].

### 4.2 Sequence Alignment of CCR2 and CCR5 Receptors

As CCR2 and CCR5 receptors belong to the same subfamily (CC) of chemokine receptors their amino acid sequences are highly homologous, mainly in the transmembrane (TM) domains [62, 115]. Both receptors contain two conserved disulfide bridges Cys32-Cys277 and Cys113-Cys190 in the CCR2 receptor and Cys20-Cys269 and Cys101-Cys178 in the CCR5 receptor. Comparative analysis revealed 66 % sequence identity in general between CCR2 and CCR5 receptors and 82 % identity in the active site. Receptor homology modeling studies predicted a ligand binding pocket of the CCR2 receptor formed by TM2, 3, 5, 6, and 7. In the CCR5 receptor TM1, 2, 3, 5, 6, and 7 form the binding pocket for CCR5 inhibitors, which is located at the extracellular region and is partly covered by the extracellular loop. Both CCR2 and CCR5 receptors possess glutamic acid in the anchor site in TM 7, Glu291 in CCR2 receptors, and Glu283 in CCR5 receptors. Glu291/Glu283 is essential for the interaction with protonated tertiary amines or quaternary ammonium ions. Superposition of CCR2/CCR5 binding sites revealed that all residues are

**Table 29** Dual CCR2/CCR5 antagonists, TAK-779, and TAK-652

| Compounds | | CCR2 $IC_{50}$ (nM)[a] | CCR5 $IC_{50}$ (nM)[b] |
|---|---|---|---|
| **15** (TAK-779) | 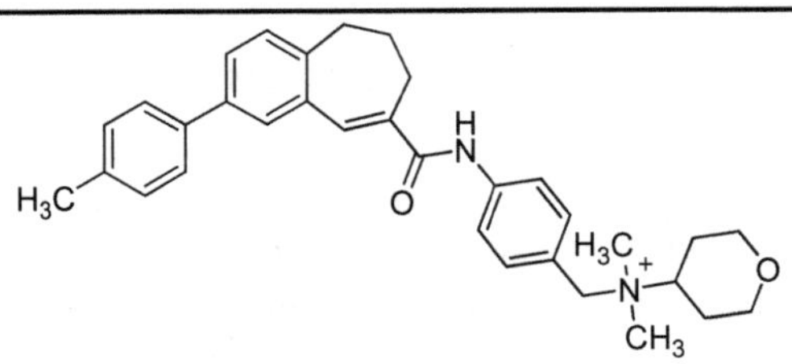 | 27 | 1.4 |
| **43** (TAK-652) | 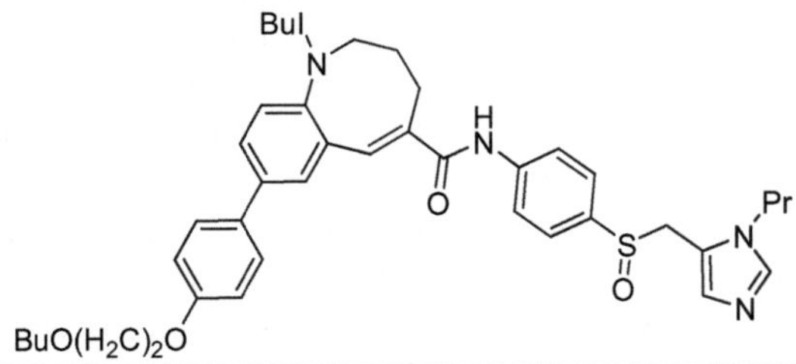 | 5.9 | 3.1 |

[a]Inhibitory effect on binding of CCL2 to human CCR5 receptor
[b]Inhibitory effect on binding of [$^{125}$I]-CCL5 to human CCR5 receptor

identical except three: Ser101/Tyr89, His121/Phe109, and Arg206/Ile198, which differ considerably in their electronic and hydrophobic properties.

Contemporary modeling studies performed on the basis of the 2010 crystallized CXCR4 receptor show higher sequence homology to CCR2/CCR5 than prior used templates based on bovine rhodopsin or the β2-adrenergic receptor [28]. The recently reported X-ray crystal structure of the CCR5 receptor/maraviroc complex [95] will allow a deep insight in the binding site and sophisticated modeling studies.

## *4.3 TAK-779 and TAK-652*

The dual CCR2/CCR5 antagonist TAK-779 (**15**) (Table 29) is the most extensively investigated compound regarding binding site experiments and computational predictions. Using computational calculations, low-energy three-dimensional receptor conformations of human CCR2 and CCR5 receptors were created, and the binding sites of **15** within the CCR2 and CCR5 receptor were predicted. Mutation experiments in which single amino acids were replaced within the receptor structure were performed, and after transient expression in the L1.2 cells chemotactic and competitive binding experiments to CCR2 and CCR5 receptors were carried out. Based on these data, it was postulated that Trp98/Thr292 in the CCR2 receptor (Fig. 3) and Trp86/Tyr108 in the CCR5 receptor (Fig. 5) were significantly associated with the efficacy of TAK-779. His121 in the CCR2 receptor was also important for antagonistic efficacy and was replaced by Tyr108 in the CCR5 receptor. An altered rotational orientation of TM3 is responsible for a different positioning of these aromatic residues in both CCR2 and CCR5 receptors. A comparison of quaternary ammonium salt TAK-779 (**15**) with antagonists including a tertiary amine showed differences in binding of these interacting residues. Depending on Glu291 in the CCR2 receptor and Glu283 in the CCR5 receptor facing either TM1 and TM2 or TM3 and TM6, different orientations of TM7 are possible: The first receptor conformation leads to a receptor activation network formed between TM 1, 2, 3, and 7, which is supposed to be required for the receptor activation by the chemokine. The second conformation is expected to be stabilized by antagonist binding [62].

The interaction of TAK-779 (**15**) with the respective binding sites of CCR2 and CCR5 receptors resulted in $IC_{50}$ values of 27 nM for the CCR2 receptor and 1.4 nM for the CCR5 receptor. **15** also inhibited the binding of CCL3 and CCL4 to the cells expressing the CCR5 receptor with an $IC_{50}$ value of 1.0 nM. No interaction was found between TAK-779 and CCR1, CCR3, CCR4, or CXCR4 receptors [71].

TAK652 (**43**) (Table 29), a benzazocine compound, was developed by Takeda Inc. as a CCR5 antagonist for anti-HIV-1 therapy in order to improve the poor oral bioavailability of the quaternary ammonium salt TAK-779.

In addition to the high CCR5 affinity ($IC_{50} = 3.1$ nM in [$^{125}$I]-CCL5 assay), TAK-652 was also found to be a potent CCR2 antagonist with binding affinity of 5.9 nM [116]. This effect was neither observed for TAK-220 (**48**) and maraviroc (**1**) nor any other CCR5 ligand [117].

## *4.4 MK0483*

As described in Sect. 2, compounds with an aminocyclopentanecarboxamide scaffold were developed as CCR2 antagonists with promising receptor affinity but also significant hERG inhibition. MK0483 (**6**) (Table 30) showed high CCR2 affinity ($IC_{50} = 4$ nM, measured as inhibition of [$^{125}$I]-CCL2 binding) and displayed an $IC_{50}$ value of 0.3 nM in chemotactic assays. MK0483 inhibited [$^{125}$I]-CCL3 binding to CCR5 receptors with an $IC_{50}$ value of 25 nM. Additionally a low ERG affinity ($IC_{50} = 33$ μM) was found. The improved lack of hERG inhibition with regard to Merck's previous compounds was shown to be associated with the 3-carboxyphenyl in position 4 of the piperidine of **6** [118].

A broad screening against different receptors, CYP enzymes and ion channels displayed high selectivity for CCR2 and CCR5 and weak interaction with muscarinic receptors $M_2$ and $M_4$ without any interaction with CYP enzymes, including CYP 3A4, 2C9, 2D6, 1A2, and 2C19. Efficacy of **6** was also evaluated in different rhesus blood experiments, suggesting that subnanomolar potency can be achieved in vivo [118].

**Table 30** Dual CCR2/CCR5 antagonists MK0483, SKB3380732, and INCB10820/PF-4178903

| Compounds | | CCR2 $IC_{50}$ (nM)[a] | CCR5 $IC_{50}$ (nM)[b] |
|---|---|---|---|
| **6** (MK0483) | 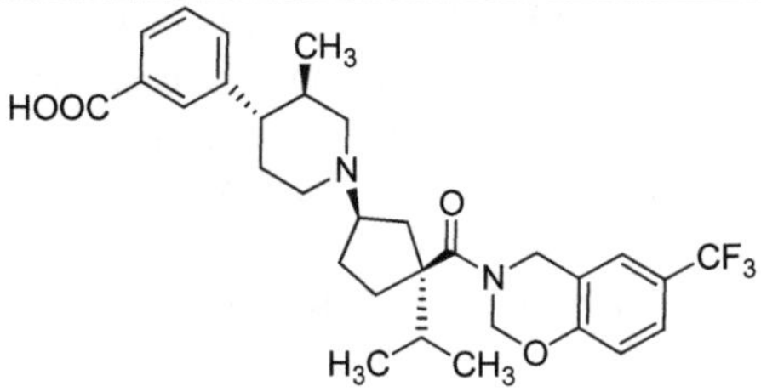 | 4 | 25 |
| **106** (SKB3380732) | 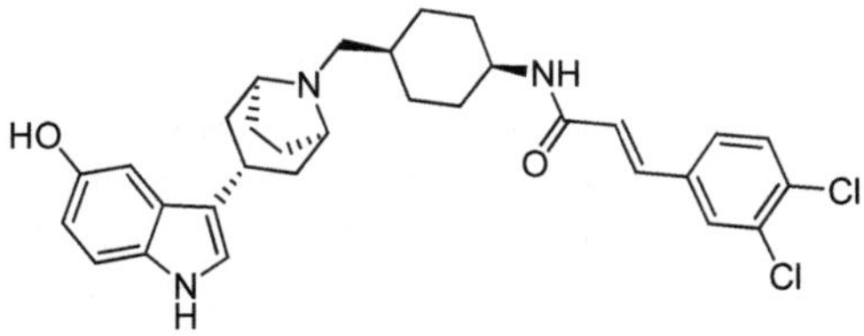 | 40 | 4,000 |
| **107** (INCB10820/PF-4178903) | 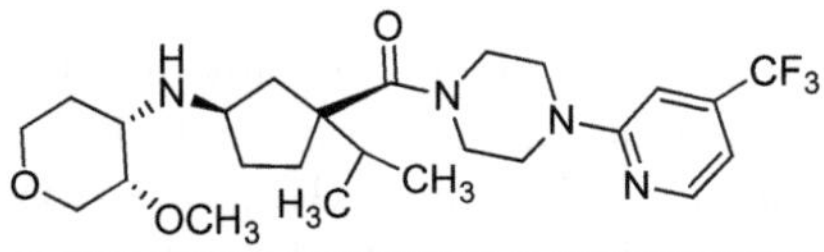 | 3 | 5.3 |

[a]Inhibitory effect on binding of CCL2 to human CCR2 receptor
[b]Inhibitory effect on binding of [$^{125}$I]-CCL5 to human CCR5 receptor

## 4.5 *SKB3380732*

The indolyltropane SKB3380732 (**106**) (Table 30), developed as a potent CCR2 ligand, displayed an $IC_{50}$ value of 40 nM for the CCR2 receptor. The development of **106** started from a potent CCR2 antagonist with an indolylpiperidine scaffold. It showed selectivity over the CCR5 receptor, but due to its high structural similarity with serotonin (5-hydroxytryptamine), this indolylpiperidine was not selective and interacted with several serotonergic and dopaminergic receptors. In order to improve the selectivity, a conformational restriction of the indolylpiperidine via a tropane moiety was performed, and the steric bulk around the basic amine was increased. Moreover, the flexible pentyl chain was exchanged by a methylcyclohexyl linker. The conformational constraint of both the piperidine ring and the pentyl alkyl chain led to 1,000-fold increased CCR2 selectivity over a number of serotonin and dopamine receptors but retained high CCR2 affinity. In contrast to previous compounds, **106** showed moderate CCR5 affinity ($IC_{50} = 4{,}000$ nM) (Table 30) [41, 119].

## 4.6 INCB10820/PF4178903

Incyte and Pfizer discovered a series of dual CCR2/CCR5 antagonists, leading to INCB10820/PF-4178903 (**107**) (Table 30) as the most potent compound. SAR studies on both the left and right part of the molecule, containing an aminocyclopentanecarboxamide with an isopropyl moiety, resulted in **107** as most promising dual antagonist.

Compound **107** displayed a CCR2 affinity of $IC_{50} = 3.0$ nM in [$^{125}$I]-CCL2 assay and $IC_{50} = 5.3$ nM to the CCR5 receptor in [$^{125}$I]-CCL4 binding assay. Compared the analog bearing a 3-trifluoromethylphenyl instead of the 3-trifluoromethylpyridin-2-yl moiety (**107**), the CCR5 affinity increased 4-fold. Regarding to chemotactic activity, **103** showed similar $IC_{50}$ values in both CCR2 ($IC_{50} = 3.2$) and CCR5 ($IC_{50} = 4.3$ nM) binding assays. A further replacement of the 3-trifluoromethylpyridin-2-yl moiety by the 4-trifluoromethylpyrimidin-2-yl residue provided a less active analog. Screening of the affinity towards various receptors, enzymes, and ion channels (>50) indicated **107** to be a selective and dual CCR2 and CCR5 antagonist. **107** did not inhibit CYP 3A4 and 2D6 enzymes, but inhibited the hERG channel ($IC_{50} = 1.7$ μM). Due to the promising in vivo properties with oral bioavailability of 84% in rats and 57% in monkeys and high metabolic stability ($t_{1/2} = 93$ min), **107** became a candidate for clinical studies [120].

The binding of **107** was also analyzed by docking into the binding site of the CCR2 and CCR5 receptor. The tertiary amine of **107** forms a salt bridge with the acidic residues Glu291 (CCR2) and Glu283 (CCR5). The trifluoromethyl substituent was also identified to interact with Arg206 of the CCR2 receptor and Ile 198 of the CCR5 receptor [28].

## 4.7 γ-Aminobutyramides

As detailed in Sect. 2, several CCR2 antagonists from various chemical classes have already been reported. Merck developed further a new class of promising CCR2 antagonists based on the γ-aminobutyramide core. The screening of Merck's sample collection led to some lead compounds, which upon further optimization resulted in ligands **108**, **109**, and **110** (Table 31) with both CCR2 and CCR5 antagonistic properties. A high structural similarity to known CCR5 antagonists was achieved by incorporation of a substituted piperidine ring. Compound **108** with the 4-phenylpiperidine moiety demonstrated moderate binding affinity to the CCR2 receptor ($IC_{50} = 150$ nM) as well as to the CCR5 receptor ($IC_{50} = 72$ nM). **108** showed improved potency compared to previously characterized unsubstituted piperidine analogs. Further variations of the phenylpiperidine moiety by 3-phenylazetidine, 3-phenylpyrrolidine, and 4-phenylazepane showed that the piperidine ring was best in terms of CCR2 and CCR5 affinity. The spiro[indene-piperidine] **109** was about 2-fold more active with $IC_{50}$ values of 80 nM (the CCR2

**Table 31** Chemical structure of dual CCR2/CCR5 antagonists with the γ-aminobutyramide substructure

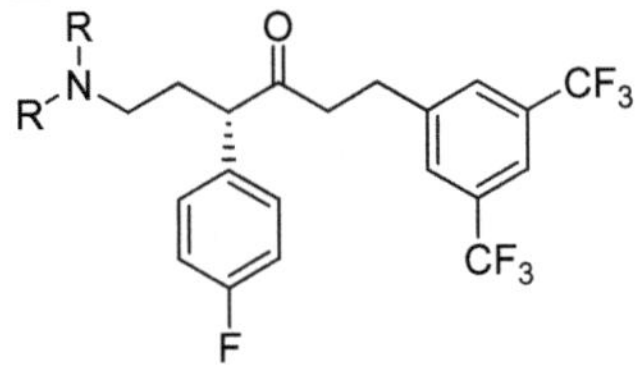


| Compounds | $NR_2$ | CCR2 $IC_{50}$ (nM)[a] | CCR5 $IC_{50}$ (nM)[b] |
|---|---|---|---|
| **108** | 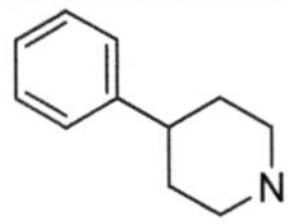 | 150 | 72 |
| **109** | 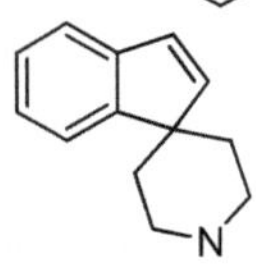 | 80 | 30 |
| **110** | 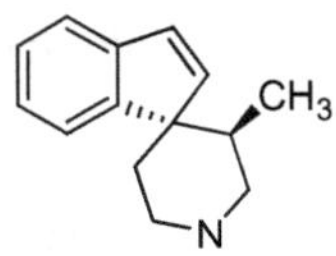 | 59 | 26% at 10 |

[a]Inhibitory effect on binding of CCL2 to human CCR5 receptor
[b]Inhibitory effect on binding of [$^{125}$I]-CCL5 to human CCR5 receptor

receptor binding affinity) and 30 nM (the CCR5 receptor binding affinity), whereas its closely related saturated spiro[indane-piperidine] analog was less potent. The introduction of a methyl group in various positions of the γ-aminobutyramide backbone and the piperidine ring decreased potency with the exception of a methyl moiety in the 3-position of the piperidine ring. However, the CCR2 and CCR5 affinity is strongly dependent on the relative and absolute configuration of the piperidine derivatives.

Compound **110** with (*R*,*R*,*S*)-configuration was the most potent ligand from this series with the $IC_{50}$ value of 59 nM in the CCR2 binding and an 26% inhibition at 10 nM in CCR5 binding. In chemotactic assays, progression from **108** to **109** and **110** was observed: **108** inhibited the CCL2 stimulated monocyte chemotaxis by only 40% at a concentration of 1 μM, while **109** with an $IC_{50}$ value of 176 nM and **110** with an $IC_{50}$ value of 41 nM. In selectivity screenings, compounds **108**, **109**, and **110** were found to be highly selective against CCR1, CCR3, CXCR3, CCR4, CXCR4, and CCR8 receptors. Only the pharmacokinetic properties of **109** were evaluated in a rat model, which showed good pharmacokinetic parameters and proper oral bioavailability at 3 mg/kg body weight [121].

## 5 Conclusions

The emerging evidence for the role of CCR2 and CCR5 receptors in human inflammatory diseases led to a growing interest in selective and dual CCR2/CCR5 antagonists. The availability of potent CCR2 and CCR5 antagonists allows the selective targeting of these receptors and the development of novel concepts for the therapy of inflammatory diseases (e.g., multiple sclerosis and atherosclerosis).

Maraviroc (**1**) is up to now the only commercially available CCR5 antagonist for the treatment of HIV-1 infections, but the growing number of reports on maraviroc-resistant viruses underlines the need of new drugs with improved resistance profile. During the development of new lead compounds, many issues like hERG channel affinity and metabolic stability had to be considered. The most promising CCR5 antagonists for the treatment of HIV infections are GSK163929 (**99**) and PF-232798 (**90c**), which will enter clinical studies.

The development of clinical candidates targeting the CCR2 receptor is also associated with the optimization of several aspects. Although the potent CCR2 antagonist MK0812 (**5**) has reached phase II clinical trials, the further development was terminated due to no significant improvement compared with placebo. The CCR2 antagonist JNJ17166864 (**16**) was tested in clinical trials for the local treatment of allergic rhinitis. However, the study was terminated due to lack of efficacy. In contrast to promising results in preclinical animal models of inflammation, CCR2 antagonists do not show sufficient efficacy in clinical trials of inflammatory diseases so far.

The clinical trials performed with selective CCR2 and CCR5 antagonists suggest that targeting a single receptor might not be sufficient for high efficacy. The fact that both receptors are important in the pathogenesis of cardiovascular and/or immunological diseases indicates great therapeutic potential of dual antagonists. Many compounds that were originally developed as selective antagonists of CCR2 or CCR5 receptors have shown later to address both subtypes. The systematic development of dual CCR2/CCR5 antagonists resulted in INCB10820 (**107**) as the most promising antagonist. In addition to the important central amine, **107** contains a fluoro substituent which interacts with Arg206 of the CCR2 receptor and Ile198 of the CCR5 receptor.

The recently published X-ray crystal structures of the CXCR4 and CCR5 receptors represent the basis for docking studies and virtual screening campaigns, which might lead to discovery of innovative ligands and the generation of novel selective and dual antagonists with desired pharmacological properties.

## References

1 Armour DR, De Groot MJ, Price DA, Stammen BLC, Wood A, Perros M, Burt C (2006) The discovery of tropane-derived CCR5 receptor antagonists. Chem Biol Drug Des 67:305–308

2. Dorr P, Westby M, Dobbs S, Griffin P, Irvine B, Macartney M, Mori J, Rickett G, Smith-Burchnell C, Napier C, Webster R, Armour D, Price D, Stammen B, Wood A, Perros M (2005) Maraviroc (UK-427,857), a potent, orally bioavailable, and selective small-molecule inhibitor of chemokine receptor CCR5 with broad-spectrum anti-human immunodeficiency virus type 1 activity. Antimicrob Agents Chemother 49:4721–4732
3. Galkina E, Ley K (2009) Immune and inflammatory mechanisms of atherosclerosis. Annu Rev Immunol 27:165–197
4. Hansson GK (1993) Immune and inflammatory mechanisms in the development of atherosclerosis. Br Heart J 69(1 Suppl):S38–S41
5. Hansson GK (1994) Immune and inflammatory mechanisms in the pathogenesis of atherosclerosis. J Atheroscler Thromb (1 Suppl) 1:S6–S9
6. Xia M, Sui Z (2009) Recent developments in CCR2 antagonists. Expert Opin Ther Pat 19:295–303
7. Feria M, Díaz-González F (2006) The CCR2 receptor as a therapeutic target. Expert Opin Ther Pat 16:49–57
8. D'Souza MP, Harden V (1996) Chemokines and HIV-1 second receptors. Nat Med 2:1293–1300
9. Weiss RA (2001) Gulliver's travels in HIVland. Nature 410:963–967
10. Karlsson I, Antonsson L, Shi Y, Karlsson A, Albert J, Leitner T, Olde B, Owman C, Fenyo EM (2003) HIV biological variability unveiled: frequent isolations and chimeric receptors reveal unprecedented variation of coreceptor use. Aids 17:2561–2569
11. Catano G, Chykarenko ZA, Mangano A, Anaya JM, He W, Smith A, Bologna R, Sen L, Clark RA, Lloyd A, Shostakovich-Koretskaya L, Ahuja SK (2011) Concordance of CCR5 genotypes that influence cell-mediated immunity and HIV-1 disease progression rates. J Infect Dis 203:263–272
12. Cohn SK Jr, Weaver LT (2006) The Black Death and AIDS: CCR5-Delta32 in genetics and history. QJM Mon J Assoc Phys 99:497–503
13. Szalai C, Duba J, Prohaszka Z, Kalina A, Szabo T, Nagy B, Horvath L, Csaszar A (2001) Involvement of polymorphisms in the chemokine system in the susceptibility for coronary artery disease (CAD). Coincidence of elevated Lp(a) and MCP-1–2518 G/G genotype in CAD patients. Atherosclerosis 158:233–239
14. Pai JK, Kraft P, Cannuscio CC, Manson JE, Rexrode KM, Albert CM, Hunter D, Rimm EB (2006) Polymorphisms in the CC-chemokine receptor-2 (CCR2) and −5 (CCR5) genes and risk of coronary heart disease among US women. Atherosclerosis 186:132–139
15. Sharda S, Gilmour A, Harris V, Singh VP, Sinha N, Tewari S, Ramesh V, Agrawal S, Mastana S (2008) Chemokine receptor 5 (CCR5) deletion polymorphism in North Indian patients with coronary artery disease. Int J Cardiol 124:254–258
16. Gomez-Reino JJ, Pablos JL, Carreira PE, Santiago B, Serrano L, Vicario JL, Balsa A, Figueroa M, de Juan MD (1999) Association of rheumatoid arthritis with a functional chemokine receptor, CCR5. Arthritis Rheum 42:989–992
17. Berce V, Repnik K, Potocnik U (2008) Association of CCR5-delta32 mutation with reduced risk of nonatopic asthma in Slovenian children. J Asthma 45(9):780–784
18. Silversides JA, Heggarty SV, McDonnell GV, Hawkins SA, Graham CA (2004) Influence of CCR5 delta32 polymorphism on multiple sclerosis susceptibility and disease course. Mult Scler J 10:149–152
19. Koenen RR, von Hundelshausen P (2008) The chemokine system as therapeutic target in cardiovascular disease. Drug Discov Today Dis Mech 5:e285–e292
20. Koenen RR, Weber C (2010) Therapeutic targeting of chemokine interactions in atherosclerosis. Nat Rev Drug Discov 9:141–153
21. de Silva E, Stumpf MP (2004) HIV and the CCR5-Delta32 resistance allele. FEMS Microbiol Lett 241:1–12
22. Pease JE, Horuk R (2009) Chemokine receptor antagonists: part 1. Expert Opin Ther Pat 19:39–58

23. Pease JE, Horuk R (2009) Chemokine receptor antagonists: part 2. Expert Opin Ther Pat 19:199–221
24. Struthers M, Pasternak A (2010) CCR2 antagonists. Curr Top Med Chem 10:1278–1298
25. Xue CB, Wang A, Meloni D, Zhang K, Kong L, Feng H, Glenn J, Huang T, Zhang Y, Cao G, Anand R, Zheng C, Xia M, Han Q, Robinson DJ, Storace L, Shao L, Li M, Brodmerkel CM, Covington M, Scherle P, Diamond S, Yeleswaram S, Vaddi K, Newton R, Hollis G, Friedman S, Metcalf B (2010) Discovery of INCB3344, a potent, selective and orally bioavailable antagonist of human and murine CCR2. Bioorg Med Chem Lett 20:7473–7478
26. Shin N, Baribaud F, Wang K, Yang G, Wynn R, Covington MB, Feldman P, Gallagher KB, Leffet LM, Lo YY, Wang A, Xue CB, Newton RC, Scherle PA (2009) Pharmacological characterization of INCB3344, a small molecule antagonist of human CCR2. Biochem Biophys Res Commun 387:251–255
27. Hou C, Sui Z (2012) Chapter 12 CCR2 antagonists for the treatment of diseases associated with inflammation. In: Levin JI, Laufer S (eds) Anti-inflammatory drug discovery, vol 26. The Royal Society of Chemistry, Cambridge
28. Kothandan G, Gadhe CG, Cho SJ (2012) Structural insights from binding poses of CCR2 and CCR5 with clinically important antagonists: a combined in silico study. PLoS One 7:e32864
29. Shamovsky I, Connolly S, David L, Ivanova S, Norden B, Springthorpe B, Urbahns K (2008) Overcoming undesirable HERG potency of chemokine receptor antagonists using baseline lipophilicity relationships. J Med Chem 51:1162–1178
30. Wu B, Chien EYT, Mol CD, Fenalti G, Liu W, Katritch V, Abagyan R, Brooun A, Wells P, Bi FC, Hamel DJ, Kuhn P, Handel TM, Cherezov V, Stevens RC (2010) Structures of the CXCR4 Chemokine GPCR with small-molecule and cyclic peptide antagonists. Science 330:1066–1071
31. Zhang X, Hufnagel H, Markotan T, Lanter J, Cai C, Hou C, Singer M, Opas E, McKenney S, Crysler C, Johnson D, Sui Z (2011) Overcoming hERG activity in the discovery of a series of 4-azetidinyl-1-aryl-cyclohexanes as CCR2 antagonists. Bioorg Med Chem Lett 21:5577–5582
32. Jamieson C, Moir EM, Rankovic Z, Wishart G (2006) Medicinal chemistry of hERG optimizations: highlights and hang-ups. J Med Chem 49:5029–5046
33. Rezazadeh S, Hesketh JC, Fedida D (2004) Rb+ flux through hERG channels affects the potency of channel blocking drugs: correlation with data obtained using a high-throughput Rb + efflux assay. J Biomol Screen 9:588–597
34. Hughes RO, Rogier DJ, Devraj R, Zheng C, Cao G, Feng H, Xia M, Anand R, Xing L, Glenn J, Zhang K, Covington M, Morton PA, Hutzler JM, Davis JW 2nd, Scherle P, Baribaud F, Bahinski A, Mo ZL, Newton R, Metcalf B, Xue CB (2011) Discovery of ((1S,3R)-1-isopropyl-3-((3S,4S)-3-methoxy-tetrahydro-2H-pyran-4-ylamino)cyclopent yl)(4-(5-(trifluoro-methyl)pyridazin-3-yl)piperazin-1-yl)methanone, PF-4254196, a CCR2 antagonist with an improved cardiovascular profile. Bioorg Med Chem Lett 21:2626–2630
35. Xue CB, Feng H, Cao G, Huang T, Glenn J, Anand R, Meloni D, Zhang K, Kong L, Wang A, Zhang Y, Zheng C, Xia M, Chen L, Tanaka H, Han Q, Robinson DJ, Modi D, Storace L, Shao L, Sharief V, Li M, Galya LG, Covington M, Scherle P, Diamond S, Emm T, Yeleswaram S, Contel N, Vaddi K, Newton R, Hollis G, Friedman S, Metcalf B (2011) Discovery of INCB3284, a potent, selective, and orally bioavailable hCCR2 antagonist. ACS Med Chem Lett 2011(2):450–454
36. Lu D, X-j Y, Evans R, Pappas A, Wang H, Su E, Hamdouchi C, Venkataraman C (2005) Cloning and functional characterization of the rabbit C-C chemokine receptor 2. BMC Immunol 6:15
37. Carter PH, Cherney RJ, Mangion IK (2007) Chapter 14 advances in the discovery of CC chemokine receptor 2 antagonists. Annu Rep Med Chem Elsevier Inc 42:211–227
38. Wijtmans M, Scholten DJ, de Esch IJP, Smit MJ, Leurs R (2012) Therapeutic targeting of chemokine receptors by small molecules. Chemokine Recep 9(4):e229–e236

39. Xue C-B, Wang A, Han Q, Zhang Y, Cao G, Feng H, Huang T, Zheng C, Xia M, Zhang K, Kong L, Glenn J, Anand R, Meloni D, Robinson DJ, Shao L, Storace L, Li M, Hughes RO, Devraj R, Morton PA, Rogier DJ, Covington M, Scherle P, Diamond S, Emm T, Yeleswaram S, Contel N, Vaddi K, Newton R, Hollis G, Metcalf B (2011) Discovery of INCB8761/PF-4136309, a potent, selective, and orally bioavailable CCR2 antagonist. ACS Med Chem Lett 2:913–918
40. Pease JE, Horuk R (2012) Chemokine receptor antagonists. J Med Chem 55:9363–9392
41. Horuk R (2009) Chemokine receptor antagonists: overcoming developmental hurdles. Nat Rev Drug Discov 8:23–33
42. Pasternak A, Goble SD, Vicario PP, Di Salvo J, Ayala JM, Struthers M, DeMartino JA, Mills SG, Yang L (2008) Potent heteroarylpiperidine and carboxyphenylpiperidine 1-alkyl-cyclopentane carboxamide CCR2 antagonists. Bioorg Med Chem Lett 18:994–998
43. Lanter JC, Markotan TP, Zhang X, Subasinghe N, Kang FA, Hou C, Singer M, Opas E, McKenney S, Crysler C, Johnson D, Molloy CJ, Sui Z (2011) The discovery of novel cyclohexylamide CCR2 antagonists. Bioorg Med Chem Lett 21:7496–7501
44. Trujillo JI, Huang W, Hughes RO, Rogier DJ, Turner SR, Devraj R, Morton PA, Xue CB, Chao G, Covington MB, Newton RC, Metcalf B (2011) Design and synthesis of novel CCR2 antagonists: investigation of non-aryl/heteroaryl binding motifs. Bioorg Med Chem Lett 21:1827–1831
45. Burrows JN, Cumming JG, Fillery SM, Hamlin GA, Hudson JA, Jackson RJ, McLaughlin S, Shaw JS (2005) Modulators of the human CCR5 receptor. Part 1: Discovery and initial SAR of 1-(3,3-diphenylpropyl)-piperidinyl amides and ureas. Bioorg Med Chem Lett 15:25–28
46. Cumming JG, Brown SJ, Cooper AE, Faull AW, Flynn AP, Grime K, Oldfield J, Shaw JS, Shepherd E, Tucker H, Whittaker D (2006) Modulators of the human CCR5 receptor. Part 3: SAR of substituted 1-[3-(4-methanesulfonylphenyl)-3-phenylpropyl]-piperidinyl phenylacetamides. Bioorg Med Chem Lett 16:3533–3536
47. Cumming JG, Cooper AE, Grime K, Logan CJ, McLaughlin S, Oldfield J, Shaw JS, Tucker H, Winter J, Whittaker D (2005) Modulators of the human CCR5 receptor. Part 2: SAR of substituted 1-(3,3-diphenylpropyl)-piperidinyl phenylacetamides. Bioorg Med Chem Lett 15:5012–5015
48. Mirzadegan T, Diehl F, Ebi B, Bhakta S, Polsky I, McCarley D, Mulkins M, Weatherhead GS, Lapierre JM, Dankwardt J, Morgans D Jr, Wilhelm R, Jarnagin K (2000) Identification of the binding site for a novel class of CCR2b chemokine receptor antagonists: binding to a common chemokine receptor motif within the helical bundle. J Biol Chem 275:25562–25571
49. Gao Z, Metz WA (2003) Unraveling the chemistry of chemokine receptor ligands. Chem Rev 103:3733–3752
50. Yang L, Butora G, Jiao RX, Pasternak A, Zhou C, Parsons WH, Mills SG, Vicario PP, Ayala JM, Cascieri MA, MacCoss M (2007) Discovery of 3-piperidinyl-1-cyclopentanecarboxamide as a novel scaffold for highly potent CC chemokine receptor 2 antagonists. J Med Chem 50:2609–2611
51. Butora G, Jiao R, Parsons WH, Vicario PP, Jin H, Ayala JM, Cascieri MA, Yang L (2007) 3-Amino-1-alkyl-cyclopentane carboxamides as small molecule antagonists of the human and murine CC chemokine receptor 2. Bioorg Med Chem Lett 17:3636–3641
52. Shiraishi M, Aramaki Y, Seto M, Imoto H, Nishikawa Y, Kanzaki N, Okamoto M, Sawada H, Nishimura O, Baba M, Fujino M (2000) Discovery of novel, potent, and selective small-molecule CCR5 antagonists as anti-HIV-1 agents: synthesis and biological evaluation of anilide derivatives with a quaternary ammonium moiety. J Med Chem 43:2049–2063
53. Lagu B, Gerchak C, Pan M, Hou C, Singer M, Malaviya R, Matheis M, Olini G, Cavender D, Wachter M (2007) Potent and selective CC-chemokine receptor-2 (CCR2) antagonists as a potential treatment for asthma. Bioorg Med Chem Lett 17:4382–4386
54. Smethurst CA, Bevan N, Brooks C, Emmons A, Gough PJ, Mookherjee C, Moores K, Peace S, Philp J, Piercy V, Watson SP, Zippoli M (2012) In vivo activity of an azole series of CCR2 antagonists. Bioorg Med Chem Lett 22:7252–7255

55. Zhang X, Hufnagel H, Hou C, Opas E, McKenney S, Crysler C, O'Neill J, Johnson D, Sui Z (2011) Design, synthesis and SAR of indazole and benzoisoxazole containing 4-azetidinyl-1-aryl-cyclohexanes as CCR2 antagonists. Bioorg Med Chem Lett 21:6042–6048
56. Zhang X, Hou C, Hufnagel H, Singer M, Opas E, McKenney S, Johnson D, Sui Z (2012) Discovery of a 4-Azetidinyl-1-thiazoyl-cyclohexane CCR2 antagonist as a development candidate. ACS Med Chem Lett 3:1039–1044
57. Subasinghe NL, Lanter J, Markotan T, Opas E, McKenney S, Crysler C, Hou C, O'Neill J, Johnson D, Sui Z (2013) A novel series of N-(azetidin-3-yl)-2-(heteroarylamino)acetamide CCR2 antagonists. Bioorg Med Chem Lett 23:1063–1069
58. Cai C, Kang FA, Hou C, O'Neill JC, Opas E, McKenney S, Johnson D, Sui Z (2013) Novel 2-aminooctahydrocyclopentalene-3a-carboxamides as potent CCR2 antagonists. Bioorg Med Chem Lett 23:351–354
59. Cai C, McComsey DF, Hou C, O'Neill JC, Opas E, McKenney S, Johnson D, Sui Z (2013) Discovery and SAR of 5-aminooctahydrocyclopentapyrrole-3a-carboxamides as potent CCR2 antagonists. Bioorg Med Chem Lett. doi:10.1016/j.bmcl.2013.05.024
60. Berkhout TA, Blaney FE, Bridges AM, Cooper DG, Forbes IT, Gribble AD, Groot PH, Hardy A, Ife RJ, Kaur R, Moores KE, Shillito H, Willetts J, Witherington J (2003) CCR2: characterization of the antagonist binding site from a combined receptor modeling/mutagenesis approach. J Med Chem 46:4070–4086
61. Kim J-H, Lim J-W, Lee S-W, Kim K-R, No K-T (2012) Prediction of binding mode between chemokine receptor CCR2 and its known antagonists using ligand supported homology modeling. Bull Kor Chem Soc 33:717–720
62. Hall SE, Mao A, Nicolaidou V, Finelli M, Wise EL, Nedjai B, Kanjanapangka J, Harirchian P, Chen D, Selchau V, Ribeiro S, Schyler S, Pease JE, Horuk R, Vaidehi N (2009) Elucidation of binding sites of dual antagonists in the human chemokine receptors CCR2 and CCR5. Mol Pharmacol 75:1325–1336
63. Kondru R, Zhang J, Ji C, Mirzadegan T, Rotstein D, Sankuratri S, Dioszegi M (2008) Molecular interactions of CCR5 with major classes of small-molecule anti-HIV CCR5 antagonists. Mol Pharmacol 73:789–800
64. Maeda K, Das D, Ogata-Aoki H, Nakata H, Miyakawa T, Tojo Y, Norman R, Takaoka Y, Ding J, Arnold GF, Arnold E, Mitsuya H (2006) Structural and molecular interactions of CCR5 inhibitors with CCR5. J Biol Chem 281:12688–12698
65. Lynch T, Price A (2007) The effect of cytochrome P450 metabolism on drug response, interactions, and adverse effects. Am Fam Physician 76:391–396
66. Pilgrim JL, Gerostamoulos D, Drummer OH (2011) Review: Pharmacogenetic aspects of the effect of cytochrome P450 polymorphisms on serotonergic drug metabolism, response, interactions, and adverse effects. Forensic Sci Med Pathol 7:162–184
67. Van Noord C, Eijgelsheim M, Stricker BHC (2010) Drug- and non-drug-associated QT interval prolongation. Br J Clin Pharmacol 70:16–23
68. Gharu L, Ringe R, Bhattacharya J (2011) HIV-1 clade C envelopes obtained from late stage symptomatic Indian patients varied in their ability towards relative CD4 usages and sensitivity to CCR5 antagonist TAK-779. Virus Res 158:216–224
69. Samson M, Labbe O, Mollereau C, Vassart G, Parmentier M (1996) Molecular cloning and functional expression of a new human CC-chemokine receptor gene. Biochemistry 35:3362–3367
70. Dragic T, Trkola A, Thompson DA, Cormier EG, Kajumo FA, Maxwell E, Lin SW, Ying W, Smith SO, Sakmar TP, Moore JP (2000) A binding pocket for a small molecule inhibitor of HIV-1 entry within the transmembrane helices of CCR5. Proc Natl Acad Sci 97:5639–5644
71. Baba M, Nishimura O, Kanzaki N, Okamoto M, Sawada H, Iizawa Y, Shiraishi M, Aramaki Y, Okonogi K, Ogawa Y, Meguro K, Fujino M (1999) A small-molecule, nonpeptide CCR5 antagonist with highly potent and selective anti-HIV-1 activity. Proc Natl Acad Sci 96:5698–5703

72. Seto M, Miyamoto N, Aikawa K, Aramaki Y, Kanzaki N, Iizawa Y, Baba M, Shiraishi M (2005) Orally active CCR5 antagonists as anti-HIV-1 agents. Part 3: Synthesis and biological activities of 1-benzazepine derivatives containing a sulfoxide moiety. Bioorg Med Chem 13:363–386
73. Aramaki Y, Seto M, Okawa T, Oda T, Kanzaki N, Shiraishi M (2004) Synthesis of 1-benzothiepine and 1-benzazepine derivatives as orally active CCR5 antagonists. Chem Pharm Bull 52:254–258
74. Pryde DC, Jones R, Middleton DS, Laverty BJ, Fenwick DR, Mason HJ, Corless M, Smith NN (2010) An in situ oxidation strategy towards overcoming hERG affinity. Bioorg Med Chem Lett 20:6400–6404
75. Seto M, Aramaki Y, Okawa T, Miyamoto N, Aikawa K, Kanzaki N, Niwa S, Iizawa Y, Baba M, Shiraishi M (2004) Orally active CCR5 antagonists as anti-HIV-1 agents: synthesis and biological activity of 1-benzothiepine 1,1-dioxide and 1-benzazepine derivatives containing a tertiary amine moiety. Chem Pharm Bull 52:577–590
76. Baba M, Takashima K, Miyake H, Kanzaki N, Teshima K, Wang X, Shiraishi M, Iizawa Y (2005) TAK-652 inhibits CCR5-mediated human immunodeficiency virus type 1 infection in vitro and has favorable pharmacokinetics in humans. Antimicrob Agents Chemother 49:4584–4591
77. Imamura S, Nishikawa Y, Ichikawa T, Hattori T, Matsushita Y, Hashiguchi S, Kanzaki N, Iizawa Y, Baba M, Sugihara Y (2005) CCR5 antagonists as anti-HIV-1 agents. Part 3: Synthesis and biological evaluation of piperidine-4-carboxamide derivatives. Biorg Med Chem 13:397–416
78. Imamura S, Ichikawa T, Nishikawa Y, Kanzaki N, Takashima K, Niwa S, Iizawa Y, Baba M, Sugihara Y (2006) Discovery of a piperidine-4-carboxamide CCR5 antagonist (TAK-220) with highly potent anti-HIV-1 activity. J Med Chem 49:2784–2793
79. Nishikawa M, Takashima K, Nishi T, Furuta RA, Kanzaki N, Yamamoto Y, Fujisawa J (2005) Analysis of binding sites for the new small-molecule CCR5 antagonist TAK-220 on human CCR5. Antimicrob Agents Chemother 49:4708–4715
80. Armour D, de Groot MJ, Edwards M, Perros M, Price DA, Stammen BL, Wood A (2006) The discovery of CCR5 receptor antagonists for the treatment of HIV infection: hit-to-lead studies. Chemmedchem 1:706–709
81. Martens HA, Kallenberg CG, Bijl M (2009) Role of CCR5 delta32 bp deletion in RA and SLE. Autoimmunity 42:260–262
82. Andres PG, Beck PL, Mizoguchi E, Mizoguchi A, Bhan AK, Dawson T, Kuziel WA, Maeda N, MacDermott RP, Podolsky DK, Reinecker H-C (2000) Mice with a selective deletion of the CC chemokine receptors 5 or 2 are protected from dextran sodium sulfate-mediated colitis: lack of CC chemokine receptor 5 expression results in a NK1.1+ lymphocyte-associated Th2-type immune response in the intestine. J Immunol 164:6303–6312
83. Cumming JG, Tucker H, Oldfield J, Fielding C, Highton A, Faull A, Wild M, Brown D, Wells S, Shaw J (2012) Balancing hERG affinity and absorption in the discovery of AZD5672, an orally active CCR5 antagonist for the treatment of rheumatoid arthritis. Bioorg Med Chem Lett 22:1655–1659
84. Bradbury J (1996) HIV-1-resistant individuals may lack HIV-1 coreceptor. Lancet 348:463
85. Deng H, Liu R, Ellmeier W, Choe S, Unutmaz D, Burkhart M, Di Marzio P, Marmon S, Sutton RE, Hill CM, Davis CB, Peiper SC, Schall TJ, Littman DR, Landau NR (1996) Identification of a major co-receptor for primary isolates of HIV-1. Nature 381:661–666
86. Combadiere C, Ahuja SK, Murphy PM (1995) Cloning and functional expression of a human eosinophil CC chemokine receptor. J Biol Chem 270:16491–16495
87. de Groot MJ, Ackland MJ, Horne VA, Alex AA, Jones BC (1999) Novel approach to predicting P450-mediated drug metabolism: development of a combined protein and pharmacophore model for CYP2D6. J Med Chem 42:1515–1524

88. de Groot MJ, Ackland MJ, Horne VA, Alex AA, Jones BC (1999) A novel approach to predicting P450 mediated drug metabolism. CYP2D6 catalyzed N-dealkylation reactions and qualitative metabolite predictions using a combined protein and pharmacophore model for CYP2D6. J Med Chem 42:4062–4070
89. Price DA, Armour D, de Groot M, Leishman D, Napier C, Perros M, Stammen BL, Wood A (2006) Overcoming HERG affinity in the discovery of the CCR5 antagonist maraviroc. Bioorg Med Chem Lett 16:4633–4637
90. Liu H, Ji M, Luo X, Shen J, Huang X, Hua W, Jiang H, Chen K (2002) New p-methylsulfonamido phenylethylamine analogues as class III antiarrhythmic agents: design, synthesis, biological assay, and 3D-QSAR analysis. J Med Chem 45:2953–2969
91. Garcia-Perez J, Rueda P, Staropoli I, Kellenberger E, Alcami J, Arenzana-Seisdedos F, Lagane B (2011) New insights into the mechanisms whereby low molecular weight CCR5 ligands inhibit HIV-1 infection. J Biol Chem 286:4978–4990
92. Ratcliff AN, Shi W, Arts EJ (2013) HIV-1 resistance to maraviroc conferred by a CD4 binding site mutation in the envelope glycoprotein gp120. J Virol 87:923–934
93. Tilton JC, Wilen CB, Didigu CA, Sinha R, Harrison JE, Agrawal-Gamse C, Henning EA, Bushman FD, Martin JN, Deeks SG, Doms RW (2010) A maraviroc-resistant HIV-1 with narrow cross-resistance to other CCR5 antagonists depends on both N-terminal and extra-cellular loop domains of drug-bound CCR5. J Virol 84:10863–10876
94. Westby M, Smith-Burchnell C, Mori J, Lewis M, Mosley M, Stockdale M, Dorr P, Ciaramella G, Perros M (2007) Reduced maximal inhibition in phenotypic susceptibility assays indicates that viral strains resistant to the CCR5 antagonist maraviroc utilize inhibitor-bound receptor for entry. J Virol 81:2359–2371
95. Tan Q, Zhu Y, Li J, Chen Z, Han GW, Kufareva I, Li T, Ma L, Fenalti G, Li J, Zhang W, Xie X, Yang H, Jiang H, Cherezov V, Liu H, Stevens RC, Zhao Q, Wu B (2013) Structure of the CCR5 chemokine receptor-HIV entry inhibitor maraviroc complex. Science 341:1387–1390
96. Fan X, Zhang H-S, Chen L, Long Y-Q (2010) Efficient synthesis and identification of novel propane-1,3-diamino bridged CCR5 antagonists with variation on the basic center carrier. Eur J Med Chem 45:2827–2840
97. Zhang H-S, Feng D-Z, Chen L, Long Y-Q (2010) Discovery of novel (S)-α-phenyl-γ-amino butanamide containing CCR5 antagonists via functionality inversion approach. Bioorg Med Chem Lett 20:2219–2223
98. Stupple PA, Batchelor DV, Corless M, Dorr PK, Ellis D, Fenwick DR, Galan SRG, Jones RM, Mason HJ, Middleton DS, Perros M, Perruccio F, Platts MY, Pryde DC, Rodrigues D, Smith NN, Stephenson PT, Webster R, Westby M, Wood A (2010) An imidazopiperidine series of CCR5 antagonists for the treatment of HIV: the discovery of N-{(1S)-1-(3-Fluorophenyl)-3-[(3-endo)-3-(5-isobutyryl-2-methyl-4,5,6,7-tetrahydro-1H-imidazo[4,5-c]pyridin-1-yl)-8-azabicyclo[3.2.1]oct-8-yl]propyl}acetamide (PF-232798). J Med Chem 54:67–77
99. Armand-Ugon M, Moncunill G, Gonzalez E, Mena M, Ballana E, Clotet B, Este JA (2010) Different selection patterns of resistance and cross-resistance to HIV-1 agents targeting CCR5. J Antimicrob Chemother 65:417–424
100. Delgado E, Fernandez-Garcia A, Vega Y, Cuevas T, Pinilla M, Garcia V, Sanchez M, Gonzalez M, Sanchez AM, Thomson MM, Perez-Alvarez L (2012) Evaluation of genotypic tropism prediction tests compared with in vitro co-receptor usage in HIV-1 primary isolates of diverse subtypes. J Antimicrob Chemother 67:25–31
101. Barber CG, Blakemore DC, Chiva J-Y, Eastwood RL, Middleton DS, Paradowski KA (2009) 1-Amido-1-phenyl-3-piperidinylbutanes – CCR5 antagonists for the treatment of HIV: part 2. Bioorg Med Chem Lett 19:1499–1503
102. Barber CG, Blakemore DC, Chiva JY, Eastwood RL, Middleton DS, Paradowski KA (2009) 1-Amido-1-phenyl-3-piperidinylbutanes – CCR5 antagonists for the treatment of HIV. Part 1. Bioorg Med Chem Lett 19:1075–1079

103. Kazmierski WM, Aquino C, Chauder BA, Deanda F, Ferris R, Jones-Hertzog DK, Kenakin T, Koble CS, Watson C, Wheelan P, Yang H, Youngman M (2008) Discovery of bioavailable 4,4-disubstituted piperidines as potent ligands of the chemokine receptor 5 and inhibitors of the human immunodeficiency virus-1. J Med Chem 51:6538–6546
104. Kazmierski WM, Anderson DL, Aquino C, Chauder BA, Duan M, Ferris R, Kenakin T, Koble CS, Lang DG, McIntyre MS, Peckham J, Watson C, Wheelan P, Spaltenstein A, Wire MB, Svolto A, Youngman M (2011) Novel 4,4-disubstituted piperidine-based C-C chemokine receptor-5 inhibitors with high potency against human immunodeficiency virus-1 and an improved human ether-a-go-go related gene (hERG) profile. J Med Chem 54:3756–3767
105. Tallant MD, Duan M, Freeman GA, Ferris RG, Edelstein MP, Kazmierski WM, Wheelan PJ (2011) Synthesis and evaluation of 2-phenyl-1,4-butanediamine-based CCR5 antagonists for the treatment of HIV-1. Bioorg Med Chem Lett 21:1394–1398
106. Takeuchi T, Kameda H (2012) What is the future of CCR5 antagonists in rheumatoid arthritis? Arthritis Res Ther 14:114
107. Fleishaker DL, Garcia Meijide JA, Petrov A, Kohen MD, Wang X, Menon S, Stock TC, Mebus CA, Goodrich JM, Mayer HB, Zeiher BG (2012) Maraviroc, a chemokine receptor-5 antagonist, fails to demonstrate efficacy in the treatment of patients with rheumatoid arthritis in a randomized, double-blind placebo-controlled trial. Arthritis Res Ther 14:R11
108. Charo IF, Ransohoff RM (2006) The many roles of chemokines and chemokine receptors in inflammation. New Engl J Med 354:610–621
109. Ota T (2013) CCR5: a novel player in the adipose tissue inflammation and insulin resistance? Adipocyte 2:99–103
110. Fantuzzi L, Borghi P, Ciolli V, Pavlakis G, Belardelli F, Gessani S (1999) Loss of CCR2 expression and functional response to monocyte chemotactic protein (MCP-1) during the differentiation of human monocytes: role of secreted MCP-1 in the regulation of the chemotactic response. Blood 94:875–883
111. Kaufmann A, Salentin R, Gemsa D, Sprenger H (2001) Increase of CCR1 and CCR5 expression and enhanced functional response to MIP-1 alpha during differentiation of human monocytes to macrophages. J Leukoc Biol 69:248–252
112. Cherney RJ, Mo R, Meyer DT, Pechulis AD, Guaciaro MA, Lo YC, Yang G, Miller PB, Scherle PA, Zhao Q, Cvijic ME, Barrish JC, Decicco CP, Carter PH (2012) Benzimidazoles as benzamide replacements within cyclohexane-based CC chemokine receptor 2 (CCR2) antagonists. Bioorg Med Chem Lett 22:6181–6184
113. Mantovani A (2004) The chemokine system in diverse forms of macrophage activation and polarization. Trends Immunol 25:677–686
114. Proudfoot AE (2002) Chemokine receptors: multifaceted therapeutic targets. Nat Rev Immunol 2:106–115
115. Charo IF, Myers SJ, Herman A, Franci C, Connolly AJ, Coughlin SR (1994) Molecular cloning and functional expression of two monocyte chemoattractant protein 1 receptors reveals alternative splicing of the carboxyl-terminal tails. Proc Natl Acad Sci 91:2752–2756
116. Seto M, Aikawa K, Miyamoto N, Aramaki Y, Kanzaki N, Takashima K, Kuze Y, Iizawa Y, Baba M, Shiraishi M (2006) Highly potent and orally active CCR5 antagonists as anti-HIV-1 agents: synthesis and biological activities of 1-benzazocine derivatives containing a sulfoxide moiety. J Med Chem 49:2037–2048
117. Baba M (2007) Isolation and characterization of human immunodeficiency virus type 1 resistant to the small-molecule CCR5 antagonist TAK-652. Antimicrob Agents Chemother 51:707–715
118. Pasternak A, Goble SD, Struthers M, Vicario PP, Ayala JM, Di Salvo J, Kilburn R, Wisniewski T, DeMartino JA, Mills SG, Yang L (2009) Discovery of a potent and orally bioavailable CCR2 and CCR5 dual antagonist. ACS Med Chem Lett 1:14–18
119. Witherington J, Bordas V, Cooper DG, Forbes IT, Gribble AD, Ife RJ, Berkhout T, Gohil J, Groot PHE (2001) Conformationally restricted indolopiperidine derivatives as potent CCR2B receptor antagonists. Bioorg Med Chem Lett 11:2177–2180

120. Zheng C, Cao G, Xia M, Feng H, Glenn J, Anand R, Zhang K, Huang T, Wang A, Kong L, Li M, Galya L, Hughes RO, Devraj R, Morton PA, Rogier DJ, Covington M, Baribaud F, Shin N, Scherle P, Diamond S, Yeleswaram S, Vaddi K, Newton R, Hollis G, Friedman S, Metcalf B, Xue CB (2011) Discovery of INCB10820/PF-4178903, a potent, selective, and orally bioavailable dual CCR2 and CCR5 antagonist. Bioorg Med Chem Lett 21:1442–1446
121. Pasternak A, Marino D, Vicario PP, Ayala JM, Cascierri MA, Parsons W, Mills SG, Maccoss M, Yang L (2006) Novel, orally bioavailable gamma-aminoamide CC chemokine receptor 2 (CCR2) antagonists. J Med Chem 49:4801–4804

# Index

MIX
Papier aus verantwortungsvollen Quellen
Paper from responsible sources
FSC® C105338

If you have any concerns about our products,
you can contact us on
**ProductSafety@springernature.com**

In case Publisher is established outside the EU,
the EU authorized representative is:
**Springer Nature Customer Service Center GmbH**
**Europaplatz 3, 69115 Heidelberg, Germany**

Printed by Libri Plureos GmbH
in Hamburg, Germany